Plant Retrotransposons
Genomic, Evolutionary and Biotechnological Perspectives

Editors

Deepu Pandita
Government Department of School Education
Jammu and Kashmir, India

Anu Pandita
Vatsalya Clinic, New Delhi, India

CRC Press
Taylor & Francis Group
Boca Raton London New York

CRC Press is an imprint of the
Taylor & Francis Group, an **informa** business
A SCIENCE PUBLISHERS BOOK

First edition published 2025
by CRC Press
2385 NW Executive Center Drive, Suite 320, Boca Raton FL 33431

and by CRC Press
4 Park Square, Milton Park, Abingdon, Oxon, OX14 4RN

Library of Congress Cataloging-in-Publication Data (applied for)

ISBN: 978-1-032-66345-6 (hbk)
ISBN: 978-1-032-66346-3 (pbk)
ISBN: 978-1-032-66349-4 (ebk)

DOI: 10.1201/9781032663494

Typeset in Times New Roman
by Prime Publishing Services

This book is dedicated to

"Dr. Manoj Kumar Dhar"

Professor, School of Biotechnology, University of Jammu,
Baba Saheb Ambedkar Road, Jammu, Jammu & Kashmir, India,

and

Director, Academy of Scientific & Innovative Research (AcSIR),
CSIR-HRDC Campus, Ghaziabad, Uttar Pradesh, India.

With profound gratitude, I acknowledge Professor MK Dhar, for introducing me to the fascinating world of mobile genetic elements in Kingdom Plantae.

Deepu Pandita

Preface

Previously known as Selfish or "Junk DNA", today's "Just Unexplored Novel Know-how" is considered the genomic gold as these form the significant portion of plant genes and genomes. The expression and transposition of retrotransposons play significant roles in the genomic diversity and variation of genome size, genetic variations and genome regulation. The activation of retrotransposons turns on the survival genes in both the biotic and abiotic stress scenarios and contributes to the evolution and speciation of plants. Retrotransposons used as genetic tools in plant biology help create climate smart designer crops for sustainable agricultural production in upcoming climate change scenarios.

This book volume is essential for readers familiar with the field and for newcomers in plant genetics and genomics, epigenetics, and epigenomics, or evolutionary biology related to the silver lining of Plant Retrotransposons. The present book gives an overview and role of the ubiquitous class of mobile genetic elements categorized as retrotransposons. Transposable Elements (TEs) were discovered by Barbara McClintock in *Zea mays* in the middle of the 20th century, for which she was later awarded the Nobel Prize in Physiology or Medicine in 1983. Retrotransposons belong to the Class 1 family of transposons and transpose by the "copy and paste" transposition mechanism. Retrotransposons are categorized into two Long Terminal Repeat (LTR) retrotransposons and non-LTR retrotransposons on the basis of mechanism of transposition and structure. LTR retrotransposons have LTR sequences at the 5′ and 3′ ends and are further categorized into Ty1-copia and Ty3-gypsy classes. Non-LTR retrotransposons show an absence of LTR sequences and are characterized into Long Interspersed Nuclear Elements (LINEs) and Short Interspersed Nuclear Elements (SINEs). Retrotransposons form the bulk of genomic DNA. Heterogeneous populations of retrotransposons make >76% of the nuclear DNA content of the genome in *Hordeum vulgare* and >75% in *Zea mays*. LTR sequences localize inside or in close proximity of genes and retro-transposition affects the structure and function of genes at or near the TE insertion sites. Retrotransposons play a significant role in the development, origin, evolution, and genome organization of plants and model plants such as *Zea mays*, *Triticum aestivum*, *Hordeum vulgare*, *Glycine max*, *Brassica* spp., and *Nicotiana* spp. as is listed in chapter contents.

Finally, this book also includes chapters that describe how retrotransposons can be used for biotechnological applications of crop improvement and plant stress responses and various computational approaches for retrotransposon investigations have been deliberated in detail.

Within the cells, a tale unfolds,
Of retrotransposons, secrets untold

In the nucleus's hidden halls retrotransposons roam,
Retrotransposons, silent yet tomes.

Once dormant relics, now stirred from slumber deep,
Intriguing the minds that are forsaken.

In the strands of DNA's intricate weave,
Lurk echoes of the past, secrets conceive.

With reverse transcriptase, retrotransposons rewrite the script,
In the genetic code, their footprints cryptically crypt.

Copying and pasting, a genetic rhyme,
In genome's landscape shaping evolution, through space and time.

Oh retrotransposons, silent voyagers of yore, you weave the threads of life's
grand tale,
In the symphony of genes, you never fail.

Jumping genes jumping across with a rhythmic beat, like dancers in flight,
Retrotransposons sculpt the genome with their might.

In your dance in genome's depths, we find the key,
To understanding life's mystery.

In plants retrotransposons dwell, guardians' unseen,
Shaping evolution's cryptic scene.

From ferns to flowers, retrotransposons leave their trace,
A testament to nature's timeless grace.

Editors:
Deepu Pandita
Anu Pandita

Contents

Section C: Retrotransposons and Crop Improvement

List of Contributors

Abdul Rehman
Department of Bioinformatics and Biotechnology, Government College University, Faisalabad, Pakistan.

Akash Fatima
Institute of Plant Breeding and Biotechnology MNS-University of Agriculture, Multan, Pakistan.

Alvina Gul
Department of Plant Biotechnology, Atta-ur-Rahman School of Applied Biosciences, National University of Sciences and Technology, Islamabad, Pakistan.

Anu Pandita
Vatsalya Clinic, New Delhi, India.

Ayesha Ghazanfar
Institute of Plant Breeding and Biotechnology MNS-University of Agriculture, Multan, Pakistan.

Bhawana
UIBT, Chandigarh University, Gharuan, Mohali, Punjab-140413, India.

Bushra Hafeez Kiani
Department of Biological Sciences (Female Campus), International Islamic University, Islamabad-44000, Pakistan.

Deepu Pandita
Government Department of School Education, Jammu, Jammu and Kashmir, India.

Farrukh Azeem
Department of Bioinformatics and Biotechnology, Government College University, Faisalabad, Pakistan.

Ghulam Muhammad
Institute of Plant Breeding and Biotechnology, MNS University of Agriculture, Multan, Pakistan.

Hajira Imran
Department of Plant Biotechnology, Atta-ur-Rahman School of Applied Biosciences, National University of Sciences and Technology, Islamabad, Pakistan.

Iqra Siddique
Department of Plant Biotechnology, Atta-ur-Rahman School of Applied Biosciences, National University of Sciences and Technology, Islamabad, Pakistan.

Jasdeep Singh
UIBT, Chandigarh University, Gharuan, Mohali, Punjab-140413, India.

Joginder Singh
Department of Microbiology, School of Bioengineering and Biosciences, Lovely Professional University, Phagwara, Punjab-144411, India.

Jyoti Sarwan
UIBT, Chandigarh University, Gharuan, Mohali, Punjab-140413, India.

Khushbhu Sharma
Department of Agronomy, School of Agriculture, Lovely Professional University, Phagwara, Punjab-144411, India.

MA Syed
Plant breeding division, Bangladesh Rice Research Institute (BRRI), Gazipur-1701, Bangladesh.

MB Akter
Crop Physiology Division, Bangladesh Institute of Nuclear Agriculture (BINA), Mymensingh-2202, Bangladesh.

Medeline Joseph
St. Xavier's College (Autonomous), Mumbai-400001 affiliated with the University of Mumbai.

Muhammad Shahid
Department of Bioinformatics and Biotechnology, Government College University, Faisalabad, Pakistan.

Muhammad Waqas
Department of Bioinformatics and Biotechnology, Government College University, Faisalabad, Pakistan.

MY Khan
Plant breeding division, Bangladesh Rice Research Institute (BRRI), Gazipur-1701, Bangladesh.

N Jahan
Plant breeding division, Bangladesh Rice Research Institute (BRRI), Gazipur-1701, Bangladesh.

Noor ul huda
Atta-ur-Rahman School of Applied Biosciences, National University of Sciences &Technology, Islamabad.

Palwasha Khanum
Institute of Plant Breeding and Biotechnology, MNS University of Agriculture, Multan, Pakistan.

Pooja Bhadrecha
UIBT, Chandigarh University, Gharuan, Mohali, Punjab-140413, India.

Prasann Kumar
Department of Agronomy, School of Agriculture, Lovely Professional University, Phagwara, Punjab-144411, India

Preedhi Kapoor
Department of Biochemistry, School of Bioengineering and Biosciences, Lovely Professional University, Phagwara, Punjab-144411, India.

Priya Sundarrajan
St. Xavier's College (Autonomous), Mumbai-400001 affiliated with the University of Mumbai.

Priyanka Devi
Department of Agronomy, School of Agriculture, Lovely Professional University, Phagwara, Punjab-144411, India.

Rabia Habib
Institute of Plant Breeding and Biotechnology MNS-University of Agriculture, Multan, Pakistan.

Sadaf Batool
Department of Bioinformatics and Biotechnology, Government College University, Faisalabad, Pakistan.

Sadia Shabir
Institute of Plant Breeding and Biotechnology, MNS University of Agriculture, Multan, Pakistan.

Saurabh Oswal
St. Xavier's College (Autonomous), Mumbai-400001 affiliated with the University of Mumbai.

Sehrish Ijaz
Institute of Plant Breeding and Biotechnology MNS-University of Agriculture, Multan, Pakistan.

Shahroz Rahman
Department of Bioinformatics and Biotechnology, Government College University, Faisalabad, Pakistan.

Shubham Kumar
UIBT, Chandigarh University, Gharuan, Mohali, Punjab-140413, India.

SK Debsharma
Plant breeding division, Bangladesh Rice Research Institute (BRRI), Gazipur-1701, Bangladesh.

Ummara Waheed
Institute of Plant Breeding and Biotechnology MNS-University of Agriculture, Multan, Pakistan.

Vajiha Sahar Khan
Institute of Plant Breeding and Biotechnology MNS-University of Agriculture, Multan, Pakistan.

Zulqurnain Khan
Institute of Plant Breeding and Biotechnology MNS-University of Agriculture, Multan, Pakistan.

1

Retrotransposons: A 360 Degree Overview

Deepu Pandita[1*] and *Anu Pandita*[2]

1. Introduction

Transposable elements (TEs) which are repetitive Deoxyribo-Nucleic Acid (DNA) sequences were discovered in *Zea mays* (maize) as controlling elements in 1944 by famous geneticist Barbara McClintock of Cold Spring Harbor Laboratory in New York, before the discovery of the structure of double helix of DNA in 1953 (McClintock 1951; Watson and Crick 1953; Pray 2008; Ou et al., 2018). The first sequencing of maize Ac/Ds TEs was done in 1984 (Döring et al., 1984; Pohlman and Fedoroff 1984). In 1983, Barbara McClintock was awarded the Nobel Prize in Physiology or Medicine for her discovery of transposable elements. These ubiquitous mobile units of the nuclear genome move/transpose within the host genome and among genomes of different organisms, thus also known as "jumping genes". TEs occur in both prokaryotes and eukaryotes and form a significant portion of plant genomes contributing to their diversity and variation in the size of the genome (McClintock 1951; Mita and Boeke 2016; Keidar et al., 2018) and constitute 45% of the human genome (Pandita and Pandita 2016). There are different types besides ways to catalogue TEs. TEs can be broadly categorized into two core classes with diverse transposition mechanisms i.e., which transpose and those that do not (Kazazian et al., 1988; Mi et al., 2000; Kapitonov and Koonin 2015; Chaparro et al., 2015). The former elements belonging to Class I or autonomous or retrotransposons or copy and paste transposons move via an RNA molecule and later class II non-autonomous elements or cut and paste transposons or DNA

[1] Government Department of School Education, Jammu, Jammu and Kashmir, India: ORCID iD-https://orcid.org/0000-0002-8361-1872

[2] Vatsalya Clinic, New Delhi, India

* Corresponding author: deepupandita@gmail.com; dt.anunischal46@gmail.com

transposons move directly as DNA fragments in genomes of different organisms (Gao et al., 2004; Wicker et al., 2007; Grandbastien 2008; Orozco-Arias et al., 2019) (Fig. 1a, b).

Class I retrotransposons are the most abundant TEs and are ubiquitous in plant genomes. Class I retrotransposons are reverse-transcribing mobile genetic elements and pieces of genetic material that replicate and move within the genome because of their very effective transposition mechanism of 'copy and paste' or replicating themselves by reverse transcription of their RNA transcript by use of reverse transcriptase, integration of daughter copy DNA into the host genome and transcription of the new copy (Xiong and Eickbush, 1988; Wicker et al., 2007; Grandbastien, 2015) (Fig. 1a, b; Fig. 6). In the reverse transcription process, DNA is converted into Ribonucleic acid (RNA) and then RNA again back to DNA by reverse transcriptase enzyme, and the process is called retrotransposition. Reverse transcriptase, an enzyme that converts genetic information into RNA, is a component of the central genetic core of retrotransposons. That RNA is then reverse-transcribed back into DNA and re-integrated elsewhere in the nuclear genome.

Class I retrotransposon elements are further subdivided into two subclasses of Long Terminal Repeat (LTR) and non-Long Terminal Repeat (non-LTR) retro-transposable elements, based on the occurrence or absenteeism of long terminal repeats (LTRs). In the case of LTR retrotransposons, retrotranscription takes place in cytoplasmic virus-like particles and Double-Stranded DNA (dsDNA) after import to the nucleus, inserts back into the genome by an integrase enzyme. Where in contrast, transcription in non-LTR retrotransposons occurs at the genome target site by the mechanism of the target-primed reverse transcription (Bourque et al., 2018; Lanciano and Cristofari, 2020). The mobility of Class II DNA transposons depends on direct replication or excision of parental sequence from a donor site by transposase enzyme followed by reinsertion of DNA into another genome location known as the cut and paste mechanism (Fig. 6), except rolling-circle elements (e.g., Helitrons) which replicate through peel-and-paste mechanism (Bourque et al., 2018; Wells and Feschotte, 2020).

LTR retrotransposons are non-coding mobile genetic elements flanked by two identical 5' and 3' non-coding direct repeats of LTR nucleotides with long Open Reading Frames (ORFs) encoding the proteins required for transposition. The two LTRs flank a protein-coding reverse transcriptase or integrase set of core genes. These genes play roles in the replication and transposition of these genetic elements. LTR retrotransposons are found most abundant and widespread in eukaryotic genomes which replicate and move between locations within genomes, increase copy number, and spread across species by horizontal transfer, and are responsible for the diversification of genomes by insertion or deletion of genes. There are two main types of LTR retrotransposon superfamilies, viz., Ty1-copia and Ty3-gypsy (Neumann et al., 2019). Gypsy elements are the most abundant and are found in most plant species, while Copia elements are less abundant and primarily found in dicots (Stritt et al., 2020).

The non-LTR retrotransposons are self-replicating mobile elements found in all plants (10–20% or more of plant genomes) and animal genomes that lack the long terminal repeat sequences found in LTR retrotransposons. They are responsible for several genomic rearrangements and gene duplication or loss or regulation of gene expression, genome size and organization, and evolutionary processes. Non-LTR retrotransposons are distinct from LTR retrotransposons, a type of transposable element that lack LTRs, and are further subclassified into two groups based on their structure, the Long Interspersed Nuclear Elements (LINEs) and Short Interspersed Nuclear Elements (SINEs) (Zhang et al., 2014). LINEs are common in plant genomes and contain L1 and RTE (retro-transposable element) superfamilies with two or one ORFs respectively for retrotransposition (Wicker et al., 2007). LINEs contain 5'UTR and stretch of $(A)_n$ for L1 or $(GTT)_n$ for RTE at 3'-end for reverse transcription (Heitkam et al., 2014). Plant SINE elements are small, non-coding, non-autonomous, with high sequence diversity, derived from tRNAs, a few hundred base pairs in size, rely on LINE machinery for retrotransposition and have mostly A tail 37 pb (Angio-domain in Angiosperm) at the 3'-end (Wenke et al., 2011; Seibt et al., 2020).

Retrotransposons ubiquitous in the nuclear genome of all species (plants, animals, fungi, and protists), are an important component of gene expression, regulation, and evolution of plant genomes, increase the effective size of the genome, contribute to the genetic diversity and adaptation to the environmental changes. In plants, retrotransposons are some of the most abundant elements and can account for up to 80% of the genome. This chapter reviews the current state of research on plant retrotransposons, focusing on their classification, distribution, structure, and functions or potential applications.

2. Classification and Structure of Retrotransposons

Finnegan proposed the initial classification of TEs by based on RNA (class I elements or retrotransposons) or DNA (class II elements or DNA transposons) intermediate molecules and the basic nature of their "copy-and-paste" replicative and "cut-and-paste" conservative transposition mechanisms respectively (Finnegan 1989). At present, the most widely used and accepted method of nomenclature based on the transposition mechanism is a hierarchical classification (Fig. 1a) subdividing Class I and Class II TEs into classes, subclasses, orders, and superfamilies (Wicker et al., 2007) and lineages are added according to Rexdb and GyDB nomenclature (Orozco-Arias et al., 2019) (Fig. 1b). However, because of highly diverse TE structures and mechanisms of transposition, several glitches and disputes occur in classification systems (Piégu et al., 2015; Arkhipova, 2017) and there is a need for a combined classification system for TEs.

Retrotransposon transposable elements have genetic material that can move and copy itself within a genome. These are eukaryote genomes' main components and comprise a large proportion of the total DNA. Class I TEs/Retrotransposons are the most copious TE type in the genomes of plants and with the help of reverse

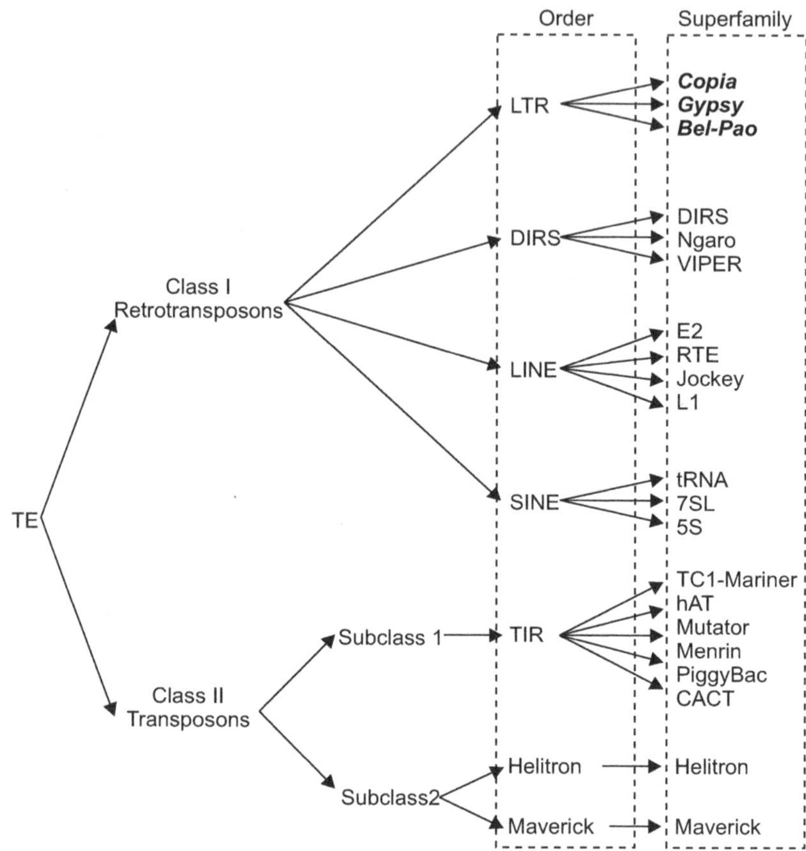

Fig. 1(a)   Classification of TE superfamilies according to Wicker's taxonomy
(*Source:* Wicker et al., 2007; Schietgat et al., 2018)

transcriptase enzymes copy and paste TEs in the genome. Class II TEs/DNA transposons use DNA polymerase as well as transposase enzymes to copy and insert TEs in the genome (Neumann et al., 2014). Retrotransposons are further divided into two subclasses Long Terminal Repeat (LTR) retrotransposons and non-Long Terminal Repeat (non-LTR) retrotransposons based on their structural features and mechanism of transposition. A detailed overview of the classification and transposition mechanism of TEs is deliberated in a chapter by Pagadala et al. (2023).

2.1   *Sub Class I retrotransposons*

Sub-class I retrotransposons also known as LTR retrotransposons (LTR-RTs) are transposable elements characterized by the presence of a pair of LTRs of around 100 to 500 base pair at the ends of their sequence and are known to transpose via "copy-and-paste" mechanism from one location in the genome to another. LTR-RTs

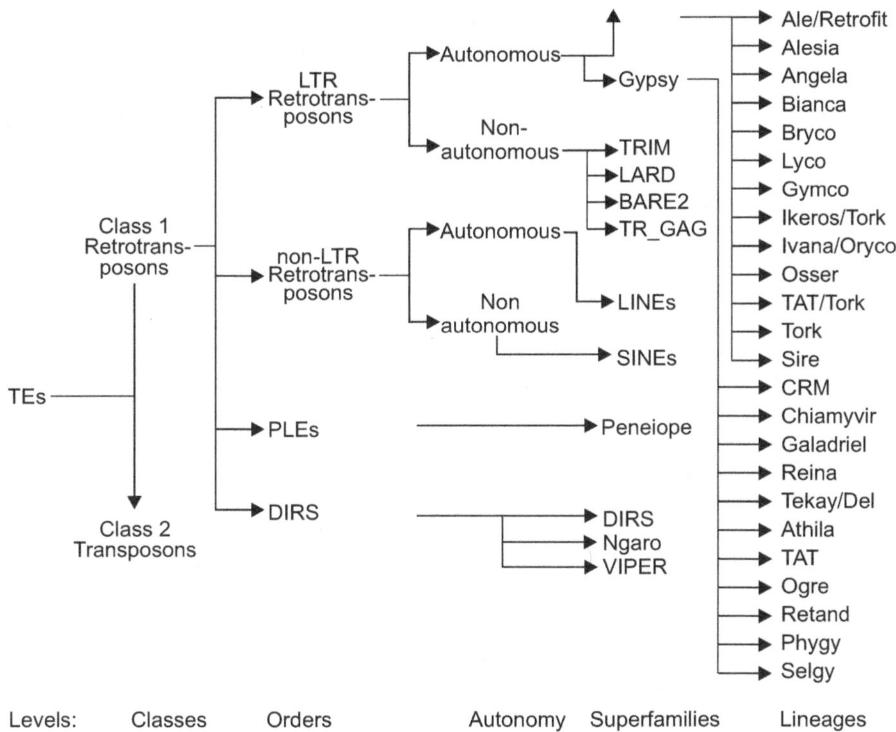

Fig. 1(b)   TE Classification according to Rexdb and GyDB nomenclature
(*Source:* Orozco-Arias et al., 2019)

replicate multiple times and generate a double-stranded DNA copy of TE. This copy then reverse-transcribes into RNA and re-inserts into the genome of a cell. LTR-RTs constitute up to 80% of genome size in *Triticum aestivum* and *Hordeum vulgare* and 40%–75% in *Zea mays* and *Saccharum officinarum* (Rahman et al., 2013; Negi et al., 2016).

LTR retrotransposons replicate and spread by a process called retro transposition. This process involves the reverse transcription of an RNA copy of retrotransposon into a DNA copy, and then the integration of DNA copy at a new site in the genome. This process can be facilitated by the presence of host proteins such as Integrase (IN) and Reverse-Transcriptase (RT), encoded by the retrotransposon. LTR retrotransposons can also spread between species through horizontal transfer. The LTR-retrotransposons have two non-coding homologous structures of a hundred to thousand bps called Long Terminal Repeats (LTRs) at both the termini of their internal coding regions (Mascagni et al., 2017a, b). The LTR retrotransposons have one or several ORFS for *Pol* gene (encodes integrase, protease, and reverse transcriptase proteins) and *GAG* gene (encode functional polyproteins) separated by stop codons and are integral for retrotransposal replication (Piednoël et al., 2013; Mascagni et al., 2015; Paz et al., 2017; Usai et al., 2017; Joly-Lopez and Bureau,

2018). LTR-retrotransposons are further subdivided into superfamilies based on their coding domain internal organization. These include Gypsy superfamily (position of Integrase (IN) after the Reverse-Transcriptase (RT) domain) and Copia superfamily (position of IN before RT domain) in plant genomes are further sub-divided into lineages and families on basis of parallels in coding region and general structures (Bennetzen et al., 2005; Wicker et al., 2007; Schnable et al., 2009; Llorens et al., 2009; Janicki et al., 2011; Gao et al., 2012a, b; Bennetzen and Wang, 2014). In plants, the Gypsy superfamily is sub-categorized into Athila, Reina, TAT, Del, Galadriel, and CRM and Copia is sub-categorized into SIRE, Oryco, Tork, Retrofit, and Bianca (Llorens et al., 2009, 2011). LTR-RTs are subdivided into superfamilies (Ty1-copia, Ty3-gypsy, and Bel-Pao) in plants based on structure and order/internal organization of the protein-coding domains (Schietgat et al., 2018). Ty1-copia and Ty3-gypsy are two key superfamilies differing in order of RT and IN domains in *Pol* ORF. The proteins encoded by Ty1-copia and Ty3-gypsy help move and integrate elements in the genome (Grandbastien 2008; Gao et al., 2012a, b). The LTR retrotransposons, Copia (Ale/Retrofit), Alesia, SIRE, Bryco, Osser, Angela, Lyco, Gymco, Bianca, Ivana/Oryco, Ikeros/Tork sto-4, Tar/Tork) clustering on the chromosomal proximal regions and Gypsy (Athila, Selgy, Clamyvir, Reina, Galadriel, Phygy, Tcn1, Tekay/Del, CRM/Centromeric Retrotransposon and TAT) with widespread and more diverse chromosomal locations in plant genomes are further sub-classed into lineages and families based on the phylogenetic analysis of RT domain (Wicker et al., 2007; Llorens et al., 2009, 2010; Schnable et al., 2009; Joly-Lopez and Bureau 2014; Orozco-Arias et al., 2019).

2.2 Sub-Class II retrotransposons

Sub-class II retrotransposons do not contain LTRs and are also termed as the non-LTR retrotransposons. Non-LTR retrotransposons show an absence of two Long Terminal Repeats (LTRs) or direct repeat sequences as found in the LTR retrotransposons. These elements use RNA intermediate sequence which after reverse-transcription into DNA inserts back into the host genome. The non-LTR retrotransposons are subdivided into SINES (Short Interspersed Nuclear Elements) and LINES (Long Interspersed Nuclear Elements) (Kapitonov and Jurka, 2008; Cordaux and Batzer, 2009). The proteins encoded by these elements are involved in the movement of the element and the formation of RNA intermediate. LINE elements carry coding sequences and SINEs are small, noncoding, few hundred base pairs in size and use LINE transposition machinery (Grandbastien, 2008). LINEs are the oldest, largest typically 6–7 kb in length, and most active non-LTR retrotransposons found in plants, having two ORFs that encode reverse transcriptase and endonuclease enzymes and are responsible for the majority of transposition events (Ma, 2013), although some suggest LINEs as highly regulated or inactive in plant genomes (Mao and Wang, 2017).

SINEs are small (80–500 bp or typically less than 1 kb in length), lack ORF for RT protein, non-autonomous elements, and are second most copious non-LTR retrotransposons with an average copy number of over 100,000 in plants (Ma, 2013). SINEs are believed to originate from the reverse transcription of tRNA or rRNA molecules and use the machinery of LINEs for replication and integration into the host genome (Ma, 2013). The SINEs shaped genomic variations of some monocot species and heterogeneity of genomes of many eukaryotic species, but besides this, less information is available about their activity in the genome of plants (Wenke et al., 2011; Sahebi et al., 2018).

2.3 *Retroviruses*

Retroviruses are characterized by the presence of an *ENV* (Envelope) gene encoding a structural protein (ENV) for envelope formation in viruses and other proteins and replicate by converting RNA intermediate into viral DNA before being inserted into the host chromosome (Gonda, 1999). ENV aids the virus in leaving the cell and infecting other cells. Retroviruses are further classified into subtypes simple retrovirus and complex retrovirus based on structure and proteins these encode for the movement of elements and formation of envelope. The simple retroviruses encode *GAG* (Group Antigens), *Pol* (reverse transcriptase), and *ENV* (envelope protein) polyproteins, while complex retroviruses encode six accessory proteins, together with polyproteins (Ryu 2017).

3. Structure of Retrotransposons

The LTR retrotransposons are subdivided into autonomous and nonautonomous elements based on the presence/encoding of domains essential for mobilization. The autonomous LTR-RTs possess two identical LTRs at terminal ends, Primer Binding Site (PBS), Poly-Purine Tract (PPT) domains, and *GAG* (structural RNA-binding capsid protein) and *Pol* (at 3′ end of *GAG*) genes at internal regions necessary for reverse transcription from RNA intermediate to complementary DNA (cDNA) strand on the minus (−) and plus (+) strands respectively (Havecker et al., 2004; Wicker et al., 2007; Eickbush and Jamburuthugoda, 2008; Bennetzen and Wang, 2014; Jedlicka et al., 2020). LTR-RTs possess promoters and terminators and short inverted repeats at ends (Wicker et al., 2007; Grandbastien 2008). *Pol* gene encodes reverse transcriptase (RT) and RNase H (RH) for the synthesis of transcript into cDNA, protease (PROT), and integrase (IN) that inserts the dsDNA daughter cDNA copy into the genome (Gao et al., 2003; Grandbastien, 2008). Few LTR retrotransposons possess an additional coding domain of ENV-like (Grandbastien, 2008) (Fig. 2).

3.1 *LTR-Retrotransposons*

In plants, LTR retrotransposons vary in size and range from 4 kb to over 23–31kb (*i.e.*, Ogre elements with >23 kb for functional and complete retroelements) (Gao et al., 2012a, 2016; Cossu et al., 2012; Bento et al., 2013; Kubat et al., 2014; Mascagni

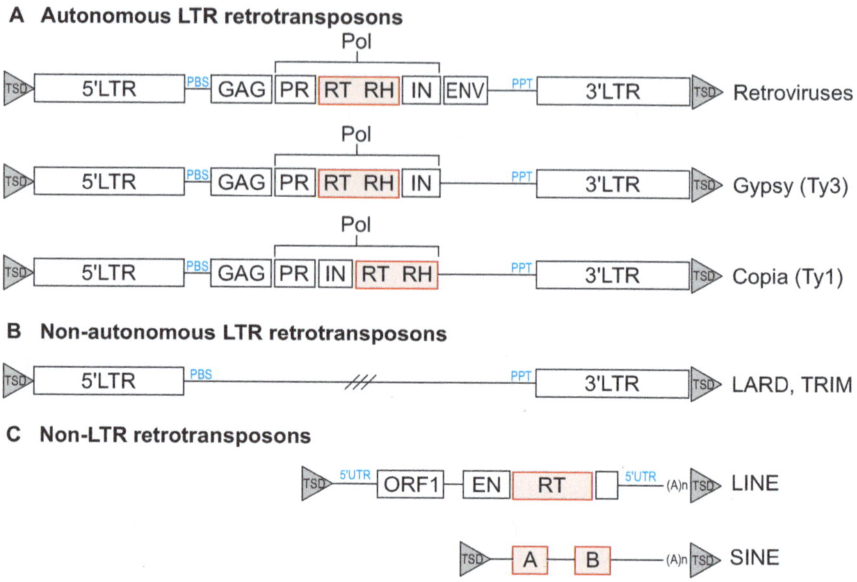

Fig. 2   Organization of different types of retrotransposons. (A) Autonomous LTR Retrotransposon (B) Non-autonomous LTR retrotransposon (C) Non-LTR retrotransposon (*Source:* Arvas et al., 2023)

et al., 2017a, b). A key characteristic of LTR retrotransposons is the existence of two homologous, non-coding, inverted repeats known as LTRs which range from a few hundred base pairs to >5kb in length and flank the ends of the central coding region (Du et al., 2018; Schietgat et al., 2018). The central protein-encoding domain of LTR-RTs shows the presence of one or more ORFs and code for *GAG* and *Pol* (polyprotein) genes with function in replication and retro-transposition and have one or more internal stop codons (Chang et al., 2013; Mascagni et al., 2017a, b; Joly-Lopez and Bureau, 2018). *GAG* domains are the most variable and encode structural protein analog of retroviral capsid or nucleocapsid for packaging of RNA and proteins (Eickbush and Jamburuthugoda, 2008; Ustyantsev et al., 2015; Zhao et al., 2016). *Pol* gene is located at the 3′ terminal end of *GAG* encoding some enzymatic domains such as PROT, RT, RNase H, and IN for the transposition machinery (Gao et al., 2003; Piednoël et al., 2013; Usai et al., 2017; Paz et al., 2017; Ustyantsev et al., 2017). All domains have specific roles in the replication cycle (Sanchez et al., 2017). Sometimes chromodomain is present which is positioned upstream of 3′ LTR and functions in targeted integration and escaping silencing by specifically targeting heterochromatic regions (Godinho et al., 2012; Novikov et al., 2012; Usai et al., 2017) (Fig. 2A).

LTRs are non-coding sequences, composed of U3, R, and U5 domains, possess start and stop signals for the retro transcription, polyadenylation and enhancers for replication, and evolve rapidly than other domains of retrotransposons (Casacuberta and Santiago 2003; Gao et al., 2012a, b; Alzohairy et al., 2014; Kriedt et al., 2014;

Giordani et al., 2016; Grover and Sharma 2017; Paz et al., 2017; Zhou et al., 2018). Because of acclimatization to variable tissue milieus and stress responses, R and U5 zones possess more conservation than U3 (Benachenhou et al., 2013; Casacuberta et al., 2013). The LTR of retrotransposons and retroviruses have analogous functions in the initiation of RNA reverse transcription (Eickbush and Jamburuthugoda 2008). RNA copy is reverse transcribed from R-to-R regions and possesses only single U5 and U3 zones and dual identical LTRs after the DNA replica is integrated back into the nuclear genome (Mascagni et al., 2018). Short-Inverted Repeat (SIR) which is a short-conserved motif of TG-5′ and 3′-CA starts and terminates LTRs (Yin et al., 2013; Mascagni et al., 2015), with some exceptions in the family Rosaceae (Yin et al., 2017). LTR possess one or two T/A-rich Goldberg–Hogness /TATA-boxes and a single AATAAA motif of polyadenylation signal and mostly AT-rich regions (Gao et al., 2012a, b; Benachenhou et al., 2013).

LTR-RTs possess primer binding sites (PBS) and Poly-Purine Tract (PPT) (Fig. 2A) as primers (Mascagni et al., 2018). The PBS is a minus (−)-strand priming site and the PPT is a plus (+)-strand priming site for the process of reverse transcription (Schulman et al., 2013; Monden et al., 2014; Mascagni et al., 2017a, b). After insertion at a site in the genome, LTR retrotransposon generates a Target

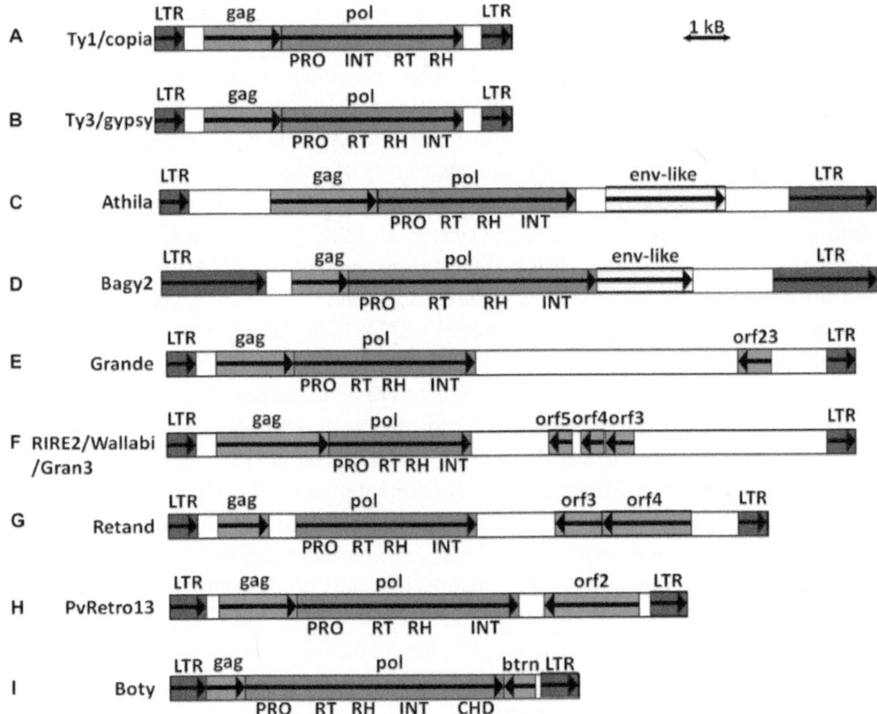

Fig. 3 Structure of LTR retrotransposons
(*Source:* Vicient and Casacuberta 2020)

Site Duplication (TSD) of 4 to 6 bps at the terminal ends, with the exception in Rosaceae (Gao et al., 2012a, b; Cossu et al., 2012; Gao et al., 2014; Roy et al., 2015; Yin et al., 2017).

LTR retrotransposons have non-autonomous forms as well (Casacuberta and Santiago 2003) known as the non-autonomous LTR retrotransposons (Fig. 2B) with deficiency of functional domains necessary for retrotransposition events and depend on the machinery of other LTR-RTs (Gao et al., 2012a, b; Yin et al., 2014; Chaparro et al., 2015; Mascagni et al., 2015; Kalendar et al., 2020). Based on structure, these are classified into very small-sized (from a few 100 bases to 4 kbp) Terminal-Repeat Retrotransposons in Miniature (TRIMs) (Witte et al., 2001; Cossu et al., 2012; Sampath and Yang, 2014; Gao et al., 2016; Kalendar et al., 2020), LARDs (of length greater than 4 kb) (Kalendar et al., 2004; Kalendar et al., 2020), TR_GAG (Chaparro et al., 2015), and BARE-2.

3.2 Non-LTR Retrotransposons

The less abundant non-LTR retrotransposons don't possess LTRs, transcribe from an internal promoter, replicate without INT domain and the RT domain starts replication of DNA from the poly-A tail of non-LTR-RT transcripts and ligates the end of new DNA copy at insertion site (Schulman et al., 2012). Based on their structure, non-LTR-RTs are sub-divided into LINEs and SINEs. LINEs and SINEs possess promoter elements and polyadenylation signals, 3′ poly(A) tail, and move within the genome by retro-transposition (Jiang 2013; Grover and Sharma, 2017). LINEs are the most copious type of non-LTR-RTs, typically 6–8 kb in length. Similar to LTR retrotransposons, LINEs possess two ORFs flanked by the Untranslated Regions (UTRs) and encode *GAG*, RT, and endonuclease (EN) (Casacuberta and Santiago 2003; Huang et al., 2012; Arvas et al., 2023). SINEs are smaller in size than LINEs, typically range from 100–500 bp, comprise tRNA-derived regions, and unrelated DNA sequences, and have only one ORF which encodes reverse transcriptase (Schulman 2013; Arvas et al., 2023) (Fig. 2C). SINEs lack the ability for self-replication and depend on LINEs (Huang et al., 2012; Schulman, 2013).

The phylogenetic trees from concatenated reverse transcriptase (RT), RNase H (RH) and integrase (IN) domain from Viridiplantae and non-Viridiplantae (Fig. 4a) show diverse locations of RH domain in Tat polyprotein (Fig. 4b) The collapsed rectangular phylogram inferred from the concatenated RT, RH and IN domains from the Viridiplantae were consistent with trees inferred from individual domains with exception of chromovirus from *Selaginella moellendorffii* (Fig. 4c) The Ty3/gypsy in plants was grouped into chromovirus and non-chromovirus (Neumann et al., 2019).

The Ty1-copia retrotransposons phylogenetic trees were calculated through the use of maximum-likelihood from alignments of the sequences of proteins of concatenated reverse transcriptase (RT), RNase H (RH), and integrase (IN) (Fig. 5a), IN (Fig. 5b), reverse transcriptase (Fig. 5c) and RNase H (Fig. 5d). The radial phylograms were generated from datasets of both the Viridiplantae and

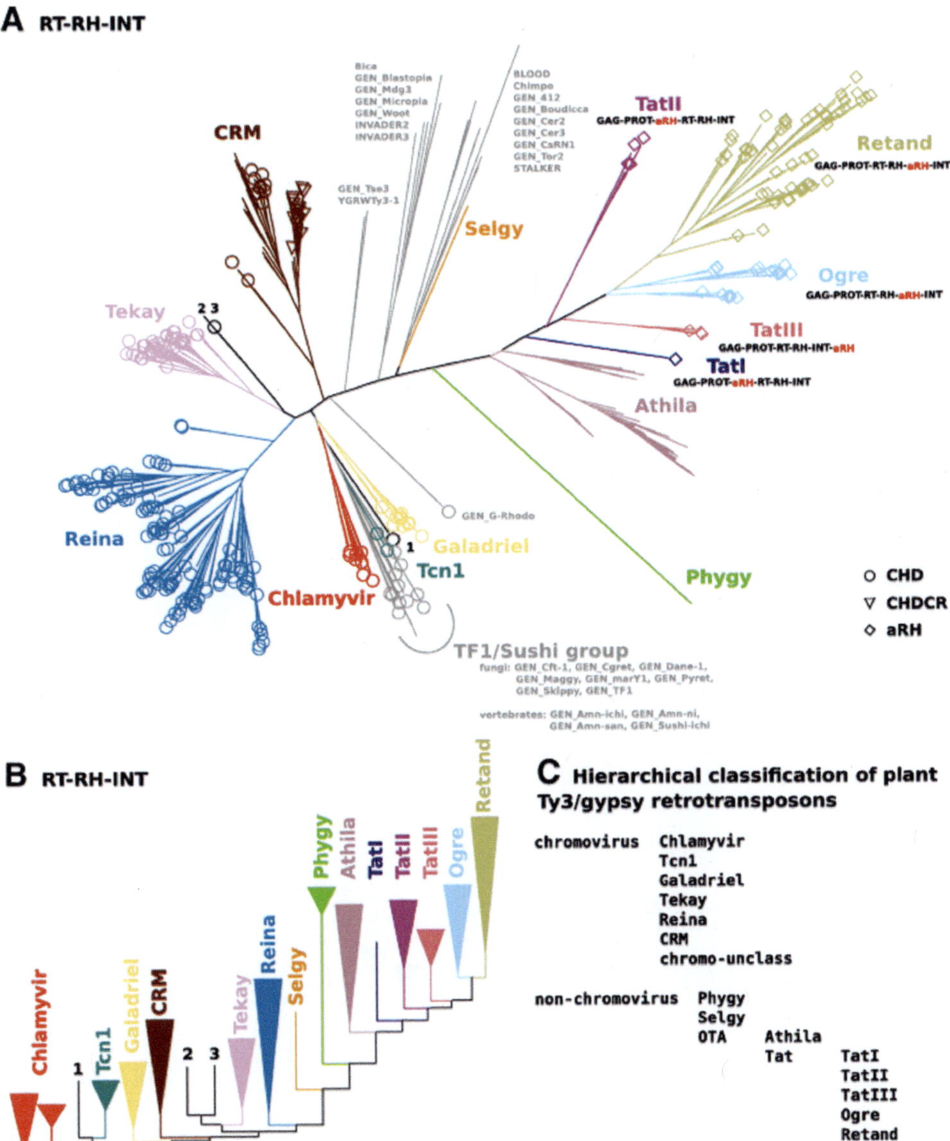

Fig. 4   Phylogenetic trees and Ty3-gypsy retrotransposon classification
(*Source:* Neumann et al., 2019)

non-Viridiplantae. The collapsed rectangular phylograms were inferred from dataset
sequences from Viridiplantae species (Neumann et al., 2019).

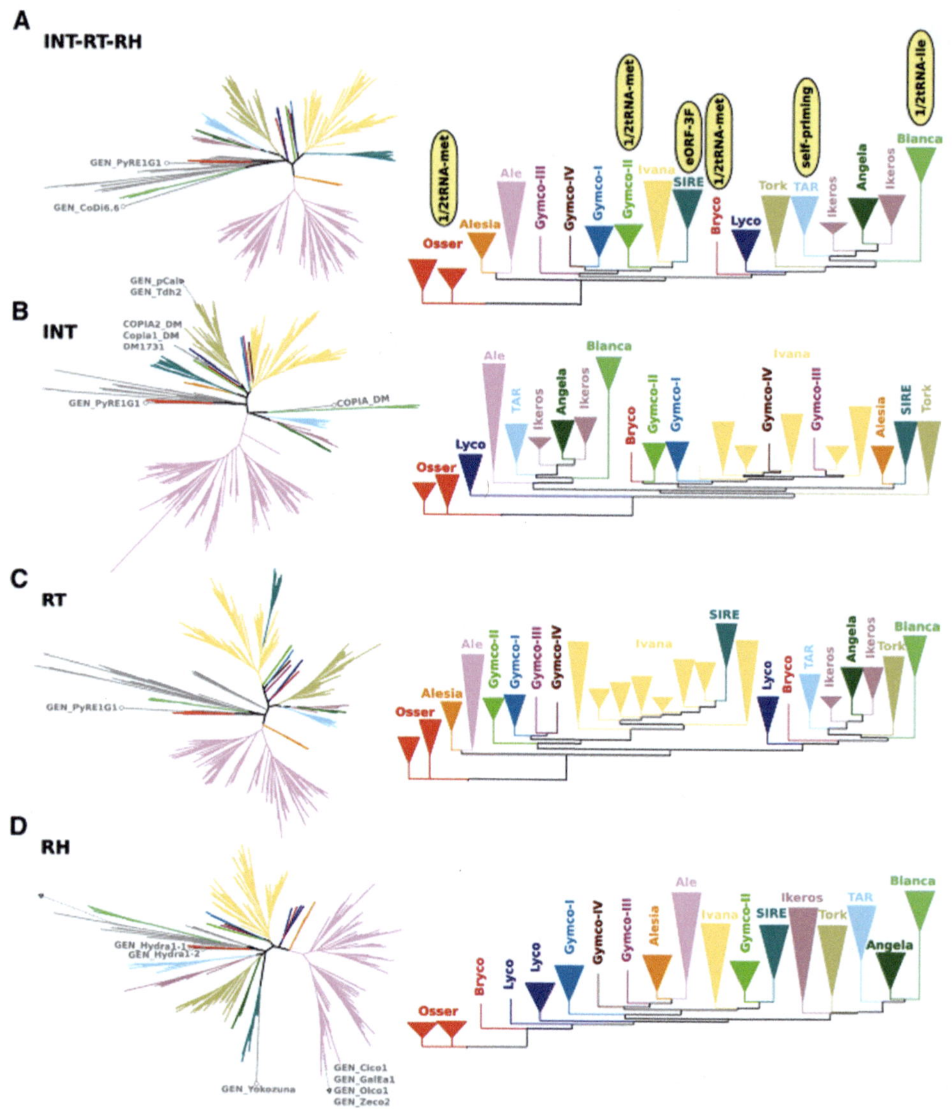

Fig. 5   Phylogenetic trees of Ty1-Copia retrotransposons
(*Source:* Neumann et al., 2019)

4.   Distribution of Retrotransposons in Plants

Retrotransposons have become widely distributed in plants, animals, fungi, and protists but not in prokaryotes and play important roles in genome evolution by introducing new genetic variations and influencing gene expression (Kidwell and

Lisch, 2001; Finnegan and Kidwell, 2008). Examples include retroviruses, LINEs and SINEs, Ty1-copia and Ty3-gypsy retro-elements, and P and hAT elements (Finnegan and Kidwell, 2008). LTR retrotransposons are most widespread in plants and LINEs in mammalian genomes (Kazazian 2000; Bennetzen and Wang, 2014). The distribution of retrotransposons in different species has been studied in detail. A direct correlation exists between the eukaryotic genome size and retrotransposon abundance but not necessarily retrotransposon type. For instance, LTR retrotransposons make up around 3% of the small *Saccharomyces cerevisiae* (yeast) genome, 37% of the larger human genome mainly LINEs and SINEs, and 75% to 90% of the maize genome (SanMiguel, 1996; Eickbush, 2001). LINEs are found in most mammalian species except platypus and SINEs are abundant in primates. P-elements are abundant in *Drosophila melanogaster* and have been found in other species of insects as well (DeBry et al., 2014).

TEs especially retrotransposons are found in most plants and determine genome size in plants besides, increases in genome-size and TE content correlate mainly in cereal crops, such as maize and sugarcane. Retrotransposons constitute 6.9% of the genome in *Arabidopsis thaliana*, 19% in *Oryza sativa* (rice), 54.52% in *Sorghum bicolor* (sorghum), 61.8% in *Solanum lycopersicum* (tomato), 75% in *Zea mays* (maize) and 40.86% in *Saccharum officinarum* (sugarcane) plants (Negi et al., 2016) (Fig. 6). In Arabidopsis, TEs form the large repertoire of 21% (~32000, mostly degenerate TE copies of 326 families) of a compact genome of 125 Mb long (TAIR 10) (Ahmed et al., 2011; Joly-Lopez and Bureau, 2014; Feng and Michaels, 2015). LTR retrotransposons increase with genome size from *Oryza sativa* (rice) which is the smallest sequenced grass genome (≈14% of its 430-Mb genome consists of LTR-RTs) (Jiang and Wessler, 2001), through *Zea mays* (maize) (≈2,500 Mb, 50–60% RTs) (SanMiguel and Bennetzen, 1998; Meyers et al., 2001), to *Hordeum vulgare* (barley) (≈4,800 Mb, >70% retrotransposons) (Vicient et al., 1999). The maize genome has 85% of retrotransposons, predominantly LTR retrotransposons (Ma, 2013). In *the Glycine max* (soybean) genome, retrotransposons have been found to account for 42.24% (Schmutz et al., 2010) from 510 families; *gypsy*-like (69%) and the remainder percentage by *copia*-like elements. According to Yi et al. (2022), 631,056 LTR elements constitute 39.95% of the whole genome in the soybean cultivar Jidou 17 (JD17). These studies demonstrate the widespread distribution of retrotransposons in plants and suggest their major role in the evolution of plant genomes.

5. Reverse Transcription and Mobilization of Retro Transposons

TEs can be distinguished into dual classes based on the transposition mechanisms, *i.e.*, active eukaryotic Class I (retrotransposons/copy and paste transposons) and active eukaryotic Class II (DNA transposons/cut and paste transposons). Further, retrotransposons can be autonomous and non-autonomous (Fig. 7a, b) and require RNA transcription to move to varied locations inside the cell genome. With the

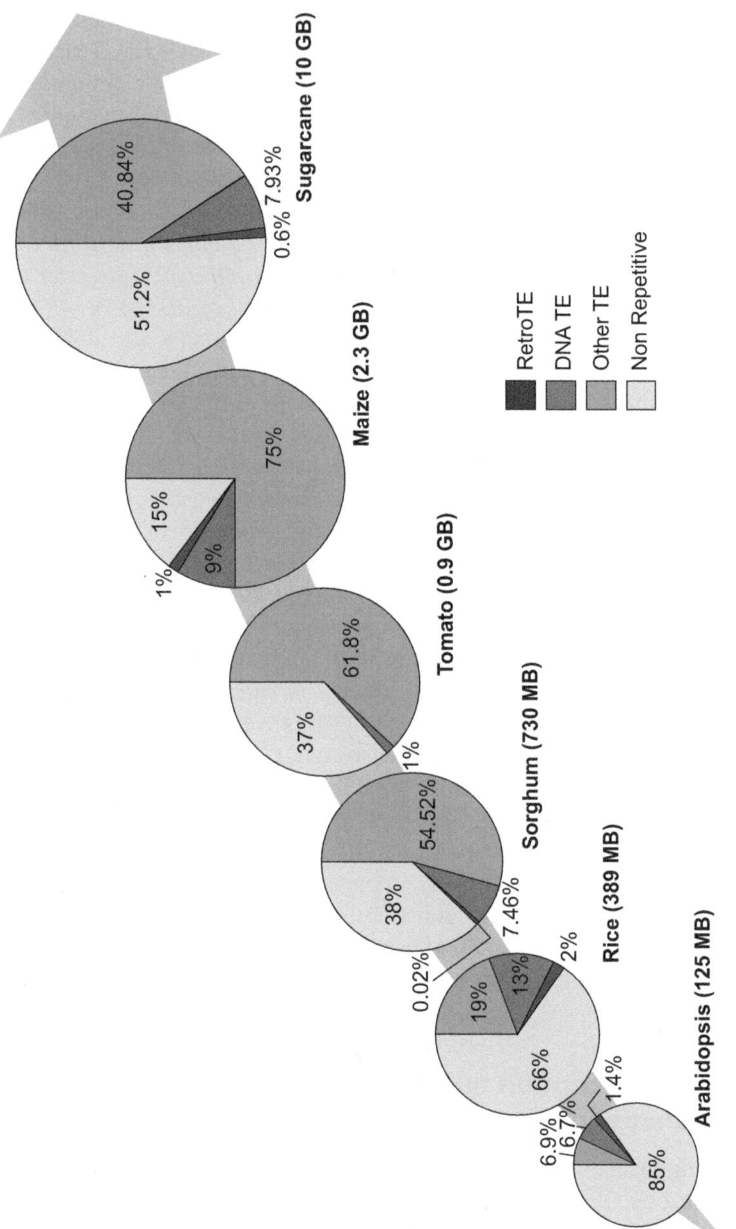

Fig. 6   TE content in various Plant Genomes
(*Source:* Negi et al., 2016)

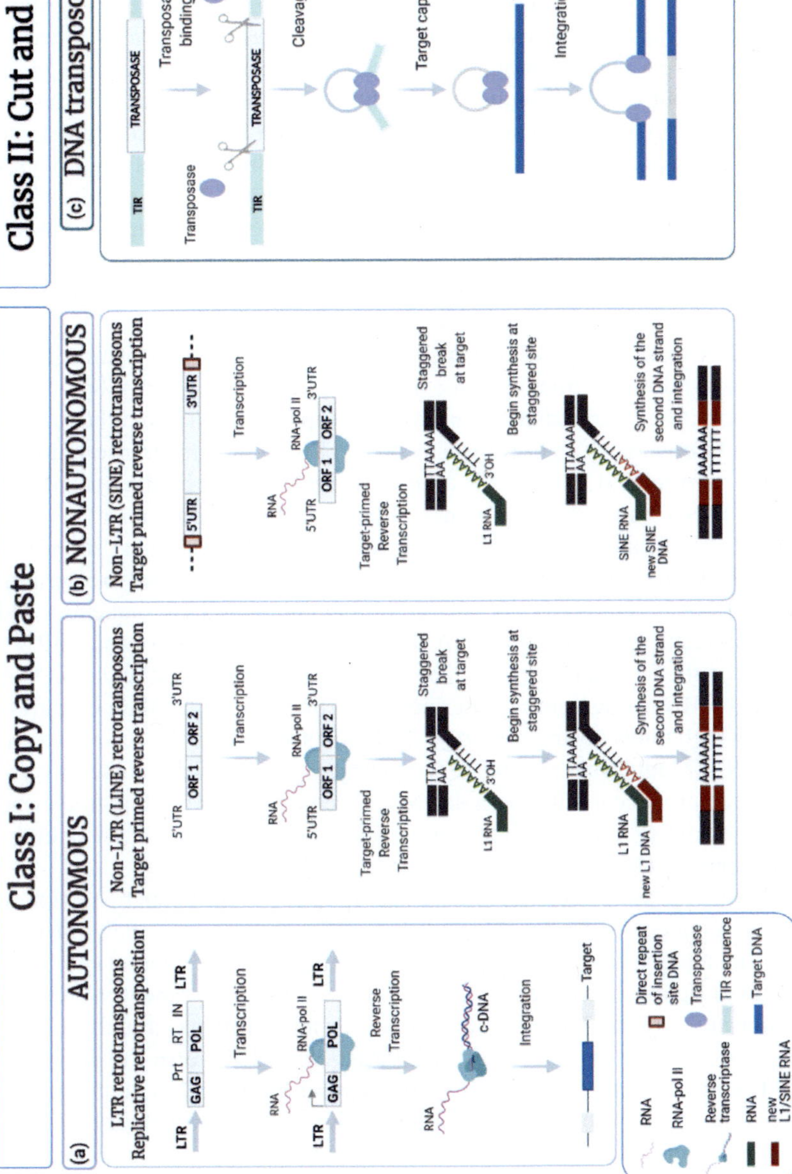

Fig. 7 Schematic Representation of Different Transposition Mechanisms
(*Source*: Colonna Romano and Fanti, 2022)

use of reverse transcriptase enzyme, retrotransposons produce cDNA sequence from the transcript (template) by "copy and paste" mechanism which randomly reinserts at a new genomic location. Autonomous RNA transposons encode all proteins necessary for the movement within the genome. Examples are Long Terminal Repeats/Endogenous Retroviruses (LTR/ERV; e.g., *Ty* element of yeast) and non-LTR LINEs. LTR-retrotransposons contain two LTRs and *GAG* (group-specific antigen), *Pol* (reverse transcriptase), IN (integrase), and *PRT* (protease) genes for producing functional proteins. The non-LTR retrotransposons have genes for transposition but lack LTR sequences and have two ORFs flanked by 5′ and 3′ untranslated regions (UTR). The non-LTR retrotransposons mobilize the Target-site Primed Reverse Transcription (TPRT) mechanism. Hydrolysis of a single DNA strand takes place at the new insertion location and then the 3′ end of this strand primes reverse transcription of new LINE cDNA by the reverse transcriptase enzyme encoded by this retro-element. Then hydrolysis of the second DNA strand releases the 3′ terminal end that primes replication of the second strand of cDNA of LINE. Then IN enzyme completes the insertion (Fig. 7a).

All LTR retrotransposons have analogous mechanisms of transposition. DNA strands are biosynthesized from opposite directions. Reverse transcription of their mRNA transposition intermediate initiates at left (upstream) LTR and the reverse transcription terminates at right (downstream) LTR. The reverse transcription is primed by 3′ termini/end of left LTR after host primer tRNA binding to mRNA sequence at tRNA base-paired sequence (PBS) nearly downstream of 5′ LTR (Boeke et al., 1985) and leads to the synthesis of first strand (minus strand) of cDNA (Matthews et al., 1997). After reverse transcriptase has extended to 5′ termini of mRNA, cDNA is relocated to 3′ termini of a template by LTR sequences and generates full-length (minus) strand DNA. The tRNA primer is removed and mRNA is destroyed by the RNase H enzyme. The priming of the second strand of cDNA synthesis is by RNA fragment at a purine-rich primer known as the polypurine tract (PPT) which binds upstream of right LTR and synthesizes the second strand (plus) of cDNA (Matthews et al., 1997). Template jump on template enables reverse transcriptase to produce a full-length dsDNA. IN enzyme after binding to the end of linear DNA, directs integration at a target site on chromosomal DNA as in the case of transposons (Ma 2013; Matthews et al., 1997) (Fig. 7a).

LINEs either possess internal promoters for initiation of upstream transcription at 3′ end of element or transcribe with target location from peripheral host promoter. The integration starts with EN cleaving (nicking) of the target site. Then a reverse transcriptase enzyme uses free 3′ termini of cDNA to prime RNA reverse transcription at 3′ termini. The polymerization of an initial (minus) strand of cDNA directly onto the insertion site is termed Target-Primed Reverse Transcription (TPRT). The mechanism of synthesis of the second (plus) cDNA strand and 5′ end element attachment to the upstream target site is unknown (Ma 2013) (Fig. 7a). The non-autonomous retrotransposons depend on the "true" (autonomous) retrotransposons for movement of sequences within various sites in the genome. SINE elements possess internal promoter for RNA polymerase III which is flanked by 5′ and 3′ UTR but lack transposition genes. SINEs use the TPRT mechanism for transposition but use LINEs for the insertion process (Fig. 7b).

Class II (DNA transposons) also known as cut and paste transposons encode transposase (TPase) enzyme flanked by the Terminal Inverted Repeats (TIRs). The TPases remove and insert TEs at the new genomic sites by two different types of machinery of "cut and paste" or "non-replicative pathway". In the first mechanism, the TEs excised from their locations and reinserted at different locations. In the "replicative pathway", TE is copied and translocated leaving the original copy in-situ (Fig. 7c) (Colonna Romano and Fanti, 2022).

6.   Functions of Retrotransposons

Previously known as JUNK, TEs are gold mines for the researchers working on crop genomics, genetic and evolutionary studies, and are proving "Just Unrevealed New-fangled Know-how" of tomorrow hidden in the genome of various organisms (Pandita and Pandita 2016; Hassan et al., 2024). The retrotransposons are major drivers and proliferators of genomic diversity, remain dormant during host development, and reactivate by mutations, expression of adjacent genes, and/ or environmental stressors turn on the survival genes and are considered as genomic gold (Hirochika et al., 2000; Grandbastien 2015; Hassan et al., 2024). Retrotransposons play pivotal roles in various biotic and abiotic stresses in plants. In *Phaeodactylum tricornutum*, activation of LTR-retrotransposons declines nitrate stress on exposure to reactive aldehydes (Maumus et al., 2009). Activation of athila LTR retrotransposons increases the number of disease-resilient gene families in *Capsicum baccatum* (Kim et al., 2017). In seedlings of *Arabidopsis thaliana* (Thale Cress) and *Vigna angularis* (Willd.) Ohwi and Ohashi) (Adzuki bean), treatment with demethylation agent zebularine increases activity and accumulation of conserved Ty1/copia-like retrotransposon known as ONSEN (Boonjing et al., 2020). In *Craterostigma plantagineum* which is a resurrection plant, dehydration induces activation of *Craterostigma Desiccation-Tolerant* (*CDT-1*) dehydration-related ABA-inducible retroelement leading to tolerance to the desiccation stress in callus (Hilbricht et al., 2008). In transgenic *Arabidopsis thaliana* plants, the *Osmotic and Alkaline Resistance 1* (*OAR1*) gene transformed from Clone L1-4 of *Boea hygrometrica* regulate osmotic and alkalinity stress resistance (Zhao et al., 2014) by improving the photochemical efficiency, integrity of cell membranes and gene expression of biomarkers. The retrotransposons have a propensity to initiate mutagenesis, affect the expression of adjacent genes, and regulate centromere structure as in maize plants (Defraia and Slotkin, 2014; Gao et al., 2015). In tomato plants, variances in volatile esters between two colored fruits are linked to retrotransposon location near the esterase family exhibiting high enzyme activity levels (Goulet et al., 2012). LTR retrotransposons are imperative for genetic variation, but can also be detrimental if un-regulated. Hence, it is vital to recognize the machinery of retroelements. The PHRE1 and PHRE2 LTR retrotransposons increase in transposition events and copy number in wild type moso bamboo

(*Phyllostachys edulis*) and *Arabidopsis thaliana* transgenics in response to high temperature stress as 5′ LTR plays the role of a promoter and regulates LTR-RTs (Papolu et al., 2021).

Non-LTR retrotransposons play a range of functions in plants. These are known to be involved in the regulation of gene expression and modulate the expression patterns of genes involved in growth and development, stress responses, flowering, gene duplication and genome evolution, genetic mutations, gene loss by insertion into genes, and disruption of their function and find use as molecular markers for genetic analysis and plant evolution. Non-LTR retrotransposons can be used to introduce transgenes into plants to manipulate traits of drought, pests, and yield of crops and to study gene regulation and identification of novel genes and gene pathways.

6.1   Role of Retrotransposons in Gene Regulation

LTR retrotransposons affect gene expression by acting as regulatory elements, promoters, enhancers, or silencers of certain genes. LTR retrotransposons have been found to have an important role in plant gene regulation (Grandbastien, 2015; Papolu et al., 2021). For instance, LTR retrotransposons can interact with transcription factors to modulate gene expression. An LTR-RT in the promoter region of a PsMYB10.2 gene promotes the regulation of anthocyanin accumulation/ biosynthesis trait for the flesh color of fruit in *Prunus salicina* (Japanese plum) (Fiol et al., 2022). The transposition of A *copia*-like retrotransposons takes place in the promoter of *Ruby* which is an MYB transcription factor after induction by cold stress. This induced MYB transcription factor is responsible for the anthocyanin production. It induces expression in the flesh of *Citrus sinensis* (Sicilian blood oranges) via *cis*-elements in the LTR (Butelli et al., 2012). In *Daucus carota* (carrots), high expression of *DcMYB7* gene of quantitative trait locus located on chromosome 3 in the purple root cultivar known as the Deep Purple contribute to anthocyanin pigmentation in the roots. *DcMYB8* generated nonfunctional tandem duplication and two transposon insertions in *DcMYB7* promoter initiate *DcMYB7* transcriptional inactivation in the roots of the non-purple carrot Kurodagosun (Xu et al., 2019).

TEs find use in insertional mutagenesis and transposon tagging. TEs in genes responsible for the anthocyanin regulation or biosynthesis contribute to appealing diversity of colors and phenotypes of flowers in *Nicotiana tabacum*, species of Ipomoea and Petunia plants (Pandita and Pandita 2023).

6.1   Role of Retrotransposons in Genome Evolution

LTR retrotransposons are mutagenic and play vital roles in the evolution of various plant genomes as revealed by genome-wide analysis (Akakpo et al., 2020; Mascagni et al., 2020; Ouyang et al., 2021; Papolu et al., 2022). The insertion of LTR-RTs into protein-coding genes can cause the production of new proteins with novel functions (Shapiro, 2014; Li et al., 2017). In addition, the insertion of LTR-RTs into non-coding regions of the genome causes fluctuations in gene expression and lead to the evolution of new traits. In *Capsicum annuum* (pepper), expression of *CaAN2* encoding an R2R3 MYB transcription factor responsible for anthocyanin biosynthesis is activated by the insertion of 4.2kb non-LTR-RT named Ca-nLTR-A in purple *C. annuum* cultivar (KC00134), which contains transcription-factor-binding sites for anthocyanin regulators at the 3' untranslated region (UTR) (Jung et al., 2019). In Vitis species, LTR-RTs contribute to the diversification of the genome (Park et al., 2021).

Conclusion

Retrotransposons are mobile genetic elements located in the plant genomes. These have been classified into LTR- and non-LTR retrotransposons. This chapter has reviewed the current state of the art of plant retrotransposons, and their structure, classification, and functions have been deliberated comprehensively. Previously known as JUNK, TEs are gold mines for the researchers working on crop genomics, genetic, and evolutionary studies and are proving as "Just Unrevealed New-fangled Know-how" of tomorrow hidden in the genome of various organisms.

References

Ahmed, I., Sarazin, A., Bowler, C., Colot, V., and Quesneville, H. (2011). Genome-wide evidence for local DNA methylation spreading from small RNA-targeted sequences in Arabidopsis. Nucleic Acids Res. 39:6919–6931.

Akakpo, R., Carpentier, M.C., Ie Hsing, Y., and Panaud, O. (2020). The impact of transposable elements on the structure, evolution and function of the rice genome. New Phytol. 226: 44–49. https://doi. org/ 10.1111/nph.16356

Alzohairy, A.M., Sabir, J.S.M., Gyulai, G., Younis, R.A.A., Jansen, R.K., and Bahieldin, A. (2014). Environmental stress activation of plant long-terminal repeat retrotransposons. *Funct. Plant Biol.* 41: 557–67.

Arkhipova, I.R. (2017). Using bioinformatic and phylogenetic approaches to classify transposable elements and understand their complex evolutionary histories. Mob. DNA 8: 19.

Arvas, Y.E., Marakli, S., Kaya, Y., and Kalendar, R. (2023). The power of retrotransposons in high-throughput genotyping and sequencing. *Front. Plant Sci.* 14:1174339. https://doi.org/10.3389/fpls.2023.1174339.

Benachenhou, F., Sperber, G.O., Bongcam-Rudloff, E., Andersson, G., Boeke, J.D., and Blomberg, J. (2013). Conserved structure and inferred evolutionary history of long terminal repeats (LTRs). Mob. DNA 4: 5.

Bennetzen, J.L., Ma, J., and Devos, K.M. (2005). Mechanisms of recent genome size variation in flowering plants. *Ann Bot.* 95(1):127– 32.

Bennetzen, J.L., Wang, H. (2014). The contributions of transposable elements to the structure, function, and evolution of plant genomes. *Annu. Rev. Plant. Biol.* 65: 505–30. https://doi.org/10.1146/annurev-arplant-050213-035811.

Bento, M., Tomás, D., Viegas, W., and Silva, M. (2013). Retrotransposons represent the most labile fraction for genomic rearrangements in polyploid plant species. *Cytogenet. Genome Res.* 140: 286–94.

Boeke, J.D., Garfinkel, D.J., Styles, C.A., and Fink, G.R. (1985). *Ty* elements transpose through an RNA intermediate. Cell. 40:491–500. doi: 10.1016/0092-8674(85)90197-7.

Boonjing, P., Masuta, Y., Nozawa, K., Kato. A., and Ito, H. (2020). The effect of zebularine on the heat-activated retrotransposon ONSEN in Arabidopsis thaliana and Vigna angularis. *Genes Genet Syst.* 95(4):165– 72.

Bourque, G., Burns, K.H., Gehring, M., Gorbunova, V., Seluanov, A., Hammell, M., et al. (2018). Ten things you should know about transposable elements. *Genome Biol.* 19:199. https://doi. org/10.1186/s13059-018-1577-z.

Butelli E., Licciardello C., Zhang Y., Liu J., Mackay S., Bailey P., et al. (2012). Retrotransposons control fruit-specific, cold-dependent accumulation of anthocyanins in blood oranges. Plant Cell 24: 1242–55. https://doi.org/10.1105/tpc.111.095232.

Casacuberta, J.M., Santiago, N. (2003). Plant LTR-retrotransposons and MITEs: control of transposition and impact on the evolution of plant genes and genomes. Gene. 311:1-11. https://doi.org/10.1016/ s0378-1119(03)00557-2.

Casacuberta, E., and González, J. (2013). The impact of transposable elements in environmental adaptation. Mol. Ecol. 22: 1503–17.

Chang, W., Jääskeläinen, M., Li, S., and Schulman, A.H. (2013). BARE Retrotransposons are translated and replicated via distinct RNA pools. PLoS ONE 8: e72270.

Chaparro, C., Gayraud, T., de Souza, R.F., Domingues, D.S., Akaffou, S., Laforga Vanzela, A.L., et al. (2015). Terminal-repeat retrotransposons with GAG domain in plant genomes: A new testimony on the complex world of transposable elements. Genome Biol. Evol. 7: 493–504.

Colonna Romano, N., and Fanti, L. (2022). Transposable Elements: Major Players in Shaping Genomic and Evolutionary Patterns. Cells. 11(6):1048. https://doi.org/10.3390/cells11061048.

Cordaux, R., and Batzer, M.A. (2009). The impact of retrotransposons on human genome evolution. Nat Rev Genet. 10(10): 691–703.

Cossu, R.M., Buti, M., Giordani, T., Natali, L., and Cavallini, A. (2012). A computational study of the dynamics of LTR retrotransposons in the Populus Trichocarpa genome. TREE Genet. Genomes 8: 61–75.

DeBry, R.W., Sharma, K., and Feschotte, C. (2014). Retrotransposons and their roles in genome evolution. Mol. Biol. Evol. 31(6): 1557–76.

Defraia, C., and Slotkin, R.K. (2014). Analysis of retrotransposon activity in plants. Methods Mol Biol. 1112:195–210.

Döring H.P., Tillmann E., and Starlinger P. (1984). DNA Sequence of the Maize Transposable Element Dissociation. *Nature.* 307: 127–130. doi: 10.1038/307127a0.

Du, D., Du, X., Mattia, M.R., Wang, Y., Yu, Q., Huang, M., Yu, Y., Grosser, J.W., and Gmitter, F.G., Jr. (2018). LTR retrotransposons from the citrus X clementina genome: Characterization and application. Tree Genet. Genomes 14: 43.

Eickbush, T.H. (2001). Retrotransposons. In (Ed(s): Sydney Brenner, Jefferey H. Miller) Encyclopedia of Genetics, Academic Press, pp 1699–1701. https://doi.org/10.1006/rwgn.2001.1111.

Eickbush, T.H., Jamburuthugoda, V.K. (2008). The diversity of retrotransposons and the properties of their reverse transcriptases. Virus Res. 134: 221–34.

Feng, W., and Michaels, S.D. (2015). Accessing the inaccessible: the organization, transcription, replication, and repair of heterochromatin in plants. Annu. Rev. Genet. 49: 439–59.

Finnegan, D.J. (1989). Eukaryotic Transposable Elements and Genome Evolution. Trends Genet. 5: 103–107. doi: 10.1016/0168-9525(89)90039-5.

Finnegan, D.J., and Kidwell, M.G. (2008). Retrotransposon dynamics in plant genomes. Mobile DNA. 1(1): 8.

Fiol, A., García, S., Dujak, C., Pacheco, I., Infante, R., and Aranzana, M.J. (2022). An LTR retrotransposon in the promoter of a *PsMYB10.2* gene associated with the regulation of fruit flesh color in Japanese plum. Hortic Res. 9. https://doi.org/10.1093/hr/uhac206.

Gao, X., Havecker, E.R., Baranov, P.V., Atkins, J.F., and Voytas, D.F. (2003). Translational recoding signals between gag and pol in diverse LTR retrotransposons. *RNA* 9: 1422–30. doi: 10.1261/ rna.5105503.

Gao, D., Jiang, N., Wing, R.A., Jiang, J., and Jackson, S.A. (2015). Transposons play an important role in the evolution and diversification of centromeres among closely related species. Front Plant Sci. 6:216.

Gao, L., McCarthy, E.M., Ganko, E.W., and McDonald, J.F. (2004). Evolutionary history of Oryza sativa LTR retrotransposons: a preliminary survey of the rice genome sequences. BMC Genomics. 5(1):18.

Gao, D., Chen, J., Chen, M., Meyers, B.C., and Jackson, S. (2012a). A highly conserved, small LTR retrotransposon that preferentially targets genes in grass genomes. PLoS ONE 7:e32010. https://doi.org/10.1371/journal.pone.0032010.

Gao, D., Jimenez-Lopez, J.C., Iwata, A., Gill, N., Jackson, S.A. (2012b). Functional and structural divergence of an unusual LTR retrotransposon family in plants. PLoS ONE 7, e48595.

Gao, D., Abernathy, B., Rohksar, D., Schmutz, J., Jackson, S.A. (2014). Annotation and sequence diversity of transposable elements in common bean (Phaseolus Vulgaris). Front. Plant Sci. 5: 339.

Gao, D., Li, Y., Do Kim, K., Abernathy, B., Jackson, S.A. (2016). Landscape and evolutionary dynamics of terminal repeat retrotransposons in miniature in plant genomes. Genome Biol. 17: 7.

Giordani, T., Cossu, R.M., Mascagni, F., Marroni, F., Morgante, M., Cavallini, A., and Natali, L. (2016). Genome-wide analysis of LTR-retrotransposon expression in leaves of populus X Canadensis water-deprived plants. Tree Genet. Genomes 12: 75.

Godinho, S., Paulo, O.S., Morais-Cecilio, L., and Rocheta, M. (2012). A new gypsy-like retroelement family in Vitis Vinifera. VITIS 51: 65–72.

Gonda, M.A. (1999). Bovine Immunodeficiency Virus (Retroviridae). In eds. (Granoff, A. and Webster, R. G.). Encyclopedia of Virology (Second Edition), Elsevier, p 184–90. https://doi.org/10.1006/rwvi.1999.0037.

Goulet, C., Mageroy, M.H., Lam, N.B., Floystad, A., Tieman, D.M., and Klee, H.J. (2012). Role of an esterase in flavor volatile variation within the tomato clade. Proc Natl Acad Sci USA. 109(46):19009–14.

Grandbastien, M.A. (2008). Retrotransposons of Plants. Elsevier. 428–36.

Grandbastien, M. A. (2015). LTR-retrotransposons, handy hitchhikers of plant regulation and stress response. Biochim. Biophys. Acta 849: 403–416. https://doi.org/10.1016/j.bbagrm.2014.07.017.

Grover, A., and Sharma, P.C. (2017). Repetitive sequences in the potato and related genomes. In eds. (Chakrabarti, S.K., Xie, C., Tiwari, J.K.). The Potato Genome, Springer: Cham, Switzerland, pp. 143–60.

Hassan, A.H., Mokhtar M.M., and El Allali, A. (2024). Transposable elements: multifunctional players in the plant genome. Front. Plant Sci. 14:1330127. https://doi.org/10.3389/fpls.2023.1330127

Havecker, E. R., Gao, X., and Voytas, D. F. (2004). The diversity of ltr retrotransposons. *Genome Biol.* 5, 1–6. doi: 10.1186/gb-2004-5-6-225.

Heitkam, T., Holtgräwe, D., Dohm, J.C., Minoche, A.E., Himmelbauer, H., Weisshaar, B., and Schmidt, T. (2014). Profiling of extensively diversified plant LINEs reveals distinct plant-specific subclades. Plant J. 79(3):385-97. doi: 10.1111/tpj.12565.

Hirochika, H., Okamoto, H., and Kakutani, T. (2000). Silencing of retrotransposons in Arabidopsis and reactivation by the ddm1 mutation. Plant Cell. 12(3): 357–69.

Huang, C.R.L., Burns, K.H., Boeke, J.D. (2012). Active transposition in genomes. Annu. Rev. Genet. 46: 651–675.

Janicki, M., Rooke, R., and Yang, G. (2011). Bioinformatics and genomic analysis of transposable elements in eukaryotic genomes. Chromosome Res. 19(6): 787– 808.

Jedlicka, P., Lexa, M., Kejnovsky, E. (2020). What can long terminal repeats tell us about the age of LTR retrotransposons, gene conversion and ectopic recombination? *Front. Plant Sci.* 11. doi: 10.3389/fpls.2020.00644.

Jiang, N. (2013). Overview of repeat annotation and de novo repeat identification. In Ed. (Peterson, T.), Plant Transposable Elements, Springer: Berlin/Heidelberg, Germany, pp. 275–87.

Jiang, N., and Wessler, S.R. (2001). Insertion preference of maize and rice miniature inverted repeat transposable elements as revealed by the analysis of nested elements. Plant Cell. 13(11): 2553–2564. doi: 10.1105/tpc.010235.

Joly-Lopez, Z,, and Bureau, T.E. (2014). Diversity and evolution of transposable elements in Arabidopsis. Chromosome Res. 22: 203–16. https://doi.org/10.1007/s10577-014-9418-8.

Joly-Lopez, Z., Bureau, T.E. (2018). Exaptation of transposable element coding sequences. Curr Opin Genet Dev. 49: 34–42.

Jung, S., Venkatesh J., Kang M.Y., Kwon J.K., and Kang B.C. (2019). A non-LTR retrotransposon activates anthocyanin biosynthesis by regulating MYB transcription factor in *Capsicum annuum*. *Plant Sci.* 287:110181. https://doi.org/10.1016/j.plantsci.2019.110181.

Kalendar, R., Vicient, C.M., Peleg, O., Anamthawat-Jonsson, K., Bolshoy, A., and Schulman, A.H. (2004). Large retrotransposon derivatives: Abundant, conserved but nonautonomous retroelements of barley and related genomes. Genetics 166: 1437–50.

Kalendar, R., Raskina, O., Belyayev, A., and Schulman, A. H. (2020). Long tandem arrays of Cassandra retroelements and their role in genome dynamics in plants. *Int. J. Mol. Sci.* 21, 2931. https://doi.org/10.3390/ijms21082931.

Kapitonov, V.V., and Jurka, J. (2008). A universal classification of eukaryotic transposable elements implemented in Repbase. Nat Rev Genet. 9(5): 411–2.

Kapitonov, V.V., Koonin, E.V. (2015). Evolution of the RAG1-RAG2 locus: Both proteins came from the same transposon. Biol. Direct 10: 20.

Kazazian, H.H., Jr., Wong, C., Youssoufian, H., Scott, A.F., Phillips, D.G., Antonarakis, S.E. (1988). Haemophilia A resulting from de novo insertion of L1 sequences represents a novel mechanism for mutation in man. *Nature* 332: 164–66.

Kazazian, H.H. Jr (2000). Genetics. L1 retrotransposons shape the mammalian genome. Science. 289: 1152–53.

Keidar, D., Doron, C., and Kashkush, K. (2018). Genome-wide analysis of a recently active retrotransposon, Au SINE, in wheat: Content, distribution within subgenomes and chromosomes, and gene associations. Plant Cell Rep. 37: 193–208.

Kidwell, M.G., and Lisch, D.R. (2001). Transposable elements and host genome evolution. Trends Ecol. Evol. 16(5): 356–61.

Kim, S., Park, J., Yeom, S.I., Kim, Y.M., Seo, E., Kim, K.T., et al. (2017). New reference genome sequences of hot pepper reveal the massive evolution of plant disease-resistance genes by retroduplication. Genome Biol. 18(1): 210.

Kriedt, R.A., Cruz, G.M.Q., Bonatto, S.L., and Freitas, L.B. (2014). Novel transposable elements in Solanaceae: Evolutionary relationships among Tnt1-related sequences in wild petunia species. Plant Mol. Biol. Rep. 32: 142–52.

Kubat, Z., Zluvova, J., Vogel, I., Kovacova, V., Cermak, T., Cegan, R., Hobza, R., Vyskot, B., and Kejnovsky, E. (2014). Possible mechanisms responsible for absence of a retrotransposon family on a plant Y chromosome. New Phytol. 202: 662–78.

Lanciano, S., Cristofari, G. (2020). Measuring and interpreting transposable element expression. Nat. Re v. Genet. 21: 721–36.

Li, S.F., Su, T., Cheng, G.Q., Wang, B.X., Li, X., Deng, C.L., and Gao, W.J. (2017). Chromosome evolution in connection with repetitive sequences and epigenetics in plants. *Genes.* 8: 290. https://doi.org/10.3390/genes8100290.

Llorens, C., Futami, R., Covelli, L., Domínguez-Escribá, L., Viu, J. M., Tamarit, D., et al. (2011). The Gypsy Database (GyDB) of mobile genetic elements: release 2.0. Nucleic Acids Res. 39: D70–D74. https://doi.org/10.1093/nar/gkq1061.

Llorens, C., Mu-oz-Pomer, A., Bernad, L., Botella, H., and Moya, A. (2009). Network dynamics of eukaryotic LTR retroelements beyond phylogenetic trees. Biol. Direct. 4: 41. https://doi.org/10.1186/1745-6150-4-41.

Llorens, C., Futami, R., Covelli, L., Domínguez-Escribá, L., Viu, J.M., Tamarit, D., Aguilar-Rodríguez, J., Vicente-Ripolles, M., Fuster, G., Bernet, G.P., et al. (2010). The Gypsy Database (GyDB) of mobile genetic elements: Release 2.0. Nucleic Acids Res. gkq1061.

Ma, J. (2013). Retrotransposons. Brenner's Encyclopedia of Genetics, 2nd edition, Volume 6 https://doi.org/10.1016/B978-0-12-374984-0.01322-X.

Mao, H., and Wang, H. (2017). Distribution, Diversity, and Long-Term Retention of Grass Short Interspersed Nuclear Elements (SINEs). Genome Biol Evol. 9(8): 2048–56.

Mascagni, F., Barghini, E., Giordani, T., Rieseberg, L.H., Cavallini, A., Natali, L. (2015). Repetitive DNA and Plant Domestication: Variation in Copy Number and Proximity to Genes of LTR-Retrotransposons among Wild and Cultivated Sunflower (Helianthus annuus) Genotypes. Genome Biol Evol. 7(12): 3368–82.

Mascagni, F., Giordani, T., Ceccarelli, M., Cavallini, A., Natali, L. (2017a). Genome-wide analysis of LTR-retrotransposon diversity and its impact on the evolution of the genus Helianthus (L.). BMC Genomics. 18(1): 634.

Mascagni, F., Cavallini, A., Giordani, T., Natali, L. (2017b). Different histories of two highly variable LTR retrotransposons in sunflower species. Gene 634: 5–14.

Mascagni, F., Vangelisti, A., Usai, G., Giordani, T., Cavallini, A., Natali, L. (2020). A computational genome-wide analysis of long terminal repeats retrotransposon expression in sunflower roots (Helianthus annuus l.). *Genetica* 148: 13–23. https://doi.org/10.1007/s10709-020-00085-4.

Mascagni, F., Usai, G., Natali, L., Cavallini, A., and Giordani, T. (2018). A Comparison of methods for LTR-retrotransposon insertion time profiling in the populus trichocarpa genome. Caryologia 71: 85–92.

Matthews, G.D., Goodwin, T.J., Butler, M.I., Berryman, T.A., and Poulter, R.T. (1997). pCal, a highly unusual Ty1/copia retrotransposon from the pathogenic yeast Candida albicans. J Bacteriol. 179(22): 7118–28. doi: 10.1128/jb.179.22.7118–7128.1997.

Maumus, F., Allen, A.E., Mhiri, C., Hu, H., Jabbari, K., Vardi, A., et al. (2009). Potential impact of stress activated retrotransposons on genome evolution in a marine diatom. BMC Genomics. 10: 624.

McClintock B. (1951). Mutable Loci in Maize. Annu. Rep. Dir. Dep. Genet. Carnegie Inst. Wash. 50: 174–81.

Meyers, B.C., Tingey, S.V., and Morgante, M. (2001). Abundance, distribution, and transcriptional activity of repetitive elements in the maize genome. Genome Res. 11(10):1660-76. https://doi.org/10.1101/gr.188201.

Mi, S., Lee, X., Li, X., Veldman, G.M., Finnerty, H., Racie, L., LaVallie, E., Tang, X.Y., Edouard, P., Howes, S., et al. (2000). Syncytin is a captive retroviral envelope protein involved in human placental morphogenesis. *Nature* 403: 785–89.

Mita, P., and Boeke, J.D. (2016). How Retrotransposons shape genome regulation. Curr. Opin. Genet. Dev. 37:90–100.

Monden, Y., Fujii, N., Yamaguchi, K., Ikeo, K., Nakazawa, Y., Waki, T., Hirashima, K., Uchimura, Y., and Tahara, M. (2014). Efficient screening of long terminal repeat retrotransposons that show high insertion polymorphism via high-throughput sequencing of the primer binding site. Genome 57: 245–52.

Negi, P., Rai, A.N., and Suprasanna, P. (2016). Moving through the stressed genome: Emerging regulatory roles for transposons in plant stress response. *Front. Plant Sci.* 7:1448. https://doi.org/10.3389/fpls.2016.01448.

Neumann, P., Novák, P., Hoštáková, N., and Macas, J. (2019). Systematic survey of plant LTR-retrotransposons elucidates phylogenetic relationships of their polyprotein domains and provides a reference for element classification. Mobile DNA 10:1. https://doi.org/10.1186/s13100-018-0144-1.

Novikov, A., Smyshlyaev, G., and Novikova, O. (2012). Evolutionary history of LTR retrotransposon chromodomains in plants. Int. J. Plant Genomics 874743.

Orozco-Arias, S., Isaza, G., and Guyot, R. (2019). Retrotransposons in Plant Genomes: Structure, Identification, and Classification through Bioinformatics and Machine Learning. Int J Mol Sci. 20(15): 3837. https://doi.org/10.3390/ijms20153837.

Ouyang, Z., Wang, Y., Ma, T., Kanzana, G., Wu, F., and Zhang, J. (2021). Genome-wide identification and development of LTR retrotransposon-based molecular markers for the melilotus genus. *Plants* 10: 890. https://doi.org/10.3390/plants10050890.

Pagadala, J.C., Pandita, D., and Pandita, A. (2023). Introduction, Classification, and Transposition of Transposable Elements (TEs). In (Eds. Pandita, D. and Pandita, A.) Plant Transposable Elements: Biology and Biotechnology (1st ed.). Apple Academic Press. https://doi.org/10.1201/9781003315193.

Pandita, D., and Pandita, A. (2016). Jumping Genes- "The Other Half of the Human Genome" and the Missing Heritability Conundrum of Human Genetic Disorders. *Biotech J. Internat.* 11(3): 1–18. https://doi.org/10.9734/BBJ/2016/13904.

Pandita, D., and Pandita, A. (Eds.). (2023). Plant Transposable Elements: Biology and Biotechnology (1st ed.). Apple Academic Press. https://doi.org/10.1201/9781003315193.

Papolu, P.K., Ramakrishnan, M., Mullasseri, S., Kalendar, R., Wei, Q., Zou, L.H., Ahmad, Z., Vinod, K.K., Yang, P., and Zhou, M. (2022). Retrotransposons: How the continuous evolutionary front shapes plant genomes for response to heat stress. *Front. Plant Sci.* 13:1064847. https://doi.org/10.3389/fpls.2022.1064847.

Papolu, P.K., Ramakrishnan, M., Wei, Q., et al. (2021). Long terminal repeats (LTR) and transcription factors regulate *PHRE1* and *PHRE2* activity in Moso bamboo under heat stress. *BMC Plant Biol* 21: 585. https://doi.org/10.1186/s12870-021-03339-1.

Park, M., Sarkhosh, A., Tsolova, V., and El-Sharkawy, I. (2021). Horizontal transfer of ltr retrotransposons contributes to the genome diversity of vitis. *Int. J. Mol. Sci.* 22: 1–16. https://doi.org/10.3390/ijms221910446.

Paz, R.C., Kozaczek, M.E., Rosli, H.G., Andino, N.P., and Sanchez-Puerta, M.V. (2017). Diversity, distribution and dynamics of full-length Copia and Gypsy LTR retroelements in Solanum lycopersicum. Genetica. 145(4-5): 417–30.

Piednoël, M., Carrete-Vega, G., and Renner, S.S. (2013). Characterization of the LTR retrotransposon repertoire of a plant clade of six diploid and one tetraploid species. Plant J. 75(4): 699–709.

Piégu, B., Bire, S., Arensburger, P., and Bigot, Y. (2015). A survey of transposable element classification systems—A call for a fundamental update to meet the challenge of their diversity and complexity. Mol. Phylogenet. Evol. 86: 90–109.

Pohlman R.F., Fedoroff N.V., and Messing J. (1984). The Nucleotide Sequence of the Maize Controlling Element Activator. Cell. 37:635–43. doi: 10.1016/0092–8674(84)90395–7.

Pray, L. (2008). Transposons: The jumping genes. *Nature Education.* 1(1): 204.

Rahman, A.Y.A., Usharraj, A.O., Misra, B.B., Thottathil, G.P., Jayasekaran, K., Feng, Y., Hou, S., Ong, S.Y., Ng, F.L., Lee, L.S., et al. (2013). Draft genome sequence of the rubber tree hevea brasiliensis. BMC Genomics 14: 75.

Rho, M., Choi, J.H., Kim, S., Lynch, M., and Tang, H. (2007). De novo identification of LTR retrotransposons in eukaryotic genomes. BMC Genomics. 8: 90.

Roy, N.S., Choi, J.Y., Lee, S.I., and Kim, N.S. (2015). Marker utility of transposable elements for plant genetics, breeding, and ecology: A review. Genes Genomics 37: 141–51.

Ryu, W.S. (2017). Retroviruses. In (Eds: Wang-Shick Ryu) Molecular Virology of Human Pathogenic Viruses, Academic Press, p. 227–46. https://doi.org/10.1016/B978-0-12-800838-6.00017-5.

Sahebi, M., Hanafi, M.M., van Wijnen, A.J., Rice, D., Rafii, M.Y., Azizi, P., et al. (2018). Contribution of transposable elements in the plant's genome. Gene. 665: 155–66.

Sampath, P., and Yang, T.J. (2014). Comparative analysis of Cassandra TRIMs in three Brassicaceae genomes. Plant Genet. Resour. Util. 12: S146–S150.

Sanchez, D.H., Gaubert, H., Drost, H.G., Zabet, N.R., and Paszkowski, J. (2017). High-frequency recombination between members of an LTR retrotransposon family during transposition bursts. Nat. Commun. 8: 1283.

SanMiguel, P., and Bennetzen, J. L. (1998). Evidence that a Recent increase in maize genome size was caused by the massive amplification of intergene retrotransposons. *Ann. Bot.* 81: 37–44. https://doi.org/10.1006/anbo.1998.0746.

SanMiguel, P., Tikhonov, A., Jin, Y.K., Motchoulskaia, N., Zakharov, D., Melake-Berhan, A., Springer, P.S., Edwards, K.J., Lee, M., Avramova, Z., and Bennetzen, J.L. (1996). Nested retrotransposons in the intergenic regions of the maize genome. Science. 274(5288): 765–768. https://doi.org/10.1126/science.

Schietgat, L., Vens, C., Cerri, R., Fischer, C.N., Costa, E., Ramon, J., Carareto, C.M.A., and Blockeel, H. (2018). A machine learning based framework to identify and classify long terminal repeat retrotransposons. PLoS Comput Biol. 14(4): e1006097. https://doi.org/10.1371/journal.pcbi.1006097.

Schmutz, J., Cannon, S.B., Schlueter, J., Ma, J., Mitros, T., Nelson, W., et al. (2010). Genome sequence of the palaeopolyploid soybean. *Nature*. 463(7278):178–83. https://doi.org/10.1038/nature08670.

Schnable, P. S., Ware, D., Fulton, R. S., Stein, J. C., Wei, F., Pasternk, S., et al. (2009). The B73 maize genome: complexity, diversity, and dynamics. *Science* 326, 1112–16. https://doi.org/10.1126/science.1178534.

Schulman, A.H. (2012). Hitching a Ride: Nonautonomous retrotransposons and parasitism as a lifestyle. In Plant Transposable Elements; Grandbastien, M.-A., Casacuberta, J.M., Eds.; Springer: Berlin/Heidelberg, Germany, pp. 71–88.

Schulman, A.H. (2013). Retrotransposon replication in plants. Curr. Opin. Virol. 3: 604–14.

Seibt, K.M., Schmidt, T., and Heitkam, T. (2020). The Conserved 3′ Angio-domain Defines a Superfamily of Short Interspersed Nuclear Elements (SINEs) in Higher Plants. Plant J. 101: 681–99.

Sgaramella, S.V., Bartels, D., Salamini, F., Furini, A. (2008). Retrotransposons and siRNA have a role in the evolution of desiccation tolerance leading to resurrection of the plant Craterostigma plantagineum. New Phytol. 179(3): 877–87.

Shapiro, J. A. (2014). Epigenetic control of mobile DNA as an interface between experience and genome change. *Front. Genet.* 5. https://doi.org/10.3389/fgene.2014.00087.

Stritt, C., Wyler, M., Gimmi, E.L., Pippel, M., and Roulin, A.C. (2020). Diversity, dynamics and effects of long terminal repeat retrotransposons in the model grass Brachypodium distachyon. New Phytol. 227(6): 1736–48. doi: 10.1111/nph.16308.

Usai, G., Mascagni, F., Natali, L., Giordani, T., and Cavallini, A. (2017). Comparative genome-wide analysis of repetitive DNA in the genus Populus L. Tree Genetics & Genomes. 13(5): 96.

Ustyantsev, K., Blinov, A., and Smyshlyaev, G. (2017). Convergence of retrotransposons in oomycetes and plants. Mob. DNA 8: 4.

Ustyantsev, K., Novikova, O., Blinov, A., and Smyshlyaev, G. (2015). Convergent evolution of ribonuclease H in LTR retrotransposons and retroviruses. Mol. Biol. Evol. 32, 1197–1207.

Vicient, C.M. and Casacuberta, J.M. (2020). Additional ORFs in Plant LTR-Retrotransposons. *Front. Plant Sci.* 11:555. https://doi.org/10.3389/fpls.2020.00555,

Vicient, C.M., Suoniemi, A., Anamthawat-Jónsson, K., Tanskanen, J., Beharav, A., Nevo, E., and Schulman, A.H. (1999). Retrotransposon BARE-1 and Its Role in Genome Evolution in the Genus Hordeum. Plant Cell. 11(9): 1769–84. https://doi.org/10.1105/tpc.11.9.1769.

Watson, J. D., and Crick, F. H. C. (1953). A structure for deoxyribose nucleic acid. *Nature* 171: 737–38.

Wells, J.N., and Feschotte, C. (2020). A Field Guide to Eukaryotic Transposable Elements. Annu. Rev. Genet. 54: 539–61.

Wenke, T., Döbel, T., Sörensen, T.R., Junghans, H., Weisshaar, B., and Schmidt, T. (2011). Targeted identification of short interspersed nuclear element families shows their widespread existence and extreme heterogeneity in plant genomes. Plant Cell. 23(9): 3117–28.

Wicker, T., Sabot, F., Hua-Van, A., Bennetzen, J.L., Capy, P., Chalhoub, B., Flavell, A., Leroy, P., Morgante, M., and Panaud, O., et al. (2007). A Unified Classification System for Eukaryotic Transposable Elements. Nat. Rev. Genet. 8: 973–82.

Witte, C.P., Le, Q.H., Bureau, T., and Kumar, A. (2001). Terminal-Repeat Retrotransposons in Miniature (TRIM) are involved in restructuring plant genomes. Proc. Natl. Acad. Sci. USA 98: 13778–83.

Xiong, Y., and Eickbush, T.H. (1988). Similarity of reverse transcriptase-like sequences of viruses, transposable elements, and mitochondrial introns. Mol Biol Evol. 5(6): 675–90.

Xu, Z.S., Yang, Q.Q., Feng, K., and Xiong, A.S. (2019). Changing carrot color: insertions in dcmyb7 alter the regulation of anthocyanin biosynthesis and modification. *Plant Physiol.* 181: 195–207. https://doi.org/10.1104/pp.19.00523.

Yi, X., Liu, J., Chen, S., Wu, H., Liu, M., and Xu, Q. et al. (2022). Genome assembly of the JD17 soybean provides a new reference genome for comparative genomics, G3 Genes|Genomes|Genetics, 12(4). https://doi.org/10.1093/g3journal/jkac017.

Yin, H., Du, J., Li, L., Jin, C., Fan, L., Li, M., Wu, J., and Zhang, S. (2014). Comparative genomic analysis reveals multiple long terminal repeats, lineage-specific amplification, and frequent interelement recombination for Cassandra retrotransposon in pear (Pyrus Bretschneideri Rehd.). Genome Biol. Evol. 6: 1423–36.

Yin, H., Liu, J., Xu, Y., Liu, X., Zhang, S., Ma, J., and Du, J. (2013). TARE1, a mutated copia-like LTR retrotransposon followed by recent massive amplification in tomato. PLoS ONE 8: e68587.

Yin, H., Wu, X., Shi, D., Chen, Y., Qi, K., Ma, Z., and Zhang, S. (2017). TGTT and AACA: Two transcriptionally active LTR retrotransposon subfamilies with a specific LTR structure and horizontal transfer in four Rosaceae species. Mob. DNA 8: 14.

Zhang, L., Yan, L., Jiang, J., Wang, Y., Jiang, Y., and Yan, T., et al. (2014). The structure and retrotransposition mechanism of LTR-retrotransposons in the asexual yeast Candida albicans. Virulence. 5(6): 655–64.

Zhao, Y., Xu, T., Shen, C.Y., Xu, G.H., Chen, S.X., and Song, L.Z., et al. (2014). Identification of a retroelement from the resurrection plant Boea hygrometrica that confers osmotic and alkaline tolerance in Arabidopsis thaliana. PLoS One. 9(5): e98098.

Zhao, D., Ferguson, A.A., and Jiang, N. (2016). What makes up plant genomes: The vanishing line between transposable elements and genes. Biochim. Biophys. Acta Gene Regul. Mech. 1859: 366–80.

Zhou, M., Liang, L., and Hänninen, H. (2018). A transposition-active phyllostachys edulis long terminal repeat (LTR) retrotransposon. J. Plant Res. 131: 203–10.

2

Origin, Evolution and Genomic Organization of Retro Transposons in the Plant Genome

MA Syed,[1*] *SK Debsharma,*[1] *N Jahan,*[1] *MY Khan*[1] *and MB Akter*[2]

1. Introduction

Transposable elements (TEs) are arranged in tandem arrays and dispensed throughout the genome at random intervals. TEs are important elements of eukaryotic genomes and can move around within them (Lisch, 2013; Bourque et al., 2018). In the middle of the 20th century, Barbara McClintock first identified TEs in maize and gave them the term jumping genes (Ravindran, 2012; Goodier, 2016). TEs are indeed sources of spontaneous mutations. The expression and activity of these genes can also boost the plant body to respond more positively to biotic and abiotic conditions (Ramakrishnan et al., 2021). Furthermore, it is now known that TE specificity aids in plant response to a variety of these stressors. There are several identified major kinds of retroelements, including retroviruses, long terminal repeat (LTR)-retrotransposons, short interspersed nuclear elements (SINE), and long interspersed nuclear elements (LINE) (Kumar and Bennetzen, 1999; Schmidt, 1999). Before reinsertion into the genome, retroposons move from one chromosome location to another through an intermediary RNA which is transformed into extrachromosomal DNA with imprinted reverse transcriptase/RNaseH enzymes (Bingham and Zachar, 1989; Boeke and Corces, 1989). Plant genomes are significantly enlarged due to this mechanism of transposition that replicates rapidly, allowing for a large

[1] Plant breeding division, Bangladesh Rice Research Institute (BRRI), Gazipur-1701, Bangladesh e-mail: sanjoybau@gmail.com; njahansau@gmail.com; yeakubkhansau@gmail.com

[2] Crop Physiology Division, Bangladesh Institute of Nuclear Agriculture (BINA), Mymensingh -2202, Bangladesh; e-mail: riponkachua@gmail.com

* Corresponding author: msyedso@yahoo.com; http://orcid.org/0000-0002-4601-1325

increase in element copy numbers (Kumar, 1996; SanMiguel and Bennetzen, 1998). Retrotransposons can cause mutations through insertion into or close to genes, just like DNA transposable elements. The sequence at the insertion site is kept because retrotransposon-induced mutations transpose via replication, which makes them relatively stable. Transposable elements appear as parasitic or self-centered DNAs based on many of their characteristics (Doolittle and Sapienza, 1980; Orgel and Crick, 1980). In addition to being expressed in the genome, retrotransposons also follow many of the same inheritance principles as the genes of the genomic "host" to which they belong. This makes retrotransposons a potential tool to create a beneficial biological process. This chapter has highlighted the origin, evolution, and genomic structure of retrotransposons in the plant genome.

2.   Role of Retrotransposon to the Evolution of Plant Genome

Retrotransposons occupy a big portion of plant genomes which contribute to structural changes, and gene regulation and therefore control their expression and evolution. Despite their availability, most of the LTR retrotransposons are dependent on autonomous elements, while others always produce defective copies (Bennetzen 2002; Sharma et al., 2008). In addition, the plant cell utilizes different superimposed mechanisms of suppression against transposable element duplication, confirming the LTR retrotransposon function (Slotkin and Martienssen 2007). The alteration of different gene families has been worked as a source of evolutionary events in plants. Repeated sequence causes positive and harmful mutations, controls gene activity, and keeps the genome stable (Belyayev et al., 2010). Due to their distribution and availability, TEs are the driving force of evolution that figure out genome makeup through recombination, genomic reorganization, gene expression, progress, and rearrangements, or by supplying primary substances for structuring the centromeres and introns (Bennetzen and Wang, 2014, Sharma et al., 2013, Vitte et al., 2014). They have potential gene-controlling factors that are responsible for mutagenic and regulatory effects of their insertion within or adjacent genes (Ahmed et al., 2011, Hayashi and Yoshida, 2009, Rebollo et al., 2012, Wang et al., 2013). In addition, it can directly disrupt gene function through alternative splicing (Cossu et al., 2012), chimeric TE-gene transduction (Xiao et al., 2008), promoter control (Kashkush et al., 2002), and gene silencing (Kashkush and Khasdan, 2007), or by the expansion of methylated TEs in flanking regions (Hollister et al., 2011, Hollister and Gaut, 2009).

Based on the sequence, reverse transcriptase, and coding regions; retrotransposons are separated into long terminal repeat retrotransposons (LTR retrotransposons), non-LTR retrotransposons, and Dictyostelium intermediate repeat sequence (DIRS) (Bourque et al., 2018). On the other hand, LTR retrotransposons are categorized into two major groups, namely; Ty1-copia and Ty3-gypsy (Wicker et al., 2007), which are the most prevalent TE components of the plant genomes (Bennetzen and Wang, 2014) and typically more active than their non-LTR partners (Paterson et al., 2009). However, most LTR retrotransposons are static and cannot be regulated for precise genomic insertion (Ramakrishnan et al., 2023). Furthermore, introductions of LTR retrotransposons into and surrounding the genes help in epigenetic activity, gene splicing, natural processes, repetition, and recombination. The copies of retrotransposon do not exit the host cell that normally shifts out of the nucleus with the combination of newly formed cDNA into a fresh locus of the same genome (Havecker et al., 2004).

Several studies revealed that LTR-RTs work as a major controlling agent for genome evolution through useful genes by structural changes and enzymatic proteins (Galindo-González et al., 2017; Grover et al., 2010; Wicker et al., 2018; Xia et al., 2020) that influence related gene expression as well as the phenotype (Chen et al., 2017; Tang et al., 2020). In addition, non-LTR retrotransposons including both LINEs (long interspersed nuclear elements) and SINEs (short interspersed nuclear elements) are specially inserted into repetitive DNA sequences into ribosomal RNA genes. For example, Menolird18, a member of LINEs of non-LTR retrotransposons, is involved in the genome evolutionary systems of melon and cucumber (Setiawan et al., 2020). However, LTR retrotransposons work in favorable epigenetic systems, such as chromatin variation, DNA methylation, siRNA control, and integration (Grandbastien, 2015; Schorn et al., 2017). In addition, LTR retrotransposons regulate genes at the transcriptional, post-transcriptional, and epigenetic levels to help different organisms tolerate stress (Galindo-González et al., 2017). Furthermore, heat-activated LTR retrotransposons demonstrated significant function in genome formation during evolution (Masuta et al., 2018). The gathering of Ty3/Gypsy heat responsive retrotransposons regulates the heat response systems in Cryptomeria japonica displayed differential expression (Ujino-Ihara, 2020).

LTR retrotransposons (LTRs) can be activated or silenced in plants to produce positive epigenetic alterations that support the integrity of the genome, the introduction of new genes, gene imprinting, the generation of genetic variety, and other processes. Besides, genomic stress, faulty gene function, mutagenesis effects, genetic rearrangements, excessive copy numbers, and other factors deceive how the genome evolves (Ramakrishnan et al., 2021; Zhang et al., 2018a). Moreover, they are extremely polymorphic and diverse at the insertion location (Liu et al., 2022). In this connection, the transposition of Tnt1 proved as a genetic engineering tool in tobacco mesophyll protoplasts (Grandbastien et al., 1989) where it creates a somaclonal variation and induced mutagenesis of tissue culture (Pouteau et al., 1991). Thus, newly developed plants using CRISPR/Cas9 derived from callus culture where homozygous plants missed Tos17 in the next generation (Saika et al., 2019), which elucidates the function of retrotransposons in genome evolution. In contrast, LTR retrotransposon, ATCOPIA93 performed as an immune-responsive gene during pathogen defense in transgenic Arabidopsis plants, which makes the connection between the sensitivity and retrotransposons to biotic stress (Zervudacki et al., 2018). Transcriptional intervention is also a mechanism by which retrotransposons might affect gene expression (White et al., 1994; Whitelaw and Martin, 2001). Therefore, chemical action and mutations are assessed for their consequence on retrotransposon activity and affected sequence in Arabidopsis

(Lindroth et al., 2001; Zilberman et al., 2003). Recently, the incorporation of LTR retrotransposon, ZmRE-1, might be correlated with changing plant height, density, and grain yield production (Li et al., 2020). In another report, LORE1 was found silent in transgenic plants which are involved in epigenetic processes (Fukai et al., 2010) and regulate the evolution in host plants through transcription of ONSEN under heat stress conditions (Masuta et al., 2017). Hence, a comprehensive investigation of LTR retrotransposons is essential for a better knowledge of plant genome evolution mechanisms and genetic diversity for further crop improvement.

3.   Retrotransposon Origin, Dispersal, and Organization in Plant Genome

Retrotransposons are genetic elements that move around and are reverse-transcribed from an intermediary RNA. They constitute a large portion of the nuclear genome in plants (typically more than 50% of the total DNA) and are present in all eukaryotic species. Their movements have an impact on gene function and expression, which eventually has an impact on evolution and genome adaptation (Shahid et al., 2020). The two classes of Transposable components are class I retrotransposons, which transpose by copying and pasting, and class II DNA transposons, which transpose by cutting and pasting (Wicker et al., 2007). All eukaryotic genomes contain long terminal repeat retrotransposons also known as LTR-RTs, which are common class I retrotransposons (Gao et al., 2004). Most plants' rapid LTR-RT proliferation is a factor in the genome's expansion (Kumar et al.1999; Schnable et al., 2009).

Plants differ significantly from certain other eukaryotes in the genomic arrangement of retrotransposons owing to their frequently high number of copies, highly diverse populations, and chromosomal dispersion patterns. Recently, significant efforts have been made to understand how retrotransposons are organized in gymnosperms and angiosperms' genomes and phenomes. Retrotransposons have been extremely successful in colonizing nearly all of the plant genomes, where they can make up more than 50% of the nuclear genome, thanks to their transpositional and amplifying reproduction system (Kossack et al., 1999; Pearce et al., 1998; Moore et al 1991; Noma et al., 1997; Pearce et al., 1996a; Pearce et al., 1996b; San Miguel et al., 1996; Yoshioka et al., 1993; Leeton et al., 1993).

3.1   *Retrotransposons Origin and Dispersal in Plants*

A very limited and straightforward placement of repeated DNAs has been found in telomeres, satellite repeats, or centromeric heterochromatin are absent from 80–90% of the Arabidopsis genome. A retrotransposon was discovered roughly every 130 kb in a contiguous area of *A. thaliana* has a 1.9 Mb gene on chromosome four (Genome Size: ca 130Mbp) (Bevan et al., 1998); this suggests that these elements make up roughly 1000 of all the genome's genic regions. These would make up roughly about four percent of the Arabidopsis genome, while the extra numerous centromeric regions with retrotransposons would make up an additional 4–6%.

Approximately 377 kb, or 3.1%, of the yeast Saccharomyces cerevisiae's 4-Mbp genome, is made up of 331 retrotransposon sequences.

Magnaporthe grisea, a filamentous fungus having a genomic volume of roughly 40 Mbp. However, both LTR and non-LTR have duplicate numbers ranging from a handful to one hundred (Kempken and Kuck, 1998). The majority of the retroelements in the 100 Mbp *Caenorhabditis elegans* genome seem to be non-LTR retrotransposons, together with fifty-nine members in the family of Rte-1, where full-length are nine (3298 bp) retrotransposons (Youngman et al., 1996). The group of Ty1-copia appears to be also nonexistent or uncommon in this nematode, however, it contains a few Ty3-gypsy group retrotransposons.

Furthermore, *C. elegans* has not been implicated in occurrences of retrotransposons or mutations brought on by them. The majority of *C. elegans* known retrotransposons were discovered to be flawed (E Berezikov, personal communication). LTR- and non-LTR retrotransposons are situated in the 165 Mbp genome of *Drosophila melanogaster*. Drosophila also has active retroviruses, in contrast to yeast, *C. elegans*, and Arabidopsis (Song et al., 1994).

It's interesting to note that, SINEs are either infrequent or absent from the genomes of Drosophila *C. elegans* and Arabidopsis. Due to the presence of SINEs in these species' close relatives, these are most likely unusual occurrences. For instance, there are several SINEs in Brassica napus, a member of the Cruciferae family like Arabidopsis (Deragon et al., 1994). The lack of SINEs in these short genome eukaryotes (C. elegans, 100 Mbp; Arabidopsis, 130 Mbp and Drosophila, roughly 165 Mbp) is unclear. Retrotransposons and pseudogenes might be eliminated from these creatures' genomes using an efficient mechanism, according to one theory. Alternately, these species' (except for Drosophila) comparatively incredibly infrequent fresh SINE insertions may have been caused by low LINE activity (Eickbush, 1992).

The number of unique families of retrotransposons and the total number of retrotransposons are related to the difference in genome size. Five distinct LTR families of retrotransposons, accounting for about three percent of the 4 Mbp *S. cerevisiae* genome, were already mentioned (Kim et al., 1998). Since retrotransposons with and without LTRs have both been discovered., there are thousands of different families of LTR and non-LTR retrotransposons in maize that collectively make up around 70% to 85% of Arabidopsis' nuclear genome contained inside with 1 to 100 families of each group, including Ty1-copia, Ty3-gypsy, LINE elements, and other elements (SanMiguel et al., 1998; SanMiguel et al., 1996).

3.2 *Retrotransposon Organization in Plant Genome*

3.2.1 *Chromosomal positions of transposable elements*

According to results obtained from the sequences of Ty1-copia retrotransposons are spread throughout the euchromatin, from time to time uniformly and sometimes sporadically, depending on the type of plant and the specific element used, as shown by chromosomes in metaphase and prophase underwent in situ hybridization. In various regions, such as centromeres, interstitial and terminal heterochromatic sections, and ribosomal DNA sites, several elements are absent or present in much reduced levels (Pearce et al., 1997; Heslop-Harrison et al., 1997; Brandes et al., 1997; Moore et al., 1994). This general observation does not apply to all elements, however, as selectively abundant elements can be found in *Allium cepa's* terminal heterochromatic regions (Pearce et al., 1996b) or the heterochromatic para-centromeric area of *Cicer arietinum* and *A. thaliana* (Heslop-Harrison et al., 1997; Brandes et al., 1997). One such instance is Athila, which is two non-retrotransposon open reading frames (ORFs) flanked by two LTRs, which are located in the para-centromeric region.

Retrotransposon sequences have been discovered by several organizations of numerous cereal species in the centromeric regions (Ananiev et al., 1998, Miller et al., 1998, Presting et al., 1998). We discovered probes with Ty3-gypsy-like sequence similarity from a centromere BAC clone from sorghum. Just the sorghum chromosomes' centromeric regions and all other Gramineae examined were hybridized by the probes (Miller et al., 1998). The barley homologue of this sorghum Ty3-gypsy integrase gene was used to clone the Cereba retrotransposon which was later found to be almost exclusively restricted to the centromeric regions of many grain species, including rye, barley, and wheat (Presting et al., 1998).

Interestingly, all of the Gramineae species under investigation had centromeric regions of Ty3-gypsy retrotransposons. This suggests that either these retrotransposons were recently inserted and amplified before the Gramineae diverged, or that they independently arrived and/or amplified in each of these species before becoming preferentially localized in centromeric regions. Centromeric portions of Gramineae species have Cereba retrotransposon sequences that exhibit an astounding degree of conservation, which raises the possibility of a link to centromeric function. Some Drosophila retrotransposons are long-lasting components of the telomeric and centromeric heterochromatin in previous studies (Pimpinelli et al., 1995). It is not yet known if plant retrotransposon sequences directly affect the functions of centromeres and/or telomeres.

Recently, it was discovered that pachytene maize chromosomes' knob 180 bps full-size copies of the retrotransposons in each individual like PREM-2, Zeon-1, RE-10, Grande, and RE-15 interrupt DNA units linked to cytologically discernible heterochromatic components (Ananiev et al., 1998). Knob DNA is linked to several genetic consequences in maize, such as a female gametophyte's distorted segregation and a late flowering phase. The 180-bp repeats were formerly thought to be the only genetic factors affecting knob heterochromatin. But the accidental retrotransposon insertions might be engaged in reducing or regulating the numerous genetic consequences connected to knob DNA.

Insitu hybridization has also been used to investigate the chromosomal distributions of retrotransposons that are not LTRs, SINEs, and LINEs. Although fewer elements found significant clustering in definite chromosomal sites, both types of the distribution of elements' chromosomes are scattered. For instance, certain

LINEs are grouped more frequently in the sub-telomeric areas of the majority of chromosome arms in sugar beets (Kubis et al., 1998). Similar to these LINEs, which were Ty1-copia retrotransposons, showed essentially no hybridization signal close to the tandem 18S-5.8S-25S rDNA gene repeats (Schmidt et al., 1995; Kubis et al., 1998). *Brassica napus'* SINE element S1Bn produced a scattered chromosomal localization pattern in in-situ hybridization tests, but it also displayed co-localization with rDNA sites as a common occurrence and a centromeric hybridization signal.

It first seemed that there were generally more euchromatic areas with retrotransposons than in heterochromatic areas based on numerous insitu and genome sequencing analyses. These findings, however, were skewed by the frequent use of retrotransposons that were first discovered close to in-situ hybridization probe using genes (Edwards et al., 1996), additionally by the fact that genic areas were the focus of genomic sequencing efforts (Bevan et al., 1998; Tikhonov et al., 1999).

Database searches have revealed that many genes either contain or are close to retrotransposon sequences. Legacy sequences of retrotransposons are frequently illegible and appear to date back in time (Wessler et al., 1995; Bureau et al., 1996; White et al., 1994). However, the majority of retrotransposons are found between genes, according to an examination of their distribution across a continuous 240 kb portion of the maize genome (SanMiguel et al., 1996). To avoid often modifying genes, several retrotransposons have probably evolved to largely transfer into repetitive sequences and other relatively inactive areas of plant chromosomes, such as intergenic spacers (SanMiguel et al., 1996). A deadly number of mutations would otherwise accumulate in the host cell. Retrotransposons would be able to multiply as scattered sequences without harming the host DNA in this fashion. By reducing the generally detrimental (for example, mutagenic) effects.

Many times, plant chromosomes include several retrotransposons like LINE, SINE, Ty1-copia, and Ty3-gypsy. Each retrotransposon family, however, also has families that are grouped or lacking in particular chromosomal areas. For instance, while certain Ty1-copia retrotransposons are occasionally found in the terminal heterochromatic region of Allium cepa chromosomes or are absent from this region in rye, while Ty3-gypsy retrotransposons are concentrated in the centromeric regions of maize, rye, and wheat (Secale cereale). Plant nucleolus organizer areas have not been discovered to include retrotransposon sequences. The majority of genes are situated in euchromatic areas, where many retrotransposons are more frequently found. These euchromatic retrotransposons in maize are primarily found in intergenic areas, according to sequence analyses of large genomic clones.

3.2.2   Transposable elements in organellar genomes

Examining whether these extracellular genomes from plants (Maier et al., 1995; Unseld et al., 1997) include the completion of the sequencing of both the chloroplast and mitochondrial genomes has allowed for the detection of any retrotransposons or retrotransposon inheritances. The small and gene-rich plants of the chloroplast genome do not contain any retrotransposons or retrotransposon fragments. On the other hand, it appears that Ty3-gypsy, Ty1-copia, and LINE-like retrotransposons are prevalent in the mitochondrial genomes. There have been no intact elements found, and there is no evidence that the insertion of a retrotransposon has changed anything. The Arabidopsis mitochondrial genome contains these retrotransposon segments, which were acquired during nine or more distinct acquisition events, in amounts of at least 4 percent (Unseld et al., 1997). These fragments may have amassed in the synthesis of DNA from the nuclear genome, resulting in the relatively large and repeat-rich plant mitochondrial genomes (Blanchard et al., 1995; Schuster and Brennicke, 1987).

3.2.3   *Local arrangements*

Using genomic DNA renaturation research, the first studies of the patterns of gene interspersion with repeated DNAs in plants started more than 25 years ago (Flavell et al., 1974). The majority of the repeated sequences under investigation at the time, particularly LTR retrotransposons, were retroelements. However, the majority of low copy-number DNAs were dotted throughout the repetitive DNA areas that looked to contain genes, as demonstrated by these Cot investigations (Hake and Walbot, 1980; Flavell et al., 1974). These studies had the disadvantage that most of their results were just averages of actual individual conditions throughout the whole genome.

The results from Cot investigations have been mainly supported by genomic sequence analyses. The retrotransposons and the micro inverted-repeat transposable elements, a subclass of small DNA transposable elements (MITEs) (White et al., 1994), which are mixed in with the genes are often found in domains that are 100 kb or smaller and make up the majority of the interspersed repetitive DNAs. Functional genes contain many transposable elements, most frequently incorporating pieces of retrotransposons (White et al., 1994). These retroelements are typically seen in small genome plants as single elements or single LTRs (Bevan et al., 1998; Chavanne et al., 1998). Additionally, several sections of large plant genomes, such as those of maize and barley, that are rich in genes, have the same clear pattern (Llaca and Messing, 1998; Panstruga et al., 1998).

The biggest section of the intricate plant genome to date to have been sequenced displays a retroelement layout pattern that is drastically different in an about 225 kb region close to maize adh1 (SanMiguel et al., 1996; Tikhonov et al., 1999). In this area, LTR retrotransposons remained common and were typically discovered as pieces stacked inside one another. Other biases in the insertions included the predilection of fivefold for insertion into the external retrotransposon domains as opposed to the internal ones and the fact that 17 of 22 were in the same possible transcriptional direction (Tikhonov et al., 1999; SanMiguel et al., 1996). These cuddled LTR retrotransposons persisted to consist of six slabs, one of which is likely to damage one of the nine potential genes discovered there. However, all 3 of the LINEs that were found here were putative. It is noteworthy that the initial sequencing of DNA in the Arabidopsis centromere region also revealed a gene-poor region that is rich in nested retrotransposons (http://www. cshl.org/protarab). It will take more genomic sequencing research to ascertain if this organizational arrangement is typical for plants or unique to the maize genome.

3.2.4 *Insertion site preferences of transposable elements*

There is some degree of insertion site preference present in all mobile DNAs. The degree of specificity can differ significantly between DNA transposable elements. Even the most haphazardly inserted mutagens, such certain genes will see the insertion of the Tn5 of *E. coli*, P elements of *D. melanogaster*, or Mutator of maize more frequently than other genes more than ten times more frequently. Numerous of these elements show a predisposition to penetrate genic regions of a genome and into certain locations within or near a gene (promoters) (Bennetzen et al., 1993; Kidwell and Lisch, 1997). Additionally, some elements, like Ac of maize, like to be inserted at sites close to the beginning element (Dooner and Belachew, 1989).

Retrovirus and retrotransposon insertion specificities have been identified in several eukaryotes (Craig, 1998; Labrador and Corces, 1997), but plants have not undergone a thorough investigation of this issue. There are instances where the favored integration sites are located in sections of the chromosomes that are actively transcribing, notably at or near promoters. Recurrent heterochromatic regions can also have preferential integration sites. In yeast, the majority of all retrotransposons expressions aim for selectivity. For instance, the Ty1, Ty2, and Ty3 elements preferentially integrate upstream the tRNA, 5S, and U6 genes, which are produced by RNA polymerase III, whereas the Ty5 element mainly inserts in silenced portions of the yeast genome, such as silent mating-type cassettes and telomeric regions (Gai et al., 1998; Zou and Voytas 1997).

Targeting oriented regions or arrangements in the host genome of retrotransposons, but how and why they do so is unclear. An integration complex, made up of the element's reverse-transcribed cDNA copy, host-encoded components, and retrotransposon-encoded integrase, may choose the insertion site (Craig, 1998; Labrador and Corces, 1997). Research on yeast retrotransposons has provided the most accurate information to date regarding preferred insertions. For instance, it was discovered that the Ty3 integration complex is a specific binding site for each transcription factor that engages tRNA genes (Kirchner et al., 1995). Similar to this, Ty5 prefers to insert into quiet chromatin. This selectivity can be lost by changes in nuclear proteins that preferentially interact with silent/heterochromatic yeast genome regions, or by a single amino acid change in the integrase protein (Gai and Voytas, 1998; Zou and Voytas, 1997).

It has been shown that some retrotransposons selectively splice into specific genomic regions in plants. For instance, in the plant species that have been studied thus far, numerous euchromatic zones are where Ty1-copia retrotransposons are most frequently detected (Hirochika et al., 1996; Garber et al., 1999; Hirochika et al., 1996). On the other hand, several of the cereal Ty3-gypsy retrotransposons prefer heterochromatic areas, such as centromeres and the knob DNA in maize, where they tend to group. Zepp, a retrotransposon that resembles LINE and builds up in Chlorella's telomeric region, also preferentially integrates into other Zepp sequences (Higashiyama et al., 1995).

It is unknown if any of these cases' biased accumulation is the result of favoured insertion, a less active method of eradication from these areas, or a lesser degree of fitness selection being used against retrotransposons there. There may be many retrotransposon families that share the specificities that are discovered. For instance, the 180-bp repeated portions of knob DNA were found to include the Grande, Zeon-1, and RE-15 retrotransposons all at the same site (Ananiev et al., 1998). The euchromatic portions of the maize genome contain the Ty1-copia and Ty3-gypsy retrotransposons in intergenic sites (SanMiguel et al., 1996). Reduced copy Ty1-copia retrotransposons whereas Tos17 of rice, Phaseolus vulgaris, and Tpv2 of a common bean are primarily targeted near and within genes (Garber et al., 1999; Hirochika et al., 1996). It appears that each retrotransposon has developed its own DNA insertion pattern as a result.

4. Characterization and Phylogenetic Analysis of Retrotransposons

4.1 Characterization of Retrotransposons

Numerous effects on an organism's genomic activities can result from the existence of transposable elements in that organism. They may alter the types of expression levels in the genome and the activities of these genes depending on the portion of the chromosome they are placed on (Nagaki et al., 2012). Although Copia retrotransposons typically cluster close to the chromosomes that are found on, Gypsy retrotransposons are much more widely distributed and occupy a more varied placement on the chromosomes in crop genome sequences (Joly-Lopez et al., 2014). Nevertheless, it is important to note that LTR retrotransposons, regardless of their origins, frequently cluster in various chromosomal locations (Schnable et al., 2009). The ubiquitous existence of retrotransposons in genome sequences has provided new important information about the characteristics of retrotransposons (Xiong and Eickbush 1988). They are divided into two subcategories: LINES and SINES (Zhang et al., 2014). Based on their genome sequences, the LTR-retrotransposons have also separated into "superfamilies," including the Gypsy, Copia, Bel-Pao, and primitive retrovirus superfamilies (Janicki et al., 2011). In contrast to Copia elements, Gypsy is greater prevalent in rice crops, which is in line with reports for the Capsicum, Solanum, and Populus species (Qin et al., 2014; Natali et al., 2015; Jouffroy et al., 2016).

In general, Copia elements have higher transcriptional activity than Gypsy elements. While less sufficient components in the genome lessen the silencing mechanisms of the host, abundant elements are more likely to be muted by the host genome (Qiu and Ungerer, 2018; Meyers et al., 2018; Vukich et al., 2009). These superfamily members from the Gypsy and Oryco superfamily members from the

Copia made up 28.49% and 4.77% of the rice LTR-RTs. As a result, both their high structural integrity and their reduced copy counts in the rice genome appear to support the robust retrotransposition activity of the Copia superfamily's Oryco elements (Cho et al., 2019).

The location of the integrase protein in the genomic areas of Gypsy retrotransposons distinguishes them from Copia retrotransposons. Integrase is located before reverse transcriptase in Copia retrotransposons and after it in Gypsy retrotransposons in the genomic sequence (Wicker et al., 2007). These superfamilies are further subdivided into various lineages utilizing phylogenetic analysis and time of divergence. The TORK, Bianca, Ale, and Maximus lineages made up the Copia superfamily, while the Attila, CRM, Del, and Galadriel lineages made up the Gypsy superfamily (Bennetzen et al., 2005). Due to their exceptional capacity to demonstrate individual activity and reproduce themselves repeatedly on chromosomes, LTR-retrotransposons display enormous diversity in number, position, and distribution throughout their coding region (Du et al., 2018). At both ends of their genetic sequence, two homologous structures known as long terminal repeats can be seen, which is a distinguishing characteristic of LTR retrotransposons and the element that provides them their namesake. The length of these DNA sequences might range from 100 bps to thousands of bps (Rho et al., 2007). The inner coding sequences are bracketed by these LTRs, which are likewise a part of retroviral sequences (Mascagni et al., 2017). LTR retrotransposons come in a wide range of sizes and functional properties. They range in length from four kilobase pairs (bp in Helianthus species (Cossu et al., 2012) to over 23 Kbp in *Populus trichocarpa* (Chang et al., 2013) in plants. Several Open Reading Frames (ORFs) are used to coordinate the architectures of LTR retrotransposons (Mascagni et al., 2015). The pol and gag genes' genetic information is contained in the ORFs, which are crucial for transcription in the host organism (Joly-Lopez and Bureau, 2018). The pol gene often houses the reverse transcriptase, while the gag genes encode functional polyproteins like their retroviral counterparts. Stop codons are often used to separate these genes (Piednoel et al., 2013). Three significant proteins are encoded by the pol gene, and each of these proteins is essential for the replication of retrotransposons in the genome (Paz et al., 2017). Integrase, protease, and reverse transcriptase are the names of these proteins (Usai et al., 2017). Genomic mechanisms are taking place to inhibit the activity of retrotransposons because they reproduce correspondingly to virus replication and can induce mutations and interfere with DNA repair (Sanchez et al., 2017). LTR retrotransposons may have a different area termed the chromodomain that allows them to bypass this silencing. The development of heterochromatin close to regions of retrotransposon activity is one method the cell employs to quiet retrotransposons (Ma et al., 2004). Because heterochromatin prevents retrotransposon proteins from easily accessing the cell DNA, replication is suppressed (Ragupathy et al., 2010). By influencing this heterochromatin, the protein encoded by the chromodomain region aids the retrotransposon in evading silencing. Retrotransposons contain chromodomains that are located upstream of the genomic sequence's 3′ terminus (Curcio et al., 1991).

4.2 *Phylogenetic Analysis of Retrotransposons*

The phylogenetic relationship builds a tree to depict the association between several taxa at the molecular basis using traits like nucleotide or amino acid sequences. This methodology has evolved into a crucial tool for assessing genomic data between various species and subgroups (Hilis 1997) and can also study domain links within a single taxon (Zhang et al., 2018b). The Maximum Likelihood approach and Tamura-Nei model were used to investigate and build the background of these plant retrotransposons (Tamura and Nei, 1993). By employing the Neighbor-joining method on the distance matrix calculated using the Tamura-Nei model. The translated proteins with the first+second+third+noncoding codon locations were all included. The final dataset included 892 places in it. Relationship analysis investigations were done using the MEGA X software (Kumar et al., 2018). The bootstrap values obtained after testing the neighbor-joining tree algorithm with 1000 bootstrap repetitions were presented above the tree's nodes (Pattengale et al., 2010). To find lineage clusters, the tree branches' cutoff value was chosen at 70% (Sacks-Davis et al., 2012). The group "C" which included well-supported retrotransposon lineage branches, is the biggest of all these groupings with values just above the threshold. This group's plant sequencing was derived through bryophytes. The well-supported bootstrap branches were group "A" (*M. grandiflora* 1 and *M. polymorpha* 2), group "B" (*A. sativa* and *A. sterilis*), group "C" (*S. cooperi* and *D. truncatula*), group "D" (*M. esculenta* and *F. virosa*), and group "E" (N. tetragona and M. grandiflora 2) (Fig. 1). Similarly, moderately supported bootstrap branches (Fig. 2) were group "F" (*M. polymorpha* 3 and *M. notabilis*), group "G" (*V. speciosa* 2 and *B. papyrifera*), and group "H" (*P. patens* 2 and *L. lagopus* 2) (Efron et al., 1996).

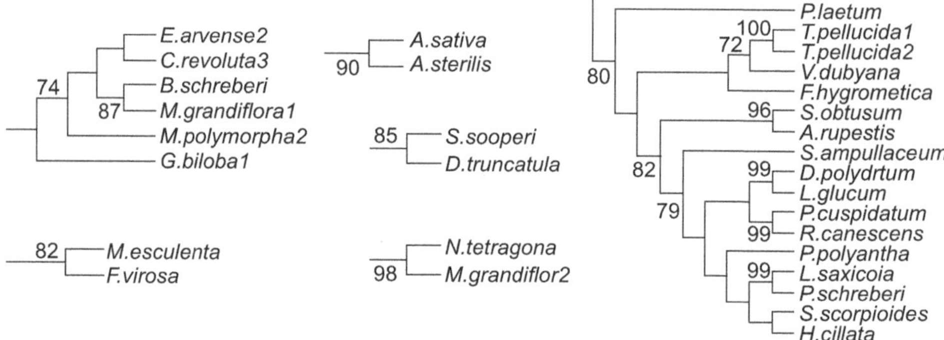

Fig. 1 Phylogenetic analysis on the basis of well-supported bootstrap branches (Brown et al., 2021).

The circular ideogram of many retrotransposons found in the kingdom Plantae's range-wide orders and families is seen in Fig. 3. This ideogram was created to effectively depict vast amounts of genetic data and secure the comprehensive presentation of large-scale data.

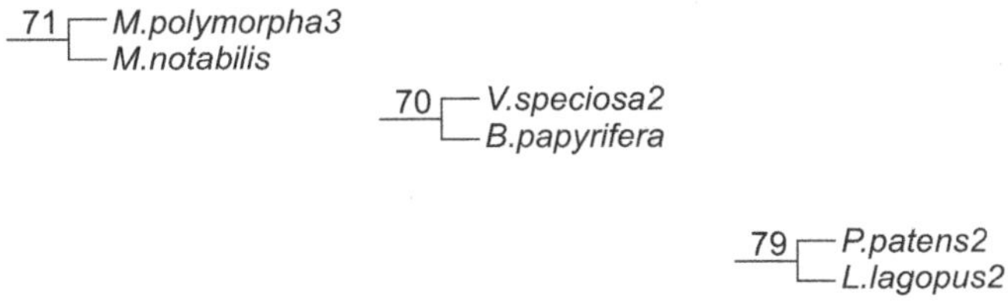

<pre>
 71 ┌─ M.polymorpha3
    └─ M.notabilis
                    70 ┌─ V.speciosa2
                       └─ B.papyrifera

                                      79 ┌─ P.patens2
                                         └─ L.lagopus2
</pre>

Fig. 2   Phylogenetic analysis based on moderately supported bootstrap branches (Brown et al., 2021).

Retrotransposons from gymnosperms made up the "red" group on the upper right, whereas retrotransposons from angiosperms made up the "blue" group. Two novel retrotransposons, Silava and Romani, exclusive to gymnosperms, were found in the "green" group. With the exception of *M. polymorpha* and *P. massoniana*, two liverworts, and gymnosperms, accordingly, the "yellow" group consists of the Gypsy family retrotransposons from angiosperms. The bryophytes oriented biggest cluster of Gypsy retrotransposons is the "orange" group. The Gypsy and Copia families oriented two gymnosperm retrotransposons make up the "purple" group. The two eudicots-oriented "pacific blue" group is a clade of non-LTR retrotransposons while the "brown" group is a clade of two gymnosperm Copia retrotransposons. The "Davidson orange" group consists largely of Gypsy retrotransposons with a few significant novel-type families, while the "ruby" group contains a collection of Copia subfamily retrotransposons (Cereba, N1, Osr30, and Silava). Cereba is unique to cereal plants, Silava is unique to gymnosperms, and Osr30 is unique to *O. sativa*.

5.   Plant Genome Modifications Through Retrotransposons Evolution

Retrotransposons act as prime modificators in the plant genome by contributing towards gene expression and regulation. Plant genome modifications through retrotransposons evolution manifested with certain putative biological functions causing diversification of genetic material. The retrotransposons-oriented modifications alter the expression level in the regulatory sequences of plant genes and improve plant adaptation to environmental changes. This type of retrotransposons-oriented modification acts differently on the type, copy number, and transpositional activity with transposition rates. The genomic architecture of plants is capable of controlling the mobilization pattern of retrotransposon that paves out tremendous variable activities over divergent taxa and species (Huang et al., 2012). The presence of a huge number of residual sequences of retrotransposon proves that genomes possess well organized post-insertion appliances for the inactivation or removal of retrotransposon (Casacuberta and Santiago, 2003). TEs activity varies within time periods ranging from very active to hardly any TE insertions taking place (Kralova et al., 2014, Grandbastien 2015, Bonchev 2016). As plants are immobile organisms, need extremely dynamic genomes (Kejnovsky

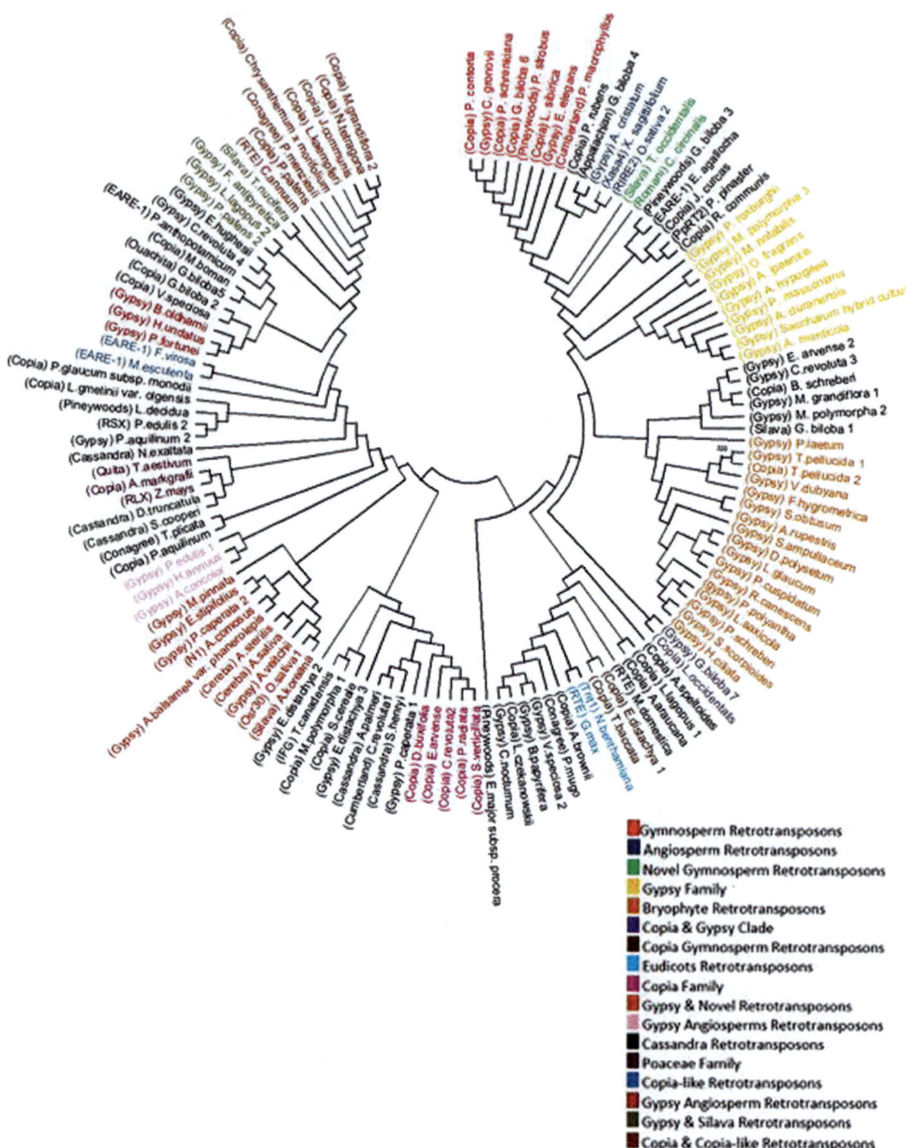

Fig. 3   Ideogram of retrotransposons in a circle pattern found in several plant species at the level of the genome (Brown et al., 2021).

et al., 2015), and provide a probable explanation for TEs activation in plants accompanied by internal and external factors (Bonchev and Parisod 2013, Negi et al., 2016, Vicient and Casacuberta, 2017). The genomic response of plants upon retrotransposon activation occurs under stressed and non-stressed conditions.

5.1 Retrotransposon Oriented Genome Redesigning in Non-stressed Conditions

Plant genomes allow extreme alterations to turn into a multicellular organism from a single-celled embryo. Numerous classes of interspersed repeats offer multiple recombinations that act as the major part of these changes. These profuse families with interspersed repeats are termed "junk DNA" that is primarily not involved in any major protein coding pathways. Recent findings indicate that this type of incessant DNA acts vitally like the functional non-coding DNA in the genome (Ariel and Manavella 2021; Pennisi 2012).

Vitte et al., (2014) opined that both retrotransposons and DNA transposons constitute the prime feature for interspersed repeats in the genome. Diversified groups for retrotransposons associated with retroviral genomic materials are highly copious (Wicker et al., 2007). Unlike the true mobility of DNA transposons, retrotransposons contrast in their transposition mechanisms and are composed of heterochromatin with the formation of different types of centromeres (Bennetzen and Wang 2014) and represent the plant genome's crucial intergenic part (Kalendar et al., 2020). An RNA intermediate assures the mobility of retrotransposon followed by transpositional Copy and Paste mechanism. Using their self (sometimes not) encoded enzymes, their RNA is reversely transcribed to originate from single-stranded RNA to the double-stranded DNA in a novel genome point. insertion site oriented such type of reverse transcription can be direct like target prime reverse transcription mechanism or indirect utilizing the extra chromosomal DNA (Wicker et al., 2007). This type of copy-paste feature creates sudden aggression of the original genomic structure and causes enormous genomic changes within a short period (Piegu et al., 2006). In the year 2006, Sabot and Schulman found that the active copies are initially numbered a few and termed transcriptionally active master copies. These copies can be static for translation and show dependence upon another transcript to accomplish their enzyme related activities. Newly formulated copies with recent insertion may relate phylogenetically for their similar sequencing pattern, but further mutational activity with their complicated lineage can recombine them. The evolutionary history of each inserted copy is unique and may acquire certain biological role and be accepted by their host plant.

Retrotransposons are highly variable in their lineage and structure in different families. Such variations at distinct phases may cause new species evolution as a major event in genomic modifications. Promoters in retrotransposon sequences attach the transcription factors that ultimately commence the synthesis of RNAs. In 2017, Vicient and Casacuberta researched "Additional ORFs in Plant LTR-Retrotransposons" where they identified and explored that LTR type of retrotransposons contain non-specific and auxiliary open reading frames. The findings prove the LTR retrotransposons as an important evolutionary appliance. Antisense open reading frames in categorized LTR retrotransposon groups of different angiosperm species suggested the putative role of these coding blocks for transposition activities. To elucidate genome modifications by these open reading frames, further investigation on transcript splicing of retrotransposon variants and their functional divergence are highly significant.

Retrotransposons are significant regulator of species diversity and genomic evolution for their unsteadiness in size, transposition process, and genomic organization (Vicient and Casacuberta, 2017). For the evolution of the wheat genome, potential retrotransposons showed severe genomic plasticity (Bariah et al., 2020). Retrotransposons have a persistent effect on genomic instability and serve as a driving force for the regulation of epigenetic evolution. In 2020, Pan et al. reported that the unstable cotton genome as an outcome of post-polyploidization caused numerous genomic structural changes related to the augmentation of LTR retrotransposon from TEs of corresponding less active A genome than from the D genome. Such finding revealed that the transposition event of retrotransposons was the key factor in genome shaping and speciation that relates to cotton genome divergence. In relation to that, the evolutionary history of cotton plants suggests a decaploidization to form the *Gossypium* genus ahead of the neo-tetraploidization to form the modern cultivated species of cotton. The variation among duplicated genes resulting functional innovation through their uncommonly high rates of evolution (Alix et al., 2017). Kalendar and Schulman (2006) performed genome wide profiling study of transposable elements regarding insertional polymorphisms. In 2021, Kalendar et al., made an experiment on transposon display. This research paves out the discovery of deletions and introgressions in the wheat genome as large genomic rearrangement events occurred by transposable elements. Insertional mutations caused by retrotransposons are generally found with truncated copy numbers inside the host plant genome. For instance, mutational groups in tobacco like *Tto1* or *Tnt1* and in rice like *Tos17* are observed with a comparatively negligible number (within a hundred) of transcript (Grandbastien 1998, Nie et al., 2019; Hirochika et al., 1996). Similarly, in maize, *Magellan* is reported with a copy number of four to eight (Purugganan and Wessler, 1994), *Hopscotch* in two to six copy number (White et al., 1994), the retrotransposon, *Bs1* with a copy number of one to five (Johns et al., 1985). On the contrary, retrotransposons that are very frequent like *Opie, Grande, Ji, Huck,* etc. (copy numbers of 10,000 or even more) were not engaged in characterized mutations in maize. Thus, it is presumed to have a relatedness for the copy number of the retrotransposon family with a tendency to insert into genes (SanMiguel et al., 1996). Superabundant groups of retrotransposons found in barley (*BARE-1)* with approximately 50 thousand copies and maize are found to show a preference in target site for previously transposed retrotransposons and intergenic regions (Suoneimi et al., 1997; SanMiguel et al., 1996). It is assumed that distribution patterns retrotransposons having high copies might be less suitable for genome modification. Regarding plant genomic modification, alteration in its size and arrangement is a crucial factor. Retrotransposons vitally determine plant genome architecture (Kumar, 1996). For instance, *Arabidopsis* (1C approximately130 Mbp), a short genomic plant lacks expansion of retrotransposon (Wright et al., 1996). In contrast, large genomic crops namely faba bean (1C is approximately 128,000 Mbp) and maize (1C is approximately 3200 Mbp) reported for presence of well-established and amplified colonization of retrotransposons (SanMiguel et al., 1996, Pearce et al., 1996b). In the maize genome, different groups of retrotransposons are detected with large copies of more than ten thousand. Among all types of maize retrotransposons, the entire copies counted around 300,000 covering about 50 to 80 percent of the maize nuclear genome (Cossu et al., 2012). The time of divergence is about 15 to 20 million years for connecting the maize and sorghum genomes. Both plants possess 10 pairs of chromosomes but nuclear genomic maize is three to four times longer than the genome of sorghum (Arumuganathan and Earle, 1991). Sequence analysis and molecular hybridization reports suggested genomic similarities between these two plants for gene content and arrangements (Hulbert et al., 1990, Bennetzen 1998, Moore et al., 1995). Even though around twenty-four retrotransposons comprising more than 70 percent neighbouring regions of *the adh1* gene in maize are not found in sorghum's orthologous *adh* portion (Avramova et al., 1996) and their detection was not feasible through gel blotting hybridization techniques in the genomic sorghum (Bennetzen et al., 1994). Genome size extension event in maize was a preliminary action opted for the enormous changing ability of retrotransposons. Genomic separation of maize with sorghum paves out the establishment for insertion of these elements through dating experiments indicating that, within the past 2 to 6 million years, amplification of retrotransposon originated in the maize genome (SanMiguel et al., 1998). Retrotransposons can redesign and alter genomic modules with or without transpositional competence such as deletions, duplications, and inversions. Coding sequence amplification can act as the unexposed material for new gene evolution. In plants, the sparse activity of retrotransposon like integration functions upon ordinary mRNAs in cells and reverse transcription has generated pseudogenes without intron (Drouin and Dover 1987, Loguercio and Wilkins, 1998). In the maize genome, *Bs1* retrotransposon captured the part of other genes followed by amplification through transposition showing how genes can be diffused in genomes (Khan et al., 2022). Retrotransposons in plants with more frequent copies and genome wide dispersion might be the sites for ectopic or unequal recombination. This kind of unbalanced recombination within the LTRs from unique retrotransposon in plants is frequently reported by independent LTRs presence (SanMiguel et al., 1996, Chen et al., 1998, Vicient et al., 1999). Reciprocal duplications and deletions may result from misplaced recombination among identical groups of retrotransposons in forward orientation. In contrast, unequal recombinations within the elements of the same chromosome with reverse arrangement will invert the sequences of chromosomes connecting two retrotransposons. Misplaced recombination on different chromosomes might create reciprocal translocations connecting two retro-elements. Genome rearrangement in yeast and *Drosophila* mediated by unequal recombination event of retrotransposon was founded by Kidwell in 2002. In plants, such chromosomal rearrangements for the structural and mechanistic arrangements have not yet been studied with proper documentation and will be a crucial thrust for future research. Effects of retrotransposon insertion events within or close to plant genes are compiled in Table 1.

Table 1   Effect of retrotransposon insertion event within or close to plant genes

Retrotransposon	Gene	Plant	Function and Insertion site	Reference
Tos17-Copia	*Oryza sativa* homeobox (*OSH15*)	Rice (*Oryza sativa*)	1. Defects in internode elongation 2. Insertion into exon 4 3. No transcript form	Sato et al., 1999
p-SINE-1	Catalase	Rice (*Oryza sativa*)	1. Insertion of *CatA* gene between exon2 and exon3 2. *pSINE*-1 retro-element act as intron in *CatA* gene	Iwamoto et al., 1999
Wis2-1A-*Copia*	Glutelin (*Glu-1*)	Wheat (*Triticum aestivum*)	1. Insertion of the *Glu1* gene inside coding sequence 2. No transcript	Moore et al., 1991
Magellan-Gypsy	*Waxy-M* (*wx*-M) locus	Maize (*Zea mays*)	1. Mutant phenotype (Waxy) 2. *wx* (exon3) gene insertion in the splice junction of intron2 and exon3	Purugganan and Wessler 1994
PREM2-Copia	Polygalacturanase (*Polgal*)	Maize (*Zea mays*)	1. No visible effect observed on gene expression 2. Three *Polgal* genes insertion at the identical 5ʳ flanking point	Rudolf et al., 2021
Bs1-Copia	Alcohol dehydrogenase (*Adh1*-S5446)	Maize (*Zea mays*)	1. Mutant phenotype derived from pollen resistant to allyl alcohol 2. Inactivation of the gene 3. Insertion into *Adh1* gene (just at exon9)	Stitzer et al., 2021
Cin4-LINE	Anthocyanin pigment gene (*A1*)	Maize (*Zea mays*)	1. Non mutant phenotype 2. *A1* gene insertion within the coding sequence	Schuster and Brennicke 1987
B5-Copia	Caffeic acid (*COMT*)	Maize (*Zea mays*)	1. Mutant pigmentation (reddish-brown) of the midrib caused by insertion near the junction of the 3ʳ coding region of the COMT gene intron 2. Size and amount of transcripts are altered	Vignols et al., 1995

Contd.

Table 1 *Contd.*

Retrotransposon	Gene	Plant	Function and Insertion site	Reference
Tst1-Copia	Starch phosphorylase	Potato (Solanum tuberosum)	1. Unimpaired transcript 2. Non mutant phenotype 3. Insertion of the element inside intron5 in reverse direction of specific gene	Camirand et al., 1990
Tnt1-Copia	Nitrate reductase (NR)	Tobacco (Nicotiana rustica)	1. NR deficient mutants 2. Insertions into exon2 and exon3 3. Transcripts of NR gene alteration	Grandbastien et al., 1989
Tna1-Gypsy	Expression of genes by pollination	Ornamental tobacco (Nicotiana alata)	1. This gene is not needed for fertility 2. Pointed at upstream (10 kb) of the S_6 RNase gene 3. Act as typical chimeric transcript 4. Expression observed in style due to Pollination event (within gene insertion)	Royo et al., 1996
Tnp2-Copia	Nitrate reductase (NR)	Tex-Mex Tobacco (Nicotiana plumbaginifolia)	1. NR deficient mutants 2. Transcripts alteration in *NR* gene through insertion within specific exon	Grandbastien, 1998
Tgmr-Copia	Fungal resistance of Soybean (Rps1-K)	Soybean (Glycine Max)	1. No known mutant phenotypes 2. Insertion in the flanking sequence of *Rps1*-k gene	Bhattacharyya et al., 1997
Copia-like	Dm3 Gene region	Lettuce (Lactuca sativa)	1. Responsible for retrotransposon-mediated rearrangements 2. Retrotransposon (*copia*-like) sequences Insertion in region of *Dm3* gene. 3. *Dm3* gene is occupied with candidate genes (*RGC2*) conferring to resistance.	Meyers et al., 1998

Contd.

Table 1   *Contd.*

Retrotransposon	*Gene*	*Plant*	*Function and Insertion site*	*Reference*
Melmoth-Copia	Self-incompatibility in S locus	Kale *(B. oleracea)*	1. In Brassica plants, S1 response is not dependent on active *SLA* gene 2. No transcript 3. Insertion within transcribed regions	Pastugilia et al., 1997
TS -SINE	*PAP2*	Bell pepper *(Capsicum annuum)*	1. Incomplete spliced transcripts originate through defective splicing 2. Non-mutant phenotype developed through 1st intron insertion in *PAP2* gene	Pozueta-Romero et al., 1998
Tms1-Copia	Gene specific to nodule formation *(Nms-25)*	Alfalfa *(Medicago sativa)*	1. Retrotransposon mediated rearrangements 2. Insertion into intron 9 which may be responsible for downstream deletions	Lloyd and Lister, 2022
Gypsy	*Hero* gene region	Tomato *(Solanum lycopersicum)*	1. In the region of *Hero* gene family, interspersed *gypsy* like retrotransposon sequences insertion	Kojima (2019)

5.2 Stress-activated Retrotransposons for Modifications in Plant Genomes

Multiple stresses are acting on plants, including biotic and abiotic stresses, such as tissue culture, wounding, heat, drought and salt stresses, freezing, polyploidization and hybridization events (Galindo-González et al., 2017, Fan et al., 2013), pathogens (Todorovska, 2007), pathogen elicitors (Wessler el al., 1995), defense-associated stresses (Casacuberta and Santiago, 2003), UV light (Schulman, 2012), and X-ray irradiation (Alzohairy et al., 2014, Grandbastien, 2015). It is assumed that retrotransposons invigoration is a common fact, stress-mediated retrotransposon activity can be precise to genotype in certain cases (Negi et al., 2016). The ultimate response to foreign stresses is confined to specific LTR sequences at pair ends in the case of LTR-RT (Wang et al., 2017). Retrotransposon's activation is precisely produced by cellular mechanisms or by external stresses allowing a quick activation of very specific LTR retrotransposons group (Casacuberta and González, 2013). Vicient and Casacuberta (2017) mentioned plant retrotransposons' ability to escape the silencing event of a host through anti-silencing factors expression. In some specified events regarding LTR-RTs insertion, absolutely identical LTR sequences are created. So, an exact mutation rate will assist in measuring the time of insertion of LTR retrotransposon utilizing the sequential difference between two LTRs (Yin et al., 2013). This computation helps in understanding the evolutionary fluctuation of all kinds of retrotransposon for the host plant genome modifications. Elbaidouri and Panaud (2012) reported that retrotransposons can interrupt crowded plant genomes, but their identified transpositional activation is very scanty and sequestered yet in plants. Table 2 presents a list of stress-responsive retrotransposons found in the genomes of plants.

Conclusion

Plant genomes contain several retrotransposons that are distinguished by specific motifs and domains. They play a crucial part in plant evolution and adaptability through their dispersion and transposition activities. The retrotransposons in plant genomes reserve harmful genes as well as genes from other life domains. They may also retain helpful genes that aid in the survival of their hosts under challenging circumstances. Retrotransposon fragments are frequently found in gene promoters and now help regulate many genes. It is generally known how retrotransposons affect the structure of genes and genomes, including how big they make them. Phylogenetic research is essential for understanding the functions, complexity, and dynamics of plant genomes. It has provided crucial knowledge about the characteristics, recurring patterns, individual profiles, variety, and phylogenetic relationships of retrotransposons throughout the whole plant kingdom. Specificities for the expression and insertion of retrotransposons have been described in numerous studies. Several research teams have started describing how elements involved in the host's developmental and stress response pathways interact to regulate LTR

Table 2   Stress-responsive retrotransposons recorded in the genomes of plants.

Retrotransposon	Foreign Stress	Species	Reference
Tos-17	Viral infection and tissue culture	Rice (*Oryza sativa*)	(Hirochika, 2001)
MAGGY	Heat shock	Rice (*Oryza sativa*)	Farman et al., 1996)
Wis2-1A	Hybridization (Interspecific)	Wheat (*Triticum aestivum*)	(Alzohairy et al., 2014)
Erika	Fungal infection	Wild wheat	(Alzohairy et al., 2014)
OARE-1	Wounding, jasmonic and salicylic acid, UV light, infection with an incompatible race of the crown rust fungus	Oat (*Avena sativa*)	(Kimura et al., 2001)
Bs-1	Barley stripe mosaic virus infection	Maize (*Zea mays*)	(Jin and Bennetzen, 1989)
ZmMI1	Cold	Maize (*Zea mays*)	(Peng and Zhang, 2009
BARE-1	Water-induced stress	Barley (*Hordeum spontaneum)*	(Vicient et al., 1999)
Tnt-1	Protoplast and tissue culture, pathogens, pathogen elicitors, compounds related to plant defense, wounding, freezing, in vitro regeneration, mechanical damage, and microbial factors	Tobacco *(Nicotiana rustica)*	(Pouteau et al., 1991)
Tto-1	Wounding, methyl jasmonate, tissue culture, fungal elicitors, chilling, cytosine demethylation, resistance to bacterial blight, and plant development	Tobacco *(Nicotiana rustica)*	(Hirochika et al., 1996)
ONSEN	Heat stress	Members of the Brassicaceae and *A. thaliana*	(Matsunaga et al., 2011)
LORE-1	Tissue culture	*Lotus japonicus*	Madsen et al. (2005)
Reme-1	UV light	Melon (*Cucumis melo*)	(Ramallo et al., 2008)
CLCoi1	Wounding and salt stress	Lemon (*Citrus limon*)	(Cavrak et al., (2014)

Contd.

Table 2   *Contd.*

Retrotransposon	Foreign Stress	Species	Reference
GBRE-1	Heat stress	*Cotton (Gossypium spp.)*	(Cao et al., 2015)
Tlc-1	Phytohormones, wounding, protoplast preparation, high salt concentration, and stress-associated signaling molecules	Wild Tomato *(Solanum chilense)*	(Tapia et al., 2005)
FaRE-1	Hormonal treatments	Strawberry *(Fragaria ananassa)*	(He et al., 2010)
Ty I /Copia	Heat shock	Arabidopsis *(A. thaliana)*	(Pecinka et al., 2010)
Ty1/ Copia and Ty3/ Gypsy	Scots pine genome upon heat-stressed conditions, the transcriptional activation of different kinds of retrotransposon was identified	Pine *(Pinus sylvestris)*	(Voronova et al., 2011)

promoters. Recent breakthroughs in next-generation sequencing (NGS) technology have changed biology and opened up new possibilities to utilize transposable elements in plant evolution and adaptability under different circumstances.

References

Ahmed, I., Sarazin, A., Bowler, C., Colot, V., and Quesneville, H. "Genome-wide evidence for local DNA methylation spreading from small RNA-targeted sequences in Arabidopsis." *Nucleic acids research* 39, no. 16 (2011): 6919–31.

Alix, K., Gérard, P.R, Schwarzacher, T., and Heslop-Harrison, J.S. "Polyploidy and interspecific hybridization: partners for adaptation, speciation and evolution in plants." *Annals of botany* 120, no. 2 (2017): 183–94.

Alzohairy, A., Sabir, J.S.M., Gyulai, G., Younis, R.A.A., Jansen, R.K., and Bahieldin, A. "Environmental stress activation of plant long-terminal repeat retrotransposons." *Functional Plant Biology* 41, no. 6 (2014): 557–67.

Ananiev, E.V., Phillips, R.L., and Rines, H.W. "Complex structure of knob DNA on maize chromosome 9: retrotransposon invasion into heterochromatin." *Genetics* 149, no. 4 (1998): 2025–37.

Ariel, F.D., and Manavella, P.A. "When junk DNA turns functional: Transposon-derived non-coding RNAs in plants." *Journal of Experimental Botany* 72, no. 11 (2021): 4132–43.

Arumuganathan, Ka, and Earle, E.D. "Nuclear DNA content of some important plant species." *Plant molecular biology reporter* 9 (1991): 208–18.

Avramova, Z., Tikhonov, A., SanMiguel, P., Jin, Y-K., Liu, C., Woo, S-S., Wing, R.A., and Bennetzen, J.K. "Gene identification in a complex chromosomal continuum by local genomic cross-referencing." *The Plant Journal* 10, no. 6 (1996): 1163–68.

Bariah, I., Keidar-Friedman, D., and Kashkush, K. "Where the wild things are: transposable elements as drivers of structural and functional variations in the wheat genome." *Frontiers in Plant Science* 11 (2020): 585515.

Belyayev, A., Kalendar, R., Brodsky, L., Nevo, E., Schulman, A.H., and Raskina, O. "Transposable elements in a marginal plant population: temporal fluctuations provide new insights into genome evolution of wild diploid wheat." *Mobile DnA* 1, no. 1 (2010): 1–16.

Bennetzen, J.L., Springer, P.S., Cresse, A.D., and Hendrickx, M. "Specificity and regulation of the Mutator transposable element system in maize." *Critical reviews in plant sciences* 12, no. 1– 2 (1993): 57–95.

Bennetzen, J.L. "Mechanisms and rates of genome expansion and contraction in flowering plants." *Genetica* 115 (2002): 29–36.

Bennetzen, J.L. "The structure and evolution of angiosperm nuclear genomes." *Current opinion in plant biology* 1, no. 2 (1998): 103–108.

Bennetzen, J.L. "Transposable element contributions to plant gene and genome evolution." *Plant molecular biology* 42 (2000): 251–69.

Bennetzen, J.L., and Wang, H. "The contributions of transposable elements to the structure, function, and evolution of plant genomes." *Annual review of plant biology* 65 (2014): 505–30.

Bennetzen, J.L., Ma, J. and Devos, k.M. "Mechanisms of recent genome size variation in flowering plants." *Annals of botany* 95, no. 1 (2005): 127–32.

Bennetzen, J.L., Schrick, K., Springer, P.S., Brown, W.E. and SanMiguel, P. "Active maize genes are unmodified and flanked by diverse classes of modified, highly repetitive DNA." *Genome* 37, no. 4 (1994): 565–76.

Bingham, P.M. and Zachar, S. Retrotransposons and the FB transposon from Drosophila melanogaster. In: *Mobile DNA*. Berg, D.E. and Howe M.M. (eds.) ASM Press, Washington, DC. (1989): 485–502.

Blanchard, J.L., and Schmidt, G.W. "Pervasive migration of organellar DNA to the nucleus in plants." *Journal of Molecular Evolution* 41 (1995): 397–406.

Boeke, J.D., and Corces, V.G. "Transcription and reverse transcription of retrotransposons." *Annual review of microbiology* 43, no. 1 (1989): 403–34.

Bonchev, G.N. "Useful parasites: The evolutionary biology and biotechnology applications of transposable elements." *Journal of genetics* 95 (2016): 1039–52.

Bonchev, G.I, and Parisod, C. "Transposable elements and microevolutionary changes in natural populations." *Molecular Ecology Resources* 13, no. 5 (2013): 765–75.

Bourque, G., Burns, K.H. Gehring, M., Gorbunova, V., Seluanov, A., Hammell, M., Imbeault, M. et al., "Ten things you should know about transposable elements." *Genome biology* 19 (2018): 1–12.

Brandes, A., Heslop-Harrison, J.S., Kamm, A., Kubis, A., Doudrick, R.L., and Schmidt, T. "Comparative analysis of the chromosomal and genomic organization of Ty1-copia-like retrotransposons in pteridophytes, gymnosperms and angiosperms." *Plant molecular biology* 33 (1997): 11–21.

Brown, A., Yllano, O.B., Arce, L., Evangelista, E.A., Esplana, F.A., Catolico, I.S.R., and Pedro, M.C.L. "Characterization, Comparative, and Phylogenetic Analyses of Retrotransposons in Diverse Plant Genomes." In *Genetic Polymorphisms-New Insights*. IntechOpen, 2021.

Bureau, T.E., Ronald, P.C., and Wessler, S.R. "A computer-based systematic survey reveals the predominance of small inverted-repeat elements in wild-type rice genes." *Proceedings of the National Academy of Sciences* 93, no. 16 (1996): 8524–29.

Camirand, A., St-Pierre, B., Marineau, C., and Brisson, N. "Occurrence of a copia-like transposable element in one of the introns of the potato starch phosphorylase gene." *Molecular and General Genetics MGG* 224 (1990): 33–39.

Cao, Y., Jiang, Y., Ding, M., He, S., Zhang, H., Lin, L., and Rong. J. "Molecular characterization of a transcriptionally active Ty1/copia-like retrotransposon in Gossypium." *Plant cell reports* 34 (2015): 1037–47.

Casacuberta, E., and González, J. "The impact of transposable elements in environmental adaptation." *Molecular ecology* 22, no. 6 (2013): 1503–17.

Casacuberta, J.M., and Santiago, N. "Plant LTR-retrotransposons and MITEs: Control of transposition and impact on the evolution of plant genes and genomes." *Gene* 311 (2003): 1–11.

Cavrak, V.V., Lettner, N., Jamge, S., Kosarewicz, A., Bayer, L.M., and Scheid, O.M. "How a retrotransposon exploits the plant's heat stress response for its activation." *PLoS genetics* 10, no. 1 (2014): e1004115.

Chang, W., Jääskeläinen, M., Li, S-P., and Schulman, A.H. "BARE retrotransposons are translated and replicated via distinct RNA pools." *PLoS One* 8, no. 8 (2013): e72270.

Chavanne, F., Zhang, D-X. Liaud, M-F., and Cerff, R. "Structure and evolution of Cyclops: a novel giant retrotransposon of the Ty3/Gypsy family highly amplified in pea and other legume species." *Plant molecular biology* 37 (1998): 363–75.

Chen, J., Zhao, H., Zheng, X., Liang, K., Guo, Y., and Sun, X. "Recent amplification of Osr4 LTR-retrotransposon caused rice D1 gene mutation and dwarf phenotype." *Plant diversity* 39, no. 2 (2017): 73–79.

Chen, M., SanMiguel, P., and Bennetzen, J.L. "Sequence organization and conservation in sh2/a1-homologous regions of sorghum and rice." *Genetics* 148, no. 1 (1998): 435–43.

Cho, J., Benoit, M., Catoni, M., Drost, H-G., Brestovitsky, A., Oosterbeek, M., and Paszkowski, J. "Sensitive detection of pre-integration intermediates of long terminal repeat retrotransposons in crop plants." *Nature plants* 5, no. 1 (2019): 26–33.

Cossu, R.M., Buti, M., Giordani, T., Natali, L., and Cavallini, A. "A computational study of the dynamics of LTR retrotransposons in the Populus trichocarpa genome." *Tree Genetics & Genomes* 8 (2012): 61–75.

Craig, N.L. "Target site selection in transposition." *Annual review of biochemistry* 66, no. 1 (1997): 437–74.

Curcio, M.J., and Garfinkel, D.J., "Regulation of retrotransposition in Saccharomyces cerevisiae." *Molecular microbiology* 5, no. 8 (1991): 1823–29.

Demirer, G.S., Zhang, H., Goh, N.S., González-Grandío, E., and Landry, M.P. "Carbon nanotube–mediated DNA delivery without transgene integration in intact plants." *Nature protocols* 14, no. 10 (2019): 2954–71.

Demirer, G.S., Zhang, H., Goh, N.S., Pinals, R.L., Chang, R., and Landry, M.P. "Carbon nanocarriers deliver siRNA to intact plant cells for efficient gene knockdown." *Science advances* 6, no. 26 (2020): eaaz0495.

Deragon, J.M., Landry, B.S., Pélissier, T., Tutois, S., Tourmente, S., and Picard, G. "An analysis of retroposition in plants based on a family of SINEs from Brassica napus." *Journal of molecular evolution* 39 (1994): 378–86.

Doolittle, W.F., and Sapienza, C. "Selfish genes, the phenotype paradigm and genome evolution." *Nature* 284, no. 5757 (1980): 601–603.

Dooner, H.K., and Belachew, A. "Transposition pattern of the maize element Ac from the bz-m2 (Ac) allele." *Genetics* 122, no. 2 (1989): 447–57.

Drouin, G., and Dover, G.A. "A plant processed pseudogene." *Nature* 328, no. 6130 (1987): 557–58.

Du, D., Du, X., Mattia, M.R., Wang, Y., Yu, Q., Huang. M., Yu, Y., Grosser, J.W., and Gmitter, F.G. "LTR retrotransposons from the Citrus x clementina genome: characterization and application." *Tree Genetics & Genomes* 14 (2018): 1–14.

Edwards, K.J., Veuskens, J., Rawles, H., Daly, A., and Bennetzen, J.L. "Characterization of four dispersed repetitive DNA sequences from *Zea mays* and their use in constructing contiguous DNA fragments using YAC clones." *Genome* 39, no. 4 (1996): 811–17.

Efron, B., Halloran, E., and Holmes, S., "Bootstrap confidence levels for phylogenetic trees." *Proceedings of the National Academy of Sciences* 93, no. 23 (1996): 13429–429.

Eickbush, T.H. "Transposing without ends: the non-LTR retrotransposable elements." *The new biologist* 4, no. 5 (1992): 430–40.

Elbaidouri, M., and Panaud, O. "Genome-wide analysis of transposition using Next Generation Sequencing technologies." *Plant Transposable Elements: Impact on Genome Structure and Function* (2012): 59–70.

Elizabeth, P. "ENCODE project writes eulogy for junk DNA." *Science* 337, no. 6099 (2012): 1159–60.

Fan, F., Wen, X., Ding, G., and Cui, B. "Isolation, identification, and characterization of genomic LTR retrotransposon sequences from masson pine (Pinus massoniana)." *Tree genetics & genomes* 9 (2013): 1237–46.

Farman, M.L., Tosa, Y., Nitta, N., Leong, S.A., and Leong, S.A. "MAGGY, a retrotransposon in the genome of the rice blast fungus Magnaporthe grisea." *Molecular and General Genetics MGG* 251 (1996): 665–74.

Flavell, R.B., Bennett, M.D., Smith, J.B., and Smith, D.B. "Genome size and the proportion of repeated nucleotide sequence DNA in plants." *Biochemical genetics* 12 (1974): 257–69.

Fukai, E., Umehara, Y., Sato, S., Endo, M., Kouchi, H., Hayashi, M., Stougaard, J., and Hirochika, H. "Derepression of the plant chromovirus LORE1 induces germline transposition in regenerated plants." *PLoS genetics* 6, no. 3 (2010): e1000868.

Gai, X., and Voytas, D.F. "A single amino acid change in the yeast retrotransposon Ty5 abolishes targeting to silent chromatin." *Molecular cell* 1, no. 7 (1998): 1051–55.

Galindo-González, L., Corinne Mhiri, C., Deyholos, M.K., and Grandbastien, M-A., "LTR-retrotransposons in plants: Engines of evolution." *Gene* 626 (2017): 14–25.

Gao, L., McCarthy, E.M., Ganko, E.W., and McDonald, J.F. "Evolutionary history of Oryza sativa LTR retrotransposons: a preliminary survey of the rice genome sequences." *Bmc Genomics* 5, no. 1 (2004): 1–18.

Garber, K., Bilic, I., Pusch, O., Tohme, J., Bachmair, A., Schweizer, D., and Jantsch, V. "The Tpv2 family of retrotransposons of Phaseolus vulgaris: structure, integration characteristics, and use for genotype classification." *Plant molecular biology* 39 (1999): 797–807.

Goodier, J.L. "Restricting retrotransposons: a review." *Mobile DNA* 7 (2016): 1–30.

Grandbastien, M-A., Spielmann, A., and Caboche, M. "Tnt1, a mobile retroviral-like transposable element of tobacco isolated by plant cell genetics." *Nature* 337, no. 6205 (1989): 376–80.

Grandbastien, M-A. "Activation of plant retrotransposons under stress conditions." *Trends in plant science* 3, no. 5 (1998): 181–87.

Grandbastien, M-A. "LTR retrotransposons, handy hitchhikers of plant regulation and stress response." *Biochimica et Biophysica Acta (BBA)-Gene Regulatory Mechanisms* 1849, no. 4 (2015): 403–416.

Grover, C.E., and Wendel, J.F. "Recent insights into mechanisms of genome size change in plants." *Journal of Botany* 2010 (2010).

Hake, S., and Walbot, V. "The genome of Zea mays, its organization and homology to related grasses." *Chromosoma* 79, no. 3 (1980): 251–70.

Havecker, E.R., Gao, X., and Voytas, D.F. "The diversity of LTR retrotransposons." *Genome biology* 5, no. 6 (2004): 1–6.

Hayashi, K., and Yoshida, H. "Refunctionalization of the ancient rice blast disease resistance gene Pit by the recruitment of a retrotransposon as a promoter." *The Plant Journal* 57, no. 3 (2009): 413–25.

He, P., Ma, Y., Zhao, G., Dai, H., Li, H., Chang, L., and Zhang, Z. "FaRE 1: a transcriptionally active Ty1-copia retrotransposon in strawberry." *Journal of plant research* 123 (2010): 707–14.

Heslop-Harrison, J. S., Brandes, A., Taketa, S., Schmidt, T., Vershinin, A.V., Alkhimova, E.G., Kamm A., et al., "The chromosomal distributions of Ty1-copia group retrotransposable elements in higher plants and their implications for genome evolution." *Genetica* 100 (1997): 197–204.

Higashiyama, T., Noutoshi, Y., Fujie, M., and Yamada, T. "Zepp, a LINE-like retrotransposon accumulated in the Chlorella telomeric region." *The EMBO Journal* 16, no. 12 (1997): 3715–23.

Hillis, D.M. Phylogenetic analysis. Current Biology. 1997; 7(3): R129–31.

Hirochika, H., Otsuki, H., Yoshikawa, M., Otsuki, Y., Sugimoto, K., and Takeda, S. "Autonomous transposition of the tobacco retrotransposon Tto1 in rice." *The plant cell* 8, no. 4 (1996): 725–34.

Hirochika, H., Sugimoto, K., Otsuki, Y., Tsugawa, H., and Kanda, M. "Retrotransposons of rice involved in mutations induced by tissue culture." *Proceedings of the National Academy of Sciences* 93, no. 15 (1996): 7783–88.

Hirochika, H. "Contribution of the Tos17 retrotransposon to rice functional genomics." *Current opinion in plant biology* 4, no. 2 (2001): 118–22.

Hollister, J.D., and Gaut, B.S. "Epigenetic silencing of transposable elements: a trade-off between reduced transposition and deleterious effects on neighboring gene expression." *Genome research* 19, no. 8 (2009): 1419–28.

Huang, C.R.L., Burns, K.H., and Boeke, J.D. "Active transposition in genomes." *Annual review of genetics* 46 (2012): 651–75.

Huang, J., Wang, Y., Liu, W., Shen, X., Fan, Q., Jian, S., and Tang, T. "EARE-1, a transcriptionally active Ty1/Copia-like retrotransposon has colonized the genome of Excoecaria agallocha through horizontal transfer." *Frontiers in plant science* 8 (2017): 45.

Hulbert, S.H., Richter, T.E., Axtell, J.D., and Bennetzen, J.L. "Genetic mapping and characterization of sorghum and related crops by means of maize DNA probes." *Proceedings of the National Academy of Sciences* 87, no. 11 (1990): 4251–55.

Iwamoto, M., Nagashima, H., Nagamine, T., Higo, H., and Higo, K. "p-SINE1-like intron of the CatA catalase homologs and phylogenetic relationships among AA-genome Oryza and related species." *Theoretical and applied genetics* 98 (1999): 853–61.

Janicki, M., Rooke, R., and Yang, G. "Bioinformatics and genomic analysis of transposable elements in eukaryotic genomes." *Chromosome research* 19 (2011): 787–808.

Jin, Y-K ., and Bennetzen, J.L. "Structure and coding properties of Bs1, a maize retrovirus-like transposon." *Proceedings of the National Academy of Sciences* 86, no. 16 (1989): 6235–39.

Johns, M. A., Mottinger, J., and Freeling, M. "A low copy number, copia-like transposon in maize." *The EMBO journal* 4, no. 5 (1985): 1093–1101.

Joly-Lopez, Z., and Bureau, T.E. "Diversity and evolution of transposable elements in Arabidopsis." *Chromosome research* 22 (2014): 203–216.

Joly-Lopez, Z., and Bureau, T.E. "Exaptation of transposable element coding sequences." *Current opinion in genetics & development* 49 (2018): 34–42.

Jouffroy, O., Saha, S., Mueller, L., Quesneville, H., and Maumus, F. "Comprehensive repeatome annotation reveals strong potential impact of repetitive elements on tomato ripening." *BMC genomics* 17, no. 1 (2016): 1–15.

Kalendar, R., and Schulman, A.H. "IRAP and REMAP for retrotransposon based genotyping and fingerprinting." *Natureprotocols* 1, (2007): 2478–84.

Kalendar, R ., Shustov, A.V., and Schulman, A.H. "Palindromic sequence-targeted (PST) PCR, version 2: an advanced method for high-throughput targeted gene characterization and transposon display." *Frontiers in plant science* 12 (2021): 691940.

Kalendar, R., Raskina, O., Belyayev, A., and Schulman, A.H. "Long tandem arrays of Cassandra retroelements and their role in genome dynamics in plants." *International Journal of Molecular Sciences* 21, no. 8 (2020): 2931.

Kashkush, K., and Khasdan, V. "Large-scale survey of cytosine methylation of retrotransposons and the impact of readout transcription from long terminal repeats on expression of adjacent rice genes." *Genetics* 177, no. 4 (2007): 1975–85.

Kashkush, K., Feldman, M., and Levy, A.V. "Transcriptional activation of retrotransposons alters the expression of adjacent genes in wheat." *Nature genetics* 33, no. 1 (2003): 102–106.

Kasuga, Salimath, S.S., Shi, J., Gijzen, M., Buzzell, R.I, and Bhattacharyya, M.K. "High resolution genetic and physical mapping of molecular markers linked to the Phytophthora resistance gene Rps1-k in soybean." *Molecular plant-microbe interactions* 10, no. 9 (1997): 1035–1044.

Kejnovsky, E., Tokan, V., and Lexa, M. "Transposable elements and G-quadruplexes." *Chromosome Research* 23 (2015): 615–623.

Kempken, F., and Kück, U. "Transposons in filamentous fungi—facts and perspectives." *BioEssays* 20, no. 8 (1998): 652–59.

Khan, A.L., Al-Harrasi, A., Wang, J-P., Sajjad Asaf, S., Riethoven, J-J.M., Shehzad, T., Liew, C-S., et al., "Genome structure and evolutionary history of frankincense producing Boswellia sacra." *Iscience* 25, no. 7 (2022): 104574.

Kidwell, M.G. "Transposable elements and the evolution of genome size in eukaryotes." *Genetica* 115 (2002): 49–63.

Kidwell, M.G., and D Lisch, D. "Transposable elements as sources of variation in animals and plants." *Proceedings of the National Academy of Sciences* 94, no. 15 (1997): 7704–11.

Kim, J.M., Vanguri, S., Boeke, J.D., Gabriel, A., and Voytas, D.F. "Transposable elements and genome organization: a comprehensive survey of retrotransposons revealed by the complete Saccharomyces cerevisiae genome sequence." *Genome research* 8, no. 5 (1998): 464–78.

Kimura, Y ., Tosa, Y., Shimada, S., Sogo, R., Kusaba, M., Sunaga, T., Betsuyaku, S., Eto, Y., Nakayashiki, H., and Mayama, S. "OARE-1, a Ty1-copia retrotransposon in oat activated by abiotic and biotic stresses." *Plant and cell Physiology* 42, no. 12 (2001): 1345–54.

Kirchner, J., Connolly, C.M., and Sandmeyer, S.B. "Requirement of RNA polymerase III transcription factors for in vitro position-specific integration of a retroviruslike element." *Science* 267, no. 5203 (1995): 1488–91.

Kojima, K.K. "Structural and sequence diversity of eukaryotic transposable elements." *Genes & Genetic Systems* 94, no. 6 (2019): 233–52.

Kossack, D.S., and Kinlaw, C.S. "IFG, a gypsy-like retrotransposon in Pinus (Pinaceae), has an extensive history in pines." *Plant Molecular Biology* 39 (1999): 417–26.

Kralova, T., Cegan, R., Kubat, Z., Vrana, J., Vyskot, B., Vogel, I., Kejnovsky, E., and Hobza, R. "Identification of a novel retrotransposon with sex chromosome-specific distribution in Silene latifolia." *Cytogenetic and Genome Research* 143, no. 1-3 (2014): 87–95.

Kubis, S.E., Heslop-Harrison, J.S., Desel, C., and Schmidt, T. "The genomic organization of non-LTR retrotransposons (LINEs) from three Beta species and five other angiosperms." *Plant molecular biology* 36 (1998): 821–31.

Kumar, A., and Bennetzen, J.L. "Plant retrotransposons." *Annual review of genetics* 33, no. 1 (1999): 479–32.

Kumar, A. "The adventures of the Ty1-copia group of retrotransposons in plants." *Trends in Genetics* 12, no. 2 (1996): 41–43.

Kumar, S.R., Stecher, G., Li, M., Knyaz, C., and Tamura. K., "MEGA X: molecular evolutionary genetics analysis across computing platforms." *Molecular biology and evolution* 35, no. 6 (2018): 1547.

Labrador, M., and Corces, V.G. "Transposable element-host interactions: regulation of insertion and excision." *Annual review of genetics* 31, no. 1 (1997): 381–404.

Leeton, P.R.J., and Smyth, D.R. "An abundant LINE-like element amplified in the genome of Lilium speciosum." *Molecular and General Genetics MGG* 237 (1993): 97–104.

Li, C., Tang, J., Hu, Z., Wang, J., Yu, T., Yi, H., and Cao, M. "A novel maize dwarf mutant generated by Ty1-copia LTR-retrotransposon insertion in Brachytic2 after spaceflight." *Plant cell reports* 39 (2020): 393–408.

Lindroth, A.M., Cao, X., Jackson, J.P., Zilberman, D., McCallum, C.M., Henikoff, S., and Jacobsen, S.E. "Requirement of CHROMOMETHYLASE3 for maintenance of CpXpG methylation." *Science* 292, no. 5524 (2001): 2077–80.

Lisch, D. "How important are transposons for plant evolution?." *Nature Reviews Genetics* 14, no. 1 (2013): 49–61.

Liu, J-X, Jing Liu, J., Yang, T-Y., Liu, L-M., Qiu, L-H., Gao, Y-J., Duan, W-X., et al., "Isolation and analysis of reverse transcriptase of Ty1-copia-like retrotransposons in sugarcane." *Sugar Tech* 24, no. 5 (2022): 1510–29.

Llaca, V., and Messing, J. "Amplicons of maize zein genes are conserved within genic but expanded and constricted in intergenic regions." *The Plant Journal* 15, no. 2 (1998): 211–20.

Lloyd, J.P.B., Ly, F., Gong, P., Pflueger, J., Swain, T., Pflueger, C., Fourie, E., Khan, M.A., Kidd, B.N., and Lister, R. "Synthetic memory circuits for stable cell reprogramming in plants." *Nature Biotechnology* (2022): 1–11.

Loguercio, L. L., and Wilkins, T.A. "Structural analysis of a hmg-coA-reductase pseudogene: insights into evolutionary processes affecting the hmgr gene family in allotetraploid cotton (Gossypium hirsutum L.)." *Current genetics* 34 (1998): 241–49.

Ma, J., Devos, K.M., and Bennetzen, J.L. "Analyses of LTR-retrotransposon structures reveal recent and rapid genomic DNA loss in rice." *Genome research* 14, no. 5 (2004): 860–69.

Madsen, L.H., Eigo Fukai, E., Radutoiu, S., Yost, C.K., Sandal, N., Schauser, L., and Stougaard, J. "LORE1, an active low-copy-number TY3-gypsy retrotransposon family in the model legume Lotus japonicus." *The Plant Journal* 44, no. 3 (2005): 372–81.

Maier, R.M., Neckermann, K., Igloi, G.L., and Kössel, H. "Complete sequence of the maize chloroplast genome: gene content, hotspots of divergence and fine tuning of genetic information by transcript editing." *Journal of molecular biology* 251, no. 5 (1995): 614–28.

Mascagni, F., Barghini, E., Giordani, T., Rieseberg, L.H., Cavallini, A., and Natali, L. "Repetitive DNA and plant domestication: variation in copy number and proximity to genes of LTR-retrotransposons among wild and cultivated sunflower (Helianthus annuus) genotypes." *Genome Biology and Evolution* 7, no. 12 (2015): 3368–82.

Mascagni, F., Giordani, T., Ceccarelli, M., Cavallini, A., and Natali, L. "Genome-wide analysis of LTR-retrotransposon diversity and its impact on the evolution of the genus Helianthus (L.)." *BMC genomics* 18, no. 1 (2017): 1–16.

Masuta, Y., Kawabe, A., Nozawa, K., Naito, K., Kato, A., and Ito, H. "Characterization of a heat-activated retrotransposon in Vigna angularis." *Breeding Science* 68, no. 2 (2018): 168–76.

Masuta, Y., Nozawa, K., Takagi, H., Yaegashi, H., Tanaka, K., Ito, T., Saito, H. et al., "Inducible transposition of a heat-activated retrotransposon in tissue culture." *Plant and Cell Physiology* 58, no. 2 (2017): 375–84.

Matsunaga, W., Kobayashi, A., Kato, A., and Ito, H. "The effects of heat induction and the siRNA biogenesis pathway on the transgenerational transposition of ONSEN, a copia-like retrotransposon in Arabidopsis thaliana." *Plant and Cell Physiology* 53, no. 5 (2012): 824–33.

Meyers, B.C., Chin, D.B., Shen, K.A., Sivaramakrishnan, S., Lavelle, D.O., Zhang, Z., and Michelmore, R.W. "The major resistance gene cluster in lettuce is highly duplicated and spans several megabases." *The Plant Cell* 10, no. 11 (1998): 1817–32.

Meyers, B.C., Tingey, S.V., and Morgante, M. "Abundance, distribution, and transcriptional activity of repetitive elements in the maize genome." *Genome Research* 11, no. 10 (2001): 1660–76.

Miller, J.T., Dong, F., Jackson, S.A., Song, J., and Jiang, J. "Retrotransposon-related DNA sequences in the centromeres of grass chromosomes." *Genetics* 150, no. 4 (1998): 1615–23.

Moore, G., Lucas, H., Batty, N., and Flavell, R. "A family of retrotransposons and associated genomic variation in wheat." *Genomics* 10, no. 2 (1991): 461–68.

Moore, G., Devos, K.M., Wang, Z., and Gale, M.D. "Cereal genome evolution: grasses, line up and form a circle." *Current biology* 5, no. 7 (1995): 737–39.

Nagaki, K., Shibata, F., Kanatani, A., Kashihara, K., and Murata, M. "Isolation of centromeric-tandem repetitive DNA sequences by chromatin affinity purification using a HaloTag7-fused centromere-specific histone H3 in tobacco." *Plant cell reports* 31 (2012): 771–79.

Natali, L., Cossu, R.M., Mascagni, F., Giordani, T., and Cavallini, A. "A survey of Gypsy and Copia LTR-retrotransposon superfamilies and lineages and their distinct dynamics in the Populus trichocarpa (L.) genome." *Tree Genetics & Genomes* 11 (2015): 1–13.

Negi, P., Rai, A.N., and Suprasanna, P. "Moving through the stressed genome: emerging regulatory roles for transposons in plant stress response." *Frontiers in plant science* 7 (2016): 1448.

Nie, Q., Qiao, G., Peng, L., and Wen, X. "Transcriptional activation of long terminal repeat retrotransposon sequences in the genome of pitaya under abiotic stress." *Plant Physiology and Biochemistry* 135 (2019): 460–68.

Noma, K., Nakajima, R., Ohtsubo, H., and Ohtsubo, E. "RIRE1, a retrotransposon from wild rice Oryza australiensis." *Genes & genetic systems* 72, no. 3 (1997): 131–40.

Orgel, L.E., and Crick, H.C. "Selfish DNA: the ultimate parasite." *Nature* 284, no. 5757 (1980): 604–607.

Pan, Y., Meng, F., and Wang, X. "Sequencing multiple cotton genomes reveals complex structures and lays foundation for breeding." *Frontiers in Plant Science* 11 (2020): 560096.

Panstruga, R., Büschges, R., Piffanelli, P., and Schulze-Lefert, P. "A contiguous 60 kb genomic stretch from barley reveals molecular evidence for gene islands in a monocot genome." *Nucleic Acids Research* 26, no. 4 (1998): 1056–62.

Pastuglia, M., Ruffio-Châble, V., Delorme, V., Gaude, T., Dumas, C., and Cock, J.M. "A functionals locus anther gene is not required for the self-incompatibility response in Brassica oleracea." *The Plant Cell* 9, no. 11 (1997): 2065–76.

Paterson, A.H., Bowers, J.E., Bruggmann, R., Dubchak, I., Grimwood, J., Gundlach, H., Haberer, G., et al., "The Sorghum bicolor genome and the diversification of grasses." *Nature* 457, no. 7229 (2009): 551–56.

Pattengale, N.D., Alipour, M., Bininda-Emonds, O.R.P., Moret, B.M.E., and Stamatakis, A. "How many bootstrap replicates are necessary?." *Journal of computational biology* 17, no. 3 (2010): 337–54.

Paz, R.C., Kozaczek, M.E., Rosli, H.G., Andino, N.P., and Sanchez-Puerta, M.V. "Diversity, distribution and dynamics of full-length Copia and Gypsy LTR retroelements in Solanum lycopersicum." *Genetica* 145 (2017): 417–430.

Pearce, S.R., Li, D., Flavell, A.J., Harrison, G., Heslop-Harrison, J.S., and Kumar, A. "The Ty1-copia group retrotransposons in Vicia species: copy number, sequence heterogeneity and chromosomal localisation." *Molecular and General Genetics MGG* 250 (1996a): 305–15.

Pearce, S.R., Harrison, G., Heslop-Harrison, P., Flavell, A.J., and Kumar, A. "Characterization and genomic organization of Ty1-copia group retrotransposons in rye (Secale cereale)." *Genome* 40, no. 5 (1997): 617–25.

Pearce, S.R., Pich, U., Harrison, G., Flavell, A.J., Heslop-Harrison, J.S., Schubert, I., and Kumar, A. "The Ty1-copia group retrotransposons of Allium cepa are distributed throughout the chromosomes but are enriched in the terminal heterochromatin." *Chromosome Research* 4 (1996b): 357–64.

Pecinka, A., Dinh, H.Q., Baubec, T., Rosa, M., Lettner, N., and Scheid, O.M.. "Epigenetic regulation of repetitive elements is attenuated by prolonged heat stress in Arabidopsis." *The Plant Cell* 22, no. 9 (2010): 3118–29.

Peng, H., and Zhang, J. "Plant genomic DNA methylation in response to stresses: Potential applications and challenges in plant breeding." *Progress in Natural Science* 19, no. 9 (2009): 1037–45.

Piednoël, M., Carrete-Vega, G., and Renner, S.S. "Characterization of the LTR retrotransposon repertoire of a plant clade of six diploid and one tetraploid species." *The Plant Journal* 75, no. 4 (2013): 699–709.

Piegu, B., Guyot, R., and Picault, N. "Doubling genome size without polyploidization: Dynamics of." *Genome Res* 21 (2011): 1201.

Pimpinelli, S., Berloco, M., Fanti, L., Dimitri, P., Bonaccorsi, S., Marchetti, E., Caizzi, R., Caggese, C., and Gatti, M. "Transposable elements are stable structural components of Drosophila melanogaster heterochromatin." *Proceedings of the National Academy of Sciences* 92, no. 9 (1995): 3804–08.

Pouteau, Sylvie, E. Huttner, Marie Angele Grandbastien, and Michel Caboche. "Specific expression of the tobacco Tnt1 retrotransposon in protoplasts." *The EMBO journal* 10, no. 7 (1991): 1911–18.

Pozueta-Romero, J., Houlné, and Schantz, R. "Identification of a short interspersed repetitive element in partially spliced transcripts of the bell pepper (Capsicum annuum) PAP gene: new evolutionary and regulatory aspects on plant tRNA-related SINEs." *Gene* 214, no. 1-2 (1998): 51–58.

Presting, G.G., Malysheva, L., Fuchs, J., and Schubert, I. "A TY3/GYPSY retrotransposon-like sequence localizes to the centromeric regions of cereal chromosomes." *The Plant Journal* 16, no. 6 (1998): 721–28.

Purugganan, M.D., and Wessler, S.R. "Molecular evolution of magellan, a maize Ty3/gypsy-like retrotransposon." *Proceedings of the National Academy of Sciences* 91, no. 24 (1994): 11674–78.

Qin, C., Yu, C., Shen, S., Fang, X., Chen, L., Min, J., Cheng., J. et al., "Whole-genome sequencing of cultivated and wild peppers provides insights into Capsicum domestication and specialization." *Proceedings of the National Academy of Sciences* 111, no. 14 (2014): 5135–40.

Qiu, F., and Ungerer, M.C. "Genomic abundance and transcriptional activity of diverse gypsy and copia long terminal repeat retrotransposons in three wild sunflower species." *BMC plant biology* 18, no. 1 (2018): 1–8.

Ragupathy, R., Banks, T., and Cloutier, S. "Molecular characterization of the Sasanda LTR copia retrotransposon family uncovers their recent amplification in Triticum aestivum (L.) genome." *Molecular Genetics and Genomics* 283 (2010): 255–71.

Ramakrishnan, M., Lakkakula S., Kalendar, R., Narayanan, M., Kandasamy, S., Sharma, A., Emamverdian, A., Wei, Q., and Zhou, M. "The dynamism of transposon methylation for plant development and stress adaptation." *International Journal of Molecular Sciences* 22, no. 21 (2021): 11387.

Ramakrishnan, M., Papolu, P.K., Mullasseri, S., Zhou, M., Sharma, A., Ahmad, Z., Viswanathan S., Kalendar, R., and Wei, Q. "The role of LTR retrotransposons in plant genetic engineering: how to control their transposition in the genome." *Plant Cell Reports* (2023): 1–13.

Ramallo, E., Kalendar, R., Schulman, A.H., and Martínez-Izquierdo, J.A. "Reme1, a Copia retrotransposon in melon, is transcriptionally induced by UV light." *Plant Molecular Biology* 66 (2008): 137–50.

Ravindran, S. "Barbara McClintock and the discovery of jumping genes." *Proceedings of the National Academy of Sciences* 109, no. 50 (2012): 20198–99.

Rebollo, R., Mark T. Romanish, and Dixie L. Mager. "Transposable elements: an abundant and natural source of regulatory sequences for host genes." *Annual review of genetics* 46 (2012): 21–42.

Rho, M., Choi, J-H., Kim, S., Lynch, M., and Tang, H. "De novo identification of LTR retrotransposons in eukaryotic genomes." *BMC genomics* 8, no. 1 (2007): 1–16.

Robbins, T.P., Walker, E.L., Kermicle, J.L., Alleman, M., and Dellaporta, S.L. "Meiotic instability of the Rr complex arising from displaced intragenic exchange and intrachromosomal rearrangement." *Genetics* 129, no. 1 (1991): 271–83.

Royo, J., Nass, N., Matton, D.P., Okamoto, S., Clarke, A.E., and Newbigin, E. "A retrotransposon-like sequence linked to the S-locus of Nicotiana alata is expressed in styles in response to touch." *Molecular and General Genetics MGG* 250 (1996): 180–88.

Rudolf, E.E., Hüther, P., Forné, I., Georgii, E., Han, Y., Hell, R., Wirtz, M., et al., "GSNOR contributes to demethylation and expression of transposable elements and stress-responsive genes." *Antioxidants* 10, no. 7 (2021): 1128.

Sabot, F., and Schulman, A.H. "Parasitism and the retrotransposon life cycle in plants: a hitchhiker's guide to the genome." *heredity* 97, no. 6 (2006): 381–88.

Sacks-Davis, R., Daraganova, G., Aitken, C., Higgs, P., Tracy, L., Bowden, S., Jenkinson, R., et al., "Hepatitis C virus phylogenetic clustering is associated with the social-injecting network in a cohort of people who inject drugs." *PloS one* 7, no. 10 (2012): e47335.

Saika, H., Mori, A., Endo, M., and Toki, S. "Targeted deletion of rice retrotransposon Tos17 via CRISPR/Cas9." *Plant cell reports* 38 (2019): 455–58.

Sanchez, D.H., Hervé G., Drost, H-G., Radu Zabet, N., and Paszkowski, J. "High-frequency recombination between members of an LTR retrotransposon family during transposition bursts." *Nature Communications* 8, no. 1 (2017): 1283.

SanMiguel, P., Tikhonov, A., Jin, Y-K., Motchoulskaia, N., Zakharov, D., Melake-Berhan, A., Springer, S. et al., "Nested retrotransposons in the intergenic regions of the maize genome." *Science* 274, no. 5288 (1996): 765–68.

Sanmiguel, P., and Bennetzen, J.L. "Evidence that a recent increase in maize genome size was caused by the massive amplification of intergene retrotransposons." *Annals of Botany* 82 (1998): 37–44.

SanMiguel, P., Gaut, B.S., Tikhonov, A., Nakajima, Y., and Bennetzen, J.L. "The paleontology of intergene retrotransposons of maize." *Nature genetics* 20, no. 1 (1998): 43–45.

Sato, Y., Sentoku, N., Miura, Y., Hirochika, Y., Kitano, Y., and Matsuoka, M. "Loss-of-function mutations in the rice homeobox gene OSH15 affect the architecture of internodes resulting in dwarf plants." *The EMBO Journal* 18, no. 4 (1999): 992–1002.

Schmidt, T., Kubis, S., and Heslop-Harrison, J.S. "Analysis and chromosomal localization of retrotransposons in sugar beet (Beta vulgaris L.): LINEs and Ty1-copia-like elements as major components of the genome." *Chromosome Research* 3 (1995): 335–45.

Schmidt, T. "LINEs, SINEs and repetitive DNA: non-LTR retrotransposons in plant genomes." *Plant molecular biology* 40 (1999): 903–10.

Schnable, P.S., Ware, D., Fulton, R.S., Stein, J.C., Wei, F., Pasternak, S., Liang, C., et al., "The B73 maize genome: complexity, diversity, and dynamics." *Science* 326, no. 5956 (2009): 1112–15.

Schorn, A.J., Gutbrod, M.J., LeBlanc, C., and Martienssen, R. "LTR-retrotransposon control by tRNA-derived small RNAs." *Cell* 170, no. 1 (2017): 61–71.

Schulman, A.H. "Hitching a ride: nonautonomous retrotransposons and parasitism as a lifestyle." *Plant transposable elements: impact on genome structure and function* (2012): 71–88.

Schuster, W. and Brennicke, A. "Plastid, nuclear and reverse transcriptase sequences in the mitochondrial genome of Oenothera: is genetic information transferred between organelles via RNA?." *The EMBO Journal* 6, no. 10 (1987): 2857–63.

Setiawan, A.B., Teo, C.H., Kikuchi, S., Sassa, H., Kato, K., and Koba, T. "Chromosomal locations of a non-LTR retrotransposon, menolird18, in Cucumis melo and Cucumis sativus, and its implication on genome evolution of cucumis species." *Cytogenetic and genome research* 160, no. 9 (2020): 554–564.

Shahid, S., and Slotkin, R.K. "The current revolution in transposable element biology enabled by long reads." *Current opinion in plant biology* 54 (2020): 49–56.

Sharma, A., Schneider, K.L., and Presting, G.G. "Sustained retrotransposition is mediated by nucleotide deletions and interelement recombinations." *Proceedings of the National Academy of Sciences* 105, no. 40 (2008): 15470–74.

Sharma, A., Wolfgruber, T.K., and Presting, G.G. "Tandem repeats derived from centromeric retrotransposons." *BMC genomics* 14, no. 1 (2013): 1–11.

Slotkin, R.K., and Martienssen, R. "Transposable elements and the epigenetic regulation of the genome." *Nature reviews genetics* 8, no. 4 (2007): 272–85.

Song, S.U., Gerasimova, T., Kurkulos, M., D. Boeke, J., and Corces, V.G. "An env-like protein encoded by a Drosophila retroelement: evidence that gypsy is an infectious retrovirus." *Genes & development* 8, no. 17 (1994): 2046–57.

Stitzer, M.C., Anderson, S.N., Springer, N.M., and Ross-Ibarra, J. "The genomic ecosystem of transposable elements in maize." *PLoS genetics* 17, no. 10 (2021): e1009768.

Suoniemi, A., Schmidt, D., and Schulman, A.H. "BARE-1 insertion site preferences and evolutionary conservation of RNA and cDNA processing sites." *Evolution and Impact of Transposable Elements* (1997): 219–30.

Suoniemi, A., Tanskanen, J., and Schulman, A.H. "Gypsy-like retrotransposons are widespread in the plant kingdom." *The Plant Journal* 13, no. 5 (1998): 699–705.

Tamura, K., and Nei, M. "Estimation of the number of nucleotide substitutions in the control region of mitochondrial DNA in humans and chimpanzees." *Molecular biology and evolution* 10, no. 3 (1993): 512–26.

Tang, M., Wu, X., Cao, Y., Qin, Y., Ding, M., Jiang, Y., Sun, C., Zhang, H., Paterson, A.H., and Rong, J. "Preferential insertion of a Ty1 LTR-retrotransposon into the A sub-genome's HD1 gene significantly correlated with the reduction in stem trichomes of tetraploid cotton." *Molecular Genetics and Genomics* 295 (2020): 47–54.

Tapia, G., Verdugo, I., Yañez, M., Ahumada, I., Theoduloz, C., Cordero, C., Poblete, F., González, E., and Ruiz-Lara, S. "Involvement of ethylene in stress-induced expression of the TLC1. 1 retrotransposon from Lycopersicon chilense Dun." *Plant Physiology* 138, no. 4 (2005): 2075–86.

Tikhonov, A.P., SanMiguel, P.J., Nakajima, Y., Gorenstein, N.M., Bennetzen, J.L., and Avramova, Z. "Colinearity and its exceptions in orthologous adh regions of maize and sorghum." *Proceedings of the National Academy of Sciences* 96, no. 13 (1999): 7409–14.

Todorovska, E. "Retrotransposons and their role in plant—genome evolution." *Biotechnology & Biotechnological Equipment* 21, no. 3 (2007): 294–305.

Ujino-Ihara, T. "Transcriptome analysis of heat stressed seedlings with or without pre-heat treatment in Cryptomeria japonica." *Molecular Genetics and Genomics* 295, no. 5 (2020): 1163–72.

Unseld, M., Marienfeld, J.R., Brandt, P., and Brennicke, A. "The mitochondrial genome of Arabidopsis thaliana contains 57 genes in 366,924 nucleotides." *Nature Genetics* 15, no. 1 (1997): 5–7 61.

Usai, G., Mascagni, Natali, L., Giordani, T., and Cavallini, A. "Comparative genome-wide analysis of repetitive DNA in the genus Populus L." *Tree genetics & genomes* 13 (2017): 1–12.

Vicient, C.M., and Casacuberta, J.M. "Impact of transposable elements on polyploid plant genomes." *Annals of Botany* 120, no. 2 (2017): 195–207.

Vicient, C., Suoniemi, A., Anamthawat-Jónsson, K., Tanskanen, J., Beharav, A., Nevo, E., and Schulman, A.H. "Retrotransposon BARE-1 and its role in genome evolution in the genus Hordeum." *The Plant Cell* 11, no. 9 (1999): 1769–84.

Vignols, F., Rigau, J., Torres, M.A., Capellades, M., and Puigdomènech, P. "The brown midrib3 (bm3) mutation in maize occurs in the gene encoding caffeic acid O-methyltransferase." *The Plant Cell* 7, no. 4 (1995): 407–16.

Vitte, C., Fustier, M-A., Alix, K., and Tenaillon, M.I. "The bright side of transposons in crop evolution." *Briefings in Functional Genomics* 13, no. 4 (2014): 276–95.

Voronova, A., Jansons, A., and Ruņģis, D. "Expression of retrotransposon-like sequences in Scots pine (Pinus sylvestris) in response to heat stress." *Environmental and Experimental Biology* 9 (2011): 121–27.

Voytas, D.F., Cummings, M.P., Koniczny, A., Ausubel, F.M., and Rodermel, S.R. "Copia-like retrotransposons are ubiquitous among plants." *Proceedings of the National Academy of Sciences* 89, no. 15 (1992): 7124–28.

Vukich, M., Giordani, T., Lucia Natali, and Andrea Cavallini. "Copia and Gypsy retrotransposons activity in sunflower (Helianthus annuus L.)." *BMC Plant Biology* 9 (2009): 1–12.

Wang, L., He, Y., Qiu, H., Guo, J., Han, M., Zhou, J., Sun, Q., and Sun, J. "Mdorycol-1, a bidirectionally transcriptional Ty1-copia retrotransposon from Malus× domestica." *Scientia Horticulturae* 220 (2017): 283–90.

Wang, Xi, Weigel, D., and Smith, L.M. "Transposon variants and their effects on gene expression in Arabidopsis." *PLoS genetics* 9, no. 2 (2013): e1003255.

Wessler, S.R., Bureau, T.E., and White, S.E. "LTR-retrotransposons and MITEs: important players in the evolution of plant genomes." *Current opinion in genetics & development* 5, no. 6 (1995): 814–21.

White, S.E., Habera, L.F, and Wessler, S.R. "Retrotransposons in the flanking regions of normal plant genes: a role for copia-like elements in the evolution of gene structure and expression." *Proceedings of the National Academy of Sciences* 91, no. 25 (1994): 11792–96.

Whitelaw, E., and Martin, D.I.K. "Retrotransposons as epigenetic mediators of phenotypic variation in mammals." *Nature Genetics* 27, no. 4 (2001): 361–65.

Wicker, T., Sabot, F., Hua-Van, A., Bennetzen, J.L., Capy, P., Chalhoub, B., Flavell, A., et al., "A unified classification system for eukaryotic transposable elements." *Nature reviews genetics* 8, no. 12 (2007): 973–82.

Wicker, T., Gundlach, H., Spannagl, M., Uauy, C., Borrill, P., Ramírez-González, R.H., De Oliveira, R., Mayer, K.F.X., Paux, E., and Choulet, F. "Impact of transposable elements on genome structure and evolution in bread wheat." *Genome biology* 19, no. 1 (2018): 1–18.

Wright, D.A., Ke, N., Smalle, J., Hauge, B.M., Goodman, H.M., and Voytas, D.F. "Multiple non-LTR retrotransposons in the genome of Arabidopsis thaliana." *Genetics* 142, no. 2 (1996): 569–78.

Xia, E., Tong, W., Hou, Y., An, Y., Chen, L., Wu, Q., Liu, Y., et al., "The reference genome of tea plant and resequencing of 81 diverse accessions provide insights into its genome evolution and adaptation." *Molecular plant* 13, no. 7 (2020): 1013–26.

Xiao, H., Jiang, N., Schaffner, E., Stockinger, E.J., and van der Knaap, E. "A retrotransposon-mediated gene duplication underlies morphological variation of tomato fruit." *Science* 319, no. 5869 (2008): 1527–30.

Xiong, Y., and Eickbush, T.H. "Similarity of reverse transcriptase-like sequences of viruses, transposable elements, and mitochondrial introns." *Molecular Biology and Evolution* 5, no. 6 (1988): 675–90.

Yin, H., Liu, J., Xu, Y., Liu, X., Zhang, S., Ma, J., and Du, J. "TARE1, a mutated Copia-like LTR retrotransposon followed by recent massive amplification in tomato." *PloS one* 8, no. 7 (2013): e68587.

Yoshioka, Y., Matsumoto, S., Kojima, S., Ohshima, K., Okada, N., and Machida, Y. "Molecular characterization of a short interspersed repetitive element from tobacco that exhibits sequence homology to specific tRNAs." *Proceedings of the National Academy of Sciences* 90, no. 14 (1993): 6562–66.

Youngman, S., van Luenen, H.G., and Plasterk, R.H.A. "Rte-1, a retrotransposon-like element in Caenorhabditis elegans." *FEBS letters* 380, no. 1–2 (1996): 1–7.

Zervudacki, J., Yu, A., DelaseAmesefe, Wang, J., Drouaud, J., Navarro, L., and Deleris, A. "Transcriptional control and exploitation of an immune-responsive family of plant retrotransposons." *The EMBO Journal* 37, no. 14 (2018): e98482.

Zhang, D., Kan, X., Huss, S.E., Jiang, L., Chen, L-Q., and Hu, Y. "Using phylogenetic analysis to investigate eukaryotic gene origin." *JoVE (Journal of Visualized Experiments)* 138 (2018b): e56684.

Zhang, H., Cao, Y., Xu, D., Goh, N.S., Demirer, G.S., Cestellos-Blanco, S., Chen, Y., Landry, M.P., and Yang, P. "Gold-nanocluster-mediated delivery of siRNA to intact plant cells for efficient gene knockdown." *Nano letters* 21, no. 13 (2021): 5859–66.

Zhang, H., Lang, Z., and Zhu, J-K. "Dynamics and function of DNA methylation in plants." *Nature reviews Molecular cell biology* 19, no. 8 (2018a): 489–506.

Zhang, L., Yan, L., Jiang, J., Wang, Y., Jiang, Y., Yan, T., and Cao, Y. "The structure and retrotransposition mechanism of LTR-retrotransposons in the asexual yeast Candida albicans." *Virulence* 5, no. 6 (2014): 655–64.

Zilberman, D., Cao, X., and Jacobsen. "ARGONAUTE4 control of locus-specific siRNA accumulation and DNA and histone methylation." *Science* 299, no. 5607 (2003): 716–719.

Zou, S., and Voytas D.F. "Silent chromatin determines target preference of the Saccharomyces retrotransposon Ty5." *Proceedings of the National Academy of Sciences* 94, no. 14 (1997): 7412–16.

3

Retrotransposons and Development of Plants

Shubham Kumar,[1] *Bhawana,*[1] *Jasdeep Singh,*[1] *Jyoti Sarwan,*[1]
and *Pooja Bhadrecha*[1*]

1. Introduction

The wide range of eukaryotic genomes are composed of mobile genetic material known as transposable elements (TEs). The structure of genes and the genome of organisms depends on the organization of these TEs. Retrotransposons and minimal DNA transposons are the most dominating TEs in shaping the eukaryotic genome. The new genome insertions occur through the duplication process of replicative transposons without changing the native form of the element (Papolu et al., 2022; Kalendar et al., 2018). TEs represent the potential reservoir of RNA level toxicity and genomic instability . TEs replicate and get located in new places due to which the somatic transpositions are generated. However, some stagnant and non-working TEs are also present. TEs are the most significant elements for the evolution of organisms as they differ in terms of size, structure, and mechanism of transposition in their genomes (Klein and Anderson 2022). The process of transposition and replication of class I and II TEs leads to novel insertions in the genome, keeping the original elements intact, by either the mechanism of 'copy & paste' or 'cut & paste', as displayed in Fig. 1.

[1] First authorship is equally shared by Bhawana and Shubham Kumar,
UIBT, Chandigarh University, Gharuan, Mohali (Punjab), 140413, India.

* Corresponding author: pbhadrecha.pb@gmail.com, pooja.e11028@cumail.in

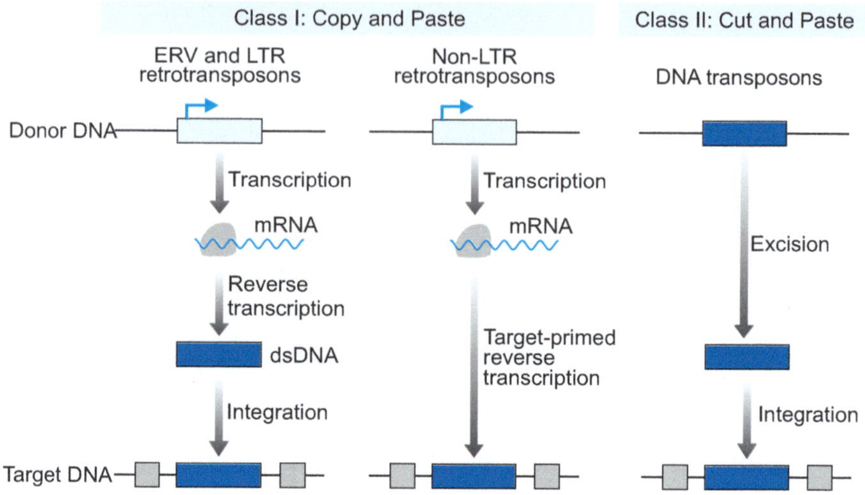

Fig. 1 Novel insertions of class I and class II TEs in the genome by copy-paste and cut-paste mechanisms (Wicker T et al., 2007)

TEs have a great impact on gene expression and regulatory patterns on long terminal repeats (LTR) retrotransposons containing the LTR's regulatory capsules on both ends so that they cannot influence the adjacent genes. LTRs can initiate antisense transcripts regulating gene expression or it is a source of regulatory sequences, also known as promoters (Sundaram and Wysocka 2020). LTRs target particular stimuli due to which the adjacent host genes influence and contribute to the organism's response by activating the stimuli and most of the LTR retrotransposons get activated during conditions of environmental stress (Ramakrishnan et al., 2022b). Figure 2 displays the basic structure of an LTR retrotransposon.

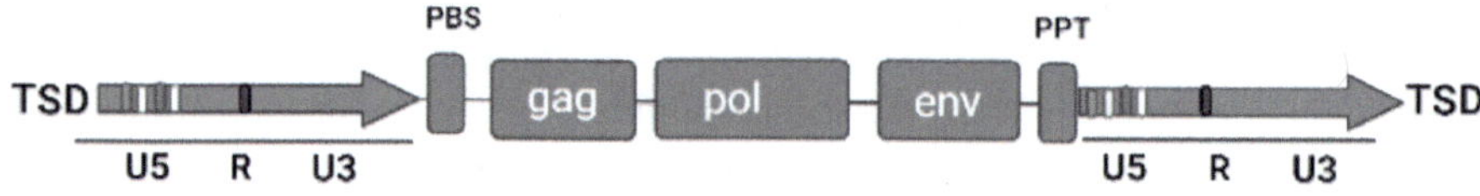

Fig. 2 Structure of LTR retrotransposon (Orozco-Arias et al., 2019)

Retrotransposons are the main components of the plant genome and are present as multiple copies. The encoding reverse transcriptase/RNaseH enzymes transform the intermediary RNA used by retrotransposons to travel to new chromosomal sites into extrachromosomal DNA before reinserted into the genome (Woodrow et al., 2012). The significant increase in the size of the genome occurs due to rapidly multiplying the genetic material by replicative mechanisms of transposition, however the elements (Ac, Tam 1, and EN/Spm) transpose by the excision mechanisms of transposition do not expand the size of plant's genome (Smith,

2015). DNA retrotransposable elements can form mutations through insertion into or close to the genes (Sahebi et al., 2018). The induced mutations of retrotransposon generally appear steady as they transfer through the process of replication, and the insertion site sequence is preserved (He et al., 2022). A retrotransposon can be used in the biological process which benefits the organisms because they exhibit genome expression and follow many of the same inheritance rules as the genes of their genomic "host". It can be better understood by an example: The telomerase function can be removed in insects by inserting the retrotransposon in Drosophila at the telomeric stage. Retrotransposons can potentially change the gene function and structure of the host genome. However, the transpositional activities are controlled by the host-encoded factors and retrotransposons, both to avoid the harmful effect on the survival of host retrotransposons (Galindo-González et al., 2017). The interactions between the host plant and retrotransposons existed prior millions of years but the evolution of their interaction can be studied, to understand the regulation of retrotransposon mechanisms such as mutational outcomes and insertion specificities for each other's survival (Goodier, 2016). In this chapter, we discuss the classification and structure of transposons, their mechanism of action, along with their role in the development of plants. Figure 3 represents the general classification of TEs, following Rexdb and GyDB nomenclature.

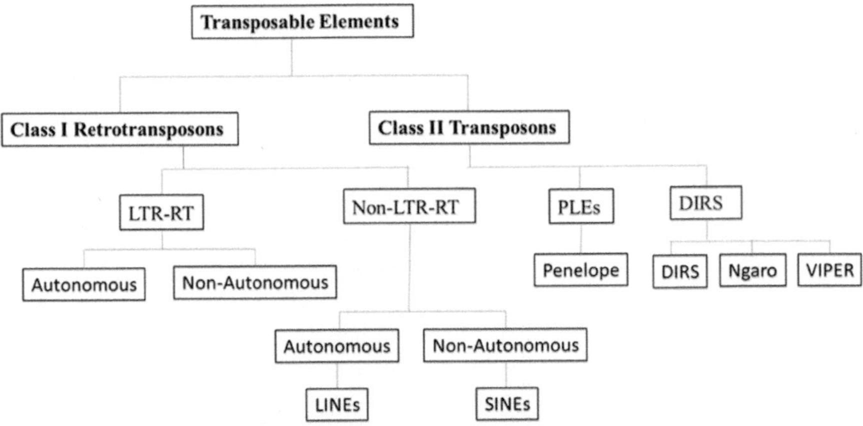

Fig. 3   Classification of Transposable Elements following Rexdb and GyDB nomenclature (Orozco-Arias et al., 2019)

2.   Origins and Evolution of Retrotransposon

The formation of retrotransposons and their entry into a given kingdom or species are both largely unknown. LTR and LINEs (Long Interspersed Nuclear Elements) retrotransposons both create conserved gene products, hence it is possible to compare their sequences, LINE elements were reported to be the primary retrotransposons in which terminal direct repeats allowed them to evolve into LTR retrotransposons (Ou and Jiang, 2018). The HeT-A element's tandem insertion in Drosophila demonstrates how this procedure could produce two flanking LTRs (McCullers and Steiniger, 2017). By acquiring an envelope function, retroviruses seem to have diverged from the retrotransposons subclass called Ty3-gypsy in animals. One theory is that the DNA-based genome that all living cells currently began to emerge from a hypothetical RNA-based cellular genome when retrotransposons first appeared (Akhlaghpour, 2022). Plant retrotransposons are extremely common, and their great sequence heterogeneity shows that they have been present in plants from ancient times (Cervera et al., 2016). Another possibility is that retrotransposons developed after the first eukaryotes were created and spread widely through a mix of vertical and horizontal transmission (Metzger et al., 2018). However, it is evident that they have multiplied quite successfully in all higher eukaryotic genomes (Tollis and Boissinot, 2012). They have made numerous and important contributions to plant genome development. The beneficial mutational modifications and selective forces can act for the advancement of organisms by using the retrotransposon sequence, just like the DBA sequence of genomic elements in plants (Bourque et al., 2018).

3.   Classification and Structure of Retrotransposons

TE is classified into two classes, one is Retrotransposons (class I) which migrate through RNA molecules and the other is Transposons (class II) which migrate through DNA molecules (Deniz et al., 2019). For example, the plant genome comprises class I and class II elements, along with Helitrons, and their insertion mechanisms are explained in Fig. 4.

They are further divided into four orders based on their types of structures and transposon's life cycle: (i) Long terminal repeat retrotransposons (LTR-RT), (ii) non-LTR retrotransposons, (iii) PLEs, (iv) DIRS. In terms of their structure, the existence and arrangement of enzyme regulatory genes, domains, and motifs, as well as their life cycles, all of these exhibit significant differences (Orozco-Arias et al., 2019). LTRs are divided into TY3-gypsy and TY-1 copia groups, which differ from one another in terms of both the degree of sequence similarity and the sequence in which the encoded gene products are found (Yadav and Prasad, 2017). Ty1-copia retrotransposons can be found in taxa ranging from single-celled algae to bryophytes, gymnosperms, and angiosperms, and they are widespread across the plant kingdom. Angiosperms and gymnosperms have been reported to contain Ty3 gy psy retrotransposons, which are widely spread throughout the plant kingdom (Lee and Kim, 2014). Furthermore, plants with large genomes frequently include both of these kinds of retrotransposons in a high number of copies. The plant species examined so far contain high copy of the non-LTR retrotransposons LINEs and SINEs as well (López-Flores and Garrido-Ramos, 2012). Similar to LTR retrotransposons, LINEs are observed in different types of angiosperms and may be prevalent throughout all plant species (He et al., 2022).

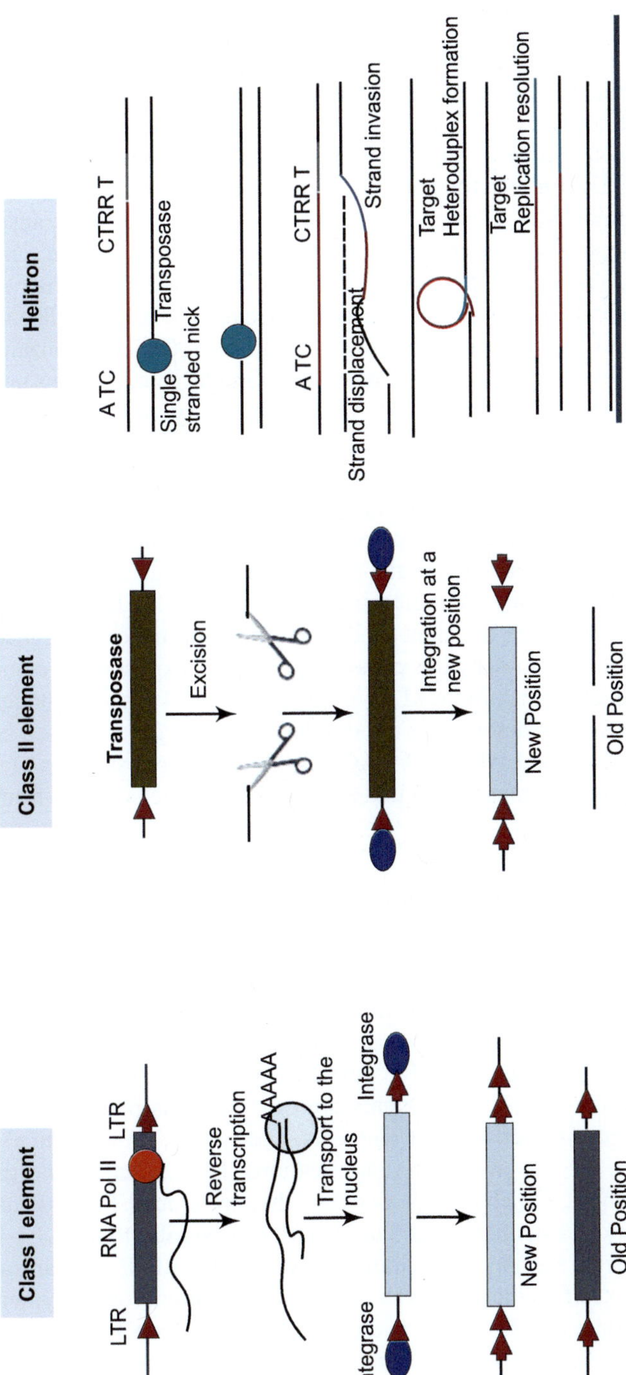

Fig. 4 Plant genomes have a small number of generic groups, which are explained below and shown in the picture (Lisch, 2013).

3.1   Long Terminal Repeat Retrotransposons (LTR)

LTR is the most abundant component in a plant's genome in which the genome size is not correlated with physiological complexity. LTR contains two main subfamilies Ty 1 *copia* and TY3-*gypsy* elements (Galindo-González et al., 2017). They can be differentiated through their internal domain organization that code for proteins essential for retrotransposons. LTR flanks the internal protein domain and contains promoter-like *cis* regulatory motifs responsible for the transcription in portable components. Therefore, retrotransposons are rejuvenated during the retro transposition process so the new copies containing the LTR on their ends are known. Consequently, LTRs are utilised as a molecular clock to regulate the time since insertion, and their sequence differences can be seen (Galindo-González et al., 2017). The Ty3-gypsy elements tend to insert in heterochromatic regions in most of the plants (González and Deyholos, 2012) However, the Ty1-*copia* element inserts in nearby regions of the genes due to the random insertions in the whole genome. This element is found nearer to the genes because they do not disrupt the gene function. When these become disruptive the purifying selections eliminate them. The disruptive elements have low copy patterns and the younger retrotransposons are most proximally associated with the gene and have a high number of families (Xu and Du, 2014). Many genomes include several LTR-RTs in very high copy counts, but most are deficient in the transposition-necessary functioning genes. Some of them possess the capacity to retrotransposon by consuming the functional mechanism produced by other LTR-RTs. These components are referred to be non-autonomous and are divided into four groups based on their structural characteristics: LARD, TR GAG, and BARE-2 are examples of TRIMs are terminal-repeat retrotransposons that are small, with sizes varying from a few hundred bases to 4 kbp.

3.2   Non LTR-Retrotransposons

LTRs are absent in non-LTR retrotransposons, although they still have an internal promoter for transcription. Even without an INT domain, these components can duplicate. Instead, the non-LTR retrotransposon transcript's poly-A tail serves as the starting point for DNA synthesis, and the newly produced DNA's end is joined to the insertion site by the RT domain (Schulman, 2012). Compared to LTR retrotransposons, these components are often far less prevalent in plants. LINEs, or long interspersed nuclear elements, and SINEs, or short interspersed nuclear elements, are the two main classifications. LINEs feature the gag and pol coding sequences, which encode domains that are crucial for structural and enzymatic activity (Casacuberta and Santiago, 2003), Similar to the LTR-RT life cycle, SINE elements are non-autonomous and rely on the LINE mechanism since they are unable to self-replicate. SINEs are made up of numerous polymerase III transcripts, including tRNA, rRNA, and others with lengths between 75 and 662 bp (Schulman, 2013). LINEs, which can be several kbp long and typically encode both RNA polymerase II produce reverse transcriptase and endonuclease genes that are located inside the same ORF. The transcription cycle frequently results in a poly-A tail at the 3′end of non-LTR retrotransposons. SINEs, like LINEs, are similarly ended by an A-rich tail, but unlike LINEs, they share the host genes' sequence. Similar to LTR retrotransposons, LINEs and SINEs generate TSDs., however, non-LTR retrotransposons produce TSDs on the insertion site that are of different sizes (Huang et al., 2012). Figure 5 displays the general structural organization of LINE and SINE.

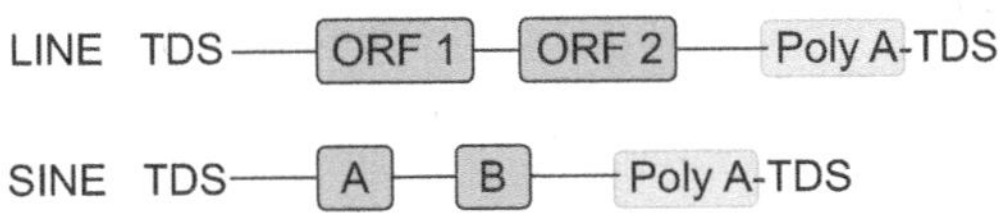

Fig. 5   Non LTR-Retrotransposons (Orozco-Arias et al., 2019)

3.2.1   *LINEs – long interspersed nuclear elements*

LTR retrotransposons are more complex, LINE elements nonetheless encode many of the same proteins. The gag and pol genes are present in LINEs, but no known integrase is present (Joan Curcio et al., 2015). DNA integration version of LINEs into chromosomal DNA may be facilitated with endonucleolytic activity (EN) protein, which is often specified by the gag loci of LINEs (Saleh et al., 2019). A low-frequency DNA repair activity may also integrate LINEs into chromosomal DNA as an alternative or extra method. LINE elements are considered to be the most ancient class of eukaryotic retrotransposons based on sequence diversity criteria and cladistic analysis suggests that the origin of the earliest LTR retrotransposons may have been the result of a LINE-acquiring LTRs (Murat et al., 2012).

3.2.2   *SINEs – short interspersed nuclear elements*

The origin of SINEs, which are short base pairs (80–500 bp) retrotransposons, is the retro transposition of different polymerase III transcripts. Typically, they lack the open reading frame (ORF) that codes for proteins but have an internal promoter that can start transcription (Price et al., 2022). The mechanism of LINE elements is hypothesised and used by SINEs to integrate and replicate into the host genome. SINEs are similar to intron free pseudogenes and the mRNA molecules before being inserted into the genome of eukaryotic organisms are frequently reverse-transcribed (Elbarbary et al., 2016). As a result, an RNA polymerase III product will contain a promoter that permits virtually normal transcription if it is only sometimes used as a template to produce DNA and subsequently integrates into a genome. Based on natural selection, the observation states that the evolution of these sequences will lead to the formation of SINEs with a growing affinity for integrating and reverse transcription mechanisms. Although the replication and integration mechanisms of SINEs are unknown, similarities between their flanking target DNAs' of various sizes and both integrated polyA tails are present, and they might utilise LINE specified functions based on how frequently their abundance is connected with LINEs and pseudogenes (Ramakrishnan et al., 2022b).

3.3 Penelope Like Element (PLEs)

PLEs are found in a wide range of organisms, including amoebae, fungi, and vertebrates except mammals. So yet, a limited number of PLEs are found in plants like Conifers. Reverse transcriptase (RT) and Endonuclease (EN) are coded by the single ORF for the number of domains, that make up PLEs (Fig. 5) (Makałowski et al., 2012). Intriguingly, the RT domain of telomerase is more similar to that of a retrotransposon compared to the LTR retrotransposons or LINEs. The GIY-YIG intron encoded endonucleases and the EN domain are linked. While some PLE elements can be oriented directly or inversely and include functional introns, they also share sequences with LTR. TSD is produced by PLEs, however, it has a variable length as opposed to LTR and non-LTR retrotransposons. It's interesting to note that the integration process of PLEs is still unknown (Poulter and Butler, 2015). Fig. 6 displays the basic structure of PLE.

Fig. 6 Structure of Penelope-like elements (PLEs) (Orozco-Arias et al., 2019)

3.4 Dictyostelium Intermediate Repeat Sequence (DIRS)

A morphologically varied collection of retrotransposons known as the DIRS order (Dictyostelium Intermediate Repeat Sequence) lacks TSDs and has a tyrosine recombinase (YR) gene in place of an INT.

The endings resemble inverted repeats or split direct repeats (SDR). These traits point to a unique integration mechanism from other retrotransposons. Almost all creatures, including plants, include DIRSs. They can also be divided into super families like VIPER, DIRS, and Ngaro (Negi et al., 2016), whose structures are presented in Fig. 7.

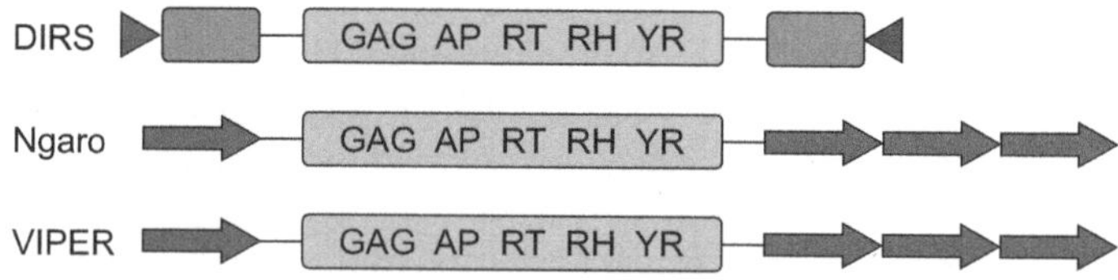

Fig. 7 Structure of Dictyostelium intermediate repeat sequences (DIRS)
(Orozco-Arias et al., 2019)

4. Mechanism of Retrotransposons

Retrotransposons are somehow beneficial for plant cells due to their properties for DNA repair (Ramakrishnan et al., 2022a). These retrotransposons also help in centromere repair (Presting, 2018). Due to their expression and repair properties, the negative impact can be minimized . They are highly adaptive in the host or parasite and they can easily interact with the host without disturbing their health. The plant transposons are adaptive and are often involved in the defence mechanisms of the host body (Seidl and Thomma, 2017). As previously known transposons have shown themselves incapable of encoding for all essential functioning by the transposons but retrotransposons are performing all the essential functioning among the host cells (Joly-Lopez and Bureau, 2014). Therefore, the retrotransposons are responsible for minimizing the negative effects of old transposons.

5.   Regulation of Expression and Transposition of Retrotransposons

The simplest strategy to manage the activity of retrotransposons would be to regulate transcriptional start because they cannot transpose without an RNA template available for reverse transcription. Numerous retrotransposons exhibit distinctive developmental and/or environmental regulatory patterns.

5.1   Regulation of Replication Mechanism

A unique role is assigned to each domain during the process of replication, as discussed by Orozco-Arias, S. et. al. (2019). Examples include digestion of larger transposon transcripts to form small proteins, as a function of the gene '*aspartic protease*'; targeted interpolation of recent LTR copies to the heterochromatin, done by identifying specific histones, directed by the gene '*chromodomain*'; transmission of retroviruses among cells as directed by the gene '*envelope*'; production of structural proteins to produce virus-alike particles, as a function of the gene '*group specific antigen*'; acceleration of cDNA retrotransposon insertion in the genome of the host as a function of the gene '*intregrase*'; the enzyme reverse transcriptase coded by the gene '*reverse transcriptase*' employs RNA in form of the template to synthesize DNA; and the degradation of RNA template in DNA-RNA hybrid, as an action of RNase H enzyme coded by the gene '*RNase H*'.

5.2   Epigenetic and Posttranscriptional Regulations

Transposons are developed in such a way that they can lessen the negative impacts of transposition and retention in both host and organisms. There are some proposed theories that the methylation of the DNA evolved in the transposons in the virus hosts (Yoth et al., 2022). There is the presence of the defence mechanism in the transposon transformations in which the involvement of hybrids between mammals i.e., *Wallabia bicolor* and *Macropus eugenii* occurs. These types of defence mechanisms are highly responsible for the DNA methylation of hybrid genomes (Serrato-Capuchina and Matute, 2018). Some studies have revealed the methylation of cytosine at 5' of CG to 5'CNG by repeated sequences of DNA in plants. The cytosine methylated genetic inactivity is majorly found in many higher animals and plants. The cytosine methylation is methylation in which de novo was first discovered the process of mutation by mutators during inactivation in higher eukaryotes and plants by transposable elements (Möller et al., 2021). These types of cytosine methylations are usually found in the plant's species like maize. Therefore, such inactivation of cytosine methylation is now observed by LTR transposon (Mhiri et al., 2022). P ast studies prove that cytosine methylation is known for the inactivation of transcription and transformation of C to T. Therefore, as a result, the transcriptional activities by high rate of transitions to mutational silencing are in the process.

5.3 *Mechanism of Retrotransposon Activation Under Environmental Stress*

In plants and eukaryotes, retrotransposons are activated by stressors and external modification is widespread. Many well-studied retrotransposons present in plants appear to be closely regulated by biochemical pathways that are triggered by stress, and their transcriptional activity is regulated by cis-regulatory sequences that are strikingly similar to those of plant defense genes. According to (Kejnovsky et al., 2015), sessile creatures with highly dynamic genomes, like plants, maybe the reason why either internal or external elicitors can activate the TEs present in plants (Negi et al., 2016; Vicient and Casacuberta, 2017). Although retrotransposon activation is a common occurrence, the stress response can occasionally become genotype-specific (Negi et al., 2016). LTR-RTs have sequences at both ends that are responsible for the response to environmental stressors (Wang et al., 2017). However, the effects of external pressures on other cellular systems rapidly activate the LTR retrotransposons family which can also activate the TEs, which is not directly caused by abiotic and biotic stress (Casacuberta and González, 2013). Nevertheless, according to some findings, retrotransposons present in plants can avoid the silencing process of the host by producing anti-silencing proteins (Vicient and Casacuberta, 2017). Some TEs that become active by transposition are isolated and have been found in plants thus, even though retrotransposons can invade and populate plant genomes in great numbers. For sessile animals like plants, a highly dynamic genome is crucial, which may help to explain why either internal or exterior elicitors can cause TEs activation in plants (Bonchev, 2016). Pathogens, pathogen elicitors, defense-associated stresses, tissue culture, wounding, heat, drought, and salt stresses, freezing, polyploidization and hybridization events, UV light, and X-ray irradiation are only a few of the biotic and abiotic factors that affect plants (Fan et al., 2013). The LTR sequences at both ends of LTR-RTs are responsible for the response to environmental stressors (Wang, et al., 2017). On the other hand, activation of TEs is not necessarily caused by external pressures directly; rather, it is caused by the impact of such stresses on other cellular systems that enable a quick activation of some particular families of LTR retrotransposons (Orozco-Arias et al., 2019). Only a small number of transpositions that ally active TEs have been found and isolated in plants, even though retrotransposons can invade and densely fill plant genomes.

Several abiotic stresses, such as protoplast isolation, cell culture, wounding, methyl jasmonate, CuCl2, and salicylic acid, significantly boost the expression of the tobacco Tnt1 and Tto1 retrotransposons (Voronova et al., 2011) Similarly, biotic stressors such as Trichoderma varied fungal extracts or pathogens such as viruses, bacteria, or fungi are used to inoculate which are used to activate the transposon present in retrotransposons. Tos17 transcription, unlike Tnt1 and Tto1, is only induced by tissue culture. The RT-PCR method was used to isolate Tobacco Tto5 as a salicylic acid-inducible retrotransposon (Grandbastien, 2015).

6. Plant Retrotransposons and Evolution

6.1 Retrotransposons and Retroviruses

LTR retrotransposons lack the sequences encoding the envelope protein (ENV), having shared striking genomic similarities with retroviruses and having similar coding sequences (Neumann et al., 2019). As a result, the replication cycle of retrotransposons is restricted to the host genome cell. Retroviruses spread from cell to cell and infect other cells through the ENV gene. The mdg-4 The Ty3-gypsy LTR retrotransposons was categorised as an endogenous retrovirus since it included a functional ENV gene (Pérez-Alegre et al., 2005). Retroviruses are thought to have originated from the Ty3-gypsy group retrotransposons can acquire the ENV gene, or the opposite, as was mentioned earlier. In plants, Ty3-gypsy retrotransposons have been found with extra sequences in the region typically inhabited by ENV, however, it is unknown if they function similarly to ENV. A Ty1-*copia* group retrotransposon with unique ENV-like sequences in the retrotransposon's typical ENV region was recently discovered in soybean (Jedlicka et al., 2019). This suggests that ENV-like and/or env-localized sequences were acquired by the Ty3-*gypsy* and Ty1-*copia* groups independently or once in one element, then they were recombined into the other groups (Galindo-González et al., 2017). Interestingly, putative ENV-like sequences can be found in plant LTR retrotransposons. In mammals, receptor-mediated endocytosis is used to transport virus particles from one cell to another through the plasma membrane. A situation like that observed in animals is unlikely to happen in plants due to the plant cell wall. Plants may experience enveloping viruses that travel between cells through plasmodesmata, or intercellular channels (Anderson et al., 2019). Plant viruses are frequently helped in their passage through plasmodesmata by virus-encoded proteins that alter the molecular size exclusion limit of those structures. Retrotransposons are not known to encode such transfer proteins (Rajput, 2015).

6.2 LTR retrotransposon's Role in Plant Evolution

The genetic and morphological changes occur due to some variations and these variations have a significant role in their development process. In the plant genome of eukaryotic organisms TEs are the main variable components, even closely related plant species' genomic architecture might differ significantly due to TEs.

Furthermore, most plant DNA is made up of TEs (Bennetzen et al., 2005). The most frequent phenotypic alterations in Gene inactivation are caused by TE. The insertion of TEs nearby genes or into the genes (Hsing et al., 2007) has been successfully used to generate new null mutations (Candela and Hake, 2008), and that is the main factor in the size of genome development. (Hawkins et al., 2006). To better understand the evolutionary characteristics of plants, mainly LTRs are deeply studied (Neumann et al., 2006; Piegu et al., 2006). LTR retrotransposons and retroviruses have similar gene architectures, however, LTR retrotransposons don't have an extracellular stage in their life cycle or an envelope gene. The transfer of the envelope gene has been considered as the mechanism by which these retroviruses separated from the LTR retrotransposon family Ty3/Gypsy (Malik and Eickbush, 2001).

7. Applications of LTR Retrotransposons

Retro-transposable and related elements are relatively common in the genomes of eukaryotic organisms, and they are often inserted into new genomic areas through a method requiring the reverse transcription of an intermediate RNA (Vuorinen et al., 2018). Meiotic prophase, internal rearrangements, and variations in the copy number of repeat elements occur on both homologous chromosomes during the induction of recombinational mechanisms. LTR retrotransposons are randomly inserted during the transposition process during a species' evolution. Genomic and species differentiation are studied to understand the development of plants, providing a lot of information. Applications of retrotransposon-based DNA markers have emerged as a crucial component of genetic diversity and variability research (Vuorinen et al., 2018; Ghonaim et al., 2020; Kalendar et al., 2021b). Their application ranges from making a map of genes to recognizing people who have similar polymorphisms (Khapilina et al., 2021a). The number of genetic diversities of agricultural plants can be analyzed by the genetic molecular marker obtained from LTR retrotransposons (Kalendar and Schulman, 2006). The impact of abiotic and biotic stress on the activation of retrotransposons is easily detected by retrotransposon-based marker systems (Kalendar et al., 2008; Belyayev et al., 2010). In addition, determining the TEs expression, which takes into account polymorphisms and the diversity of the transposon transcriptional landscape (Kalendar et al., 2011; Kalendar et al., 2018; Kalendar et al., 2021a) may provide new perspectives into host-TE interactions (Lanciano and Cristofari, 2020). The crucial role of LTR retrotransposons as markers in molecular breeding is further supported by the fact that they are related to genes which significantly comprised in potential applications of genomic variation, genomic assembly, tagging of genes, and genes analysis (Vangelisti et al., 2021).

7.1 *Retrotransposons Contribution to Genome Development*

Retrotransposons place themselves inside or close to genes to cause gene mutations. A few examples of LTR and non-LTR retrotransposons that cause alterations in plants are Ty1-copia, Ty3-gypsy, LINE, and SINE antigens (Quesneville, 2020). Many retrotransposon insertions within and close to genes have a negative impact on the expression of those genes by reducing or eliminating transcription of the gene or by adverse changes in transcript processing (Gill et al., 2021). In other cases, more complex effects on gene expression can occur from the insertion of retrotransposon sequences within or near a gene in other situations, such as when these sequences alter the temporal and spatial patterns of transcription or the structure of the resulting protein (Niu et al., 2019). The proper synthesis of a protein or a truncated protein can be occasionally restored by the splicing of retrotransposon sequences. Gross changes in a transcript's maturation can also be caused by the insertion of a retrotransposon (Anderson et al., 2019). Splicing an intron removal by using a cryptic splice site or sites within the element can occasionally result in a protein product that is different from the wild type and often has less functionality. For example, wxG, one of three maize waxy alleles that have undergone retrotransposon mutations, exhibits different tissue-specific expression patterns with thirty times greater enzymatic activity in pollen than in endosperm (Catlin and Josephs, 2022). It was shown that variations in RNA processing between different tissues have greater waxy enzymatic activity than endosperm. Therefore, the wxG allele provides a great example of a retrotransposon insertion that led to tissue-specific alternative splicing. Several plant genes that are currently regarded as wild type have been infected with retrotransposons before (Kögler, 2019). For instance, using Ty1-copia group retrotransposons as query sequences, computer-assisted database searches revealed that over 30 genes from both dicotyledonous and monocotyledonous plants have ancient, degenerate retrotransposon insertions in 50 or 30 flanking regions.

7.2 *Retrotransposons Utilization as Genomic Tools*

There are many essential benefits to choosing retrotransposons for the molecular markers. Their presence in a high number of copies is the form of retrotransposon populations, as well as they can be used because of a wide range of advantages they usually show diverse mutations including polymorphisms within the heterologous chromosomes in many plant hosts. Hence retrotransposons can produce new insertions in the genomic populations of organisms. These insertions of the genomes are led to the polymorphisms in the species. These polymorphisms by the retrotransposons are temporary insertions within the host cells to make phylogenies. Therefore, these lineages are important to study in such species and there are known species that show polymorphisms for the establishment of genetic diversity, DNA methylation, and maintaining the phylogenetic e.g., maize, barley, and pea. These retrotransposons have shown various characteristics and are highly approachable in agronomical studies. These retrotransposons can create a bridge between agricultural and mutational strategies as well wide distribution of the euchromatins domains. In fact, the retrotransposon Tnd-1 has produced a marker that is connected to tobacco's black root resistance. There are presences of a larger number of markers in the retrotransposons which are quite helpful in sequence-specific amplification polymorphism (SSAP) which is a polymorphic marker. SSAP technique is meant for the multiple amplified fragments required in the process of inserting bands on the sequence gels.

Moreover, SSAP-based markers compared with the conventional AFLP-based makers appear to be better for estimating phylogenetic relationships because their primers are specific retrotransposon-based. Therefore, a multi-retrotransposon approach has been recently used for estimating the phylogenetic relationships in species or among species (Li et al., 2019). According to the various transpositional histories of every element, this technique is specifically carrying some significant information. Examples of retroelements that can be employed for genetic linkage and intraspecific genetic diversity research include those that have recently transferred and should be highly polymorphic within a species.

In the case of SSAP based transposons in the isolation of the rapid Tyl-*copia*, molecular studies were made compulsory for all the plant species carrying LTR retrotransposons (Li et al., 2019). The SSAP transposon methods must be applicable for the subcloning by host genomes, analysis of SSAP bands, and identification of the sequences.

Conclusions and Future Prospects

Several studies revealed that plant retrotransposons are well known for the genetic mutations in the plants. It also can lead to an increase in the genomics in the plant genes. They can proliferate rapidly with their host by the means of replication with the host cells. The ability to proliferate by retrotransposons makes them a major component of plant genomes. It can be defined as the major fluid of the genomic component in plant genomes. There is a presence of transposable elements is about 90% in almost every plant genomic data. LTR plays a major role in heat stress regulations and there need to be more studies in the respective area. Hence heat stress plants can survive summers and can also beat the side effects of global warming. Smart agriculture can be possible with LTR retrotransposons. Not well-known method and contribution of the biotic and biotic stress by sRNA and inheritance of trans generation with stability of sRNA. Therefore, there is a need to understand the complete phenomena behind sRNA transfer technologies for plant breedings. In the centromere evolution, the plant retrotransposons play a vital role in the specific centromere transposition. Therefore, this somehow makes it valuable for the genomic studies of the target specific genes. There is also a need for the study of genomic sequencing from the agricultural point of view and further research on the target genes to be modified in agricultural fields. Some studies suggest that there is the importance of retrotransposons for enhancing the number of genes in plant genomes. There is a huge contribution of LTR retrotransposons as transposable elements. These LTRs are responsible for the intraspecific phenotypic variation within the plants. Hence, the presence of TE in plant genomes is required for enhancing the plant genomes and can possible the further mutations for increasing the quality of the gene and make an evolutionary study for the researchers. There is also an essential need to study the evolutionary history of transposable elements for a better understanding of the transposable elements and their roles in various applications.

References

Akhlaghpour, H. (2022). An RNA-based theory of natural universal computation. *Journal of Theoretical Biology*, 537, 110984.

Anderson, S.N., Stitzer, M.C., Zhou, P., et al. (2019). Dynamic patterns of transcript abundance of transposable element families in maize. G3 Genes, Genomes, Genet 9: 3673–82.

Belyayev, A., Kalendar, R., Brodsky, L., Nevo, E., Schulman, A.H., and Raskina, O. (2010). Transposable elements in a marginal plant population: Temporal fluctuations provide new insights into genome evolution of wild diploid wheat. *Mobile DNA* 1, 1–16. doi: 10.1186/1759-8753-1-6.

Bennetzen, J.L., Ma, J., Devos, K.M. (2005). Mechanisms of recent genome size variation in flowering plants. *Ann. Bot.* 95, 127–32. doi: 10.1093/aob/mci008.

Bonchev, G.N. (2016). Useful parasites: The evolutionary biology and biotechnology applications of transposable elements. *J. Genet. 95*, 1039–52.

Bourque, G., Burns, K.H., Gehring, M., Gorbunova, V., Seluanov, A., Hammell, M., ... and Feschotte, C. (2018). Ten things you should know about transposable elements. *Genome biology*, 19, 1–12.

Candela, H. and Hake, S. (2008). The art and design of genetic screens: Maize. *Nat. Rev. Genet.* 9, 192–203. doi: 10.1038/nrg2291.

Casacuberta, J.M. and Santiago N. (2003). Plant LTR-retrotransposons and MITEs: control of transposition and impact on the evolution of plant genes and genomes. Gene 311: 1–11.

Casacuberta, E. and González, J. (2013). The impact of transposable elements in environmental adaptation. *Molecular ecology*, 22(6), 1503–17.

Catlin, N.S. and Josephs, E.B. (2022). The important contribution of transposable elements to phenotypic variation and evolution. Curr Opin Plant Biol, Volume 65, 2022, 102140, ISSN 1369-5266, https:///10.1016/j.pbi.2021.102140.

Cervera, A., Urbina, D., and de la Peña, M. (2016). Retrozymes are a unique family of non-autonomous retrotransposons with hammerhead ribozymes that propagate in plants through circular RNAs. *Genome biology*, 17(1), 1–16.

Deniz, Ö., Frost, J.M., and Branco, M.R. (2019). Regulation of transposable elements by DNA modifications. Nature Reviews Genetics, 20(7), 417–31.

Elbarbary, R.A., Lucas, B.A., and Maquat, L.E. (2016). Retrotransposons as regulators of gene expression. *Science*, 351(6274), aac7247.

Fan, F., Wen, X., Ding, G., and Cui, B. Isolation, identification, and characterization of genomic LTR retrotransposon sequences from masson pine (*Pinus Massoniana*). *Tree Genet. Genomes* 2013, *9*, 1237–46.

Ghonaim, M., Kalendar, R., Barakat, H., Elsherif, N., Ashry, N., and Schulman, A.H. (2020). High-throughput retrotransposon-based genetic diversity of maize germplasm assessment and analysis. Mol. Biol. Rep. 47, 1589–1603. doi: 10.1007/s11033-020-05246-4.

Galindo-González, L., Mhiri, C., Deyholos, M.K., and Grandbastien M.A. (2017). LTR-retrotransposons in plants: Engines of evolution. Gene 626: 14–25.

Gill, R.A., Scossa F., King G.J., et al. (2021). On the role of transposable elements in the regulation of gene expression and subgenomic interactions in crop genomes. CRC Crit Rev Plant Sci 40: 157–89.

González, L.G. and Deyholos, M.K. (2012). Identification, characterization and distribution of transposable elements in the flax (Linum usitatissimum L.) genome. BMC Genomics 13: 1–17.

Goodier, J.L. (2016). Restricting retrotransposons: a review. Mobile DNA, 7, 1–30.

Grandbastien, M.A. (2015). LTR retrotransposons, handy hitchhikers of plant regulation and stress response. Biochim Biophys Acta (BBA)-Gene Regul Mech 1849: 403–16

He, J., Yu, Z., Jiang, J., Chen, S., Fang, W., Guan, Z., ... and Wang, H. (2022). An eruption of LTR retrotransposons in the autopolyploid genomes of Chrysanthemum nankingense (Asteraceae). Plants, 11(3), 315.

Hawkins, J.S., Kim, H., Nason, J.D., Wing, R.A., Wendel, J.F. (2006). Differential lineage-specific amplification of transposable elements is responsible for genome size variation in gossypium. *Genome Res.* 16, 1252–61. doi: 10.1101/gr.5282906.

Hsing, Y.I., Chern, C.G., Fan, M.J., Lu, P.C., Chen, K.T., Lo, S.F., et al. (2007). A rice gene activation/knockout mutant resource for high throughput functional genomics. *Plant Mol. Biol.* 63, 351–64. doi: 10.1007/s11103-006-9093-z.

Huang, C.R.L., Burns, K.H., and Boeke, J.D. (2012). Active transposition in genomes. Annual review of genetics, 46, 651–75.

Jedlicka, P., Lexa, M., Vanat, I., et al. (2019). Nested plant LTR retrotransposons target specific regions of other elements, while all LTR retrotransposons often target palindromes and nucleosome-occupied regions: in silico study. Mob DNA 10: 1–14

Joan Curcio, M., Lutz, S., and Lesage, P. (2015). The Ty1 LTR-retrotransposon of budding yeast, Saccharomyces cerevisiae. Mobile DNA III, 925–64.

Joly-Lopez, Z. and Bureau, T.E. (2014). Diversity and evolution of transposable elements in Arabidopsis. Chromosom Res 22: 203–16.

Kalendar, R., Amenov, A., and Daniyarov, A. (2018). Use of retrotransposon-derived genetic markers to analyse genomic variability in plants. *Funct. Plant Biol.* 46, 15–29. doi: 10.1071/FP18098.

Kalendar, R., Flavell, A., Ellis, T., Sjakste, T., Moisy, C., and Schulman, A.H. (2011). Analysis of plant diversity with retrotransposon-based molecular markers. *Heredity* 106, 520–30. doi: 10.1038/hdy.2010.93.

Kalendar, R., Muterko, A., and Boronnikova, S. (2021a). Retrotransposable elements: DNA fingerprinting and the assessment of genetic diversity. *Methods in molecular biology* 2222, 263–86. doi: 10.1007/978-1-0716-0997-2_15.

Kalendar, R., Sabot, F., Rodriguez, F., Karlov, G.I., Natali, L., and Alix, K. (2021b). Editorial: Mobile elements and plant genome evolution, comparative analyzes and computational tools. *Front. Plant Sci.* 12, 735134. doi: 10.3389/fpls.2021.735134.

Kalendar, R., and Schulman, A.H. (2006). IRAP and REMAP for retrotransposon-based genotyping and fingerprinting. *Nat. Protoc.* 1, 2478–84. doi: 10.1038/nprot.2006.377.

Kalendar, R., Tanskanen, J., Chang, W., Antonius, K., Sela, H., Peleg, O., et al. (2008). Cassandra Retrotransposons carry independently transcribed 5S RNA. *Proc. Natl. Acad. Sci. U.S.A.* 105, 5833–38. doi: 10.1073/pnas.0709698105.

Kejnovsky, E., Tokan, V., and Lexa, M. (2015). Transposable elements and G-quadruplexes. Chromosom Res 23: 615–23.

Khapilina, O., Raiser, O., Danilova, A., Shevtsov, V., Turzhanova, A., and Kalendar, R. (2021a). DNA Profiling and assessment of genetic diversity of relict species allium altaicum pall. on the territory of Altai. *PeerJ* 9, e10674. doi: 10.7717/peerj.10674.

Klein, S.P. and Anderson, S.N. (2022). The evolution and function of transposons in epigenetic regulation in response to the environment. Current Opinion in Plant Biology, 69, 102277.

Kögler, A. (2019). Diversity and Evolution of Short Interspersed Nuclear Elements (SINEs) in Angiosperm and Gymnosperm Species and their Application as molecular Markers for Genotyping.

Lanciano, S., and Cristofari, G. (2020). Measuring and interpreting transposable element expression. *Nat. Rev. Genet.* 21, 721–36. doi: 10.1038/s41576-020-0251-y.

Lee, S.I. and Kim, N.S. (2014). Transposable elements and genome size variations in plants. Genomics & informatics, 12(3), 87–97.

Li, S., Ramakrishnan M., Vinod, K.K, et al. (2019). Development and deployment of high-throughput retrotransposon-based markers reveal genetic diversity and population structure of Asian bamboo. Forests 11: 31.

Lisch, D. (2013). How important are transposons for plant evolution? *Nat. Rev. Genet.* 14, 49–61. doi: 10.1038/nrg3374.

López-Flores, I. and Garrido-Ramos, M. A. (2012). The repetitive DNA content of eukaryotic genomes. Repetitive DNA, 7, 1–28.

Makałowski, W., Pande. A., Gotea, V. and Makałowska, I. (2012) Transposable elements and their identification. Evol Genomics Stat Comput Methods, Vol 1 337–59

Malik, H.S., Eickbush, T.H. (2001). Phylogenetic analysis of ribonuclease H domains suggests a late, chimeric origin of LTR retrotransposable elements and retroviruses. *Genome Res.* 11, 1187–97. doi: 10.1101/gr.185101.

McCullers, T.J. and Steiniger, M. (2017). Transposable elements in Drosophila. Mob. Genet. Elements 7: 1–18.

Metzger, M.J., Paynter, A.N., Siddall, M.E., and Goff, S.P. (2018). Horizontal transfer of retrotransposons between bivalves and other aquatic species of multiple phyla. Proceedings of the National Academy of Sciences, 115(18), E4227-E4235.

Mhiri, C., Borges, F., and Grandbastien, M.A. (2022). Specificities and dynamics of transposable elements in land plants. *Biology* (Basel) 11:488.

Möller, M., Habig, M., Lorrain, C., et al. (2021). Recent loss of the Dim2 DNA methyltransferase decreases mutation rate in repeats and changes evolutionary trajectory in a fungal pathogen. PLoS Genet 17:e1009448.

Murat, F., Peer, Y.V.D., and Salse, J. (2012). Decoding plant and animal genome plasticity from differential paleo-evolutionary patterns and processes. Genome biology and evolution, 4(9), 917–28.

Negi, P., Rai, A.N., and Suprasanna, P. (2016). Moving through the stressed genome: emerging regulatory roles for transposons in plant stress response. Frontiers in plant science, 7, 1448.

Neumann, P., Koblízková, A., Navrátilová, A., and Macas, J. (2006). Significant expansion of vicia pannonica genome size mediated by amplification of a single type of giant retroelement. *Genetics* 173, 1047–1056. doi: 10.1534/genetics.106.056259.

Neumann, P., Novák, P., Hoštáková, N., and Macas, J. (2019). Systematic survey of plant LTR-retrotransposons elucidates phylogenetic relationships of their polyprotein domains and provides a reference for element classification. Mob DNA 10:1–17.

Niu, L., Zhang, H., Wu, Z., et al. (2019). Correction: Modified TCA/acetone precipitation of plant proteins for proteomic analysis. PLoS One 14:e0211612.

Orozco-Arias, S., Isaza, G., and Guyot, R. (2019). Retrotransposons in plant genomes: structure, identification, and classification through bioinformatics and machine learning. *International journal of molecular sciences*, 20(15), 3837.

Ou, S. and Jiang, N. (2018). LTR_retriever: a highly accurate and sensitive program for identification of long terminal repeat retrotransposons. Plant physiology, 176(2), 1410–1422.

Papolu, P.K., Ramakrishnan, M., Mullasseri, S., Kalendar, R., Wei, Q., Zou, L.H., ... and Zhou, M. (2022). Retrotransposons: How the continuous evolutionary front shapes plant genomes for response to heat stress. Frontiers in plant science.

Pérez-Alegre, M., Dubus, A., and Fernández, E. (2005). REM1, a new type of long terminal repeat retrotransposon in Chlamydomonas reinhardtii. Mol Cell Biol 25:10628–38.

Piegu, B., Guyot, R., Picault, N., Roulin, A., Saniyal, A., Kim, H., et al., (2006). Doubling genome size without polyploidization: dynamics of retrotransposition-driven genomic expansions in oryza australiensis, a wild relative of rice. *Genome Res.* 16, 1262–69. doi: 10.1101/gr.5290206.

Poulter, R.T. and Butler, M.I. (2015). Tyrosine recombinase retrotransposons and transposons. Mobile DNA III, 1271–1291.

Presting, G.G. (2018). Centromeric retrotransposons and centromere function. Curr Opin Genet & Dev 49:79–84.

Price, A.M., Steinbock, R.T., Lauman, R., Charman, M., Hayer, K.E. and Kumar, N. (2022). Novel viral splicing events and open reading frames revealed by long-read direct RNA sequencing of adenovirus transcripts. PLoS pathogens, 18(9), e1010797.

Quesneville, H. (2020). Twenty years of transposable element analysis in the Arabidopsis thaliana genome. Mob DNA 11: 1–13.

Rajput, M.K. (2015). Retrotransposons: the intrinsic genomic evolutionist. Genes & genomics 37: 113–123.

Ramakrishnan, M., Papolu P.K., and Mullasseri S., et al. (2022a). The role of LTR retrotransposons in plant genetic engineering: how to control their transposition in the genome. Plant Cell Rep 1–13.

Ramakrishnan, M., Satish, L., Sharma, A., et al. (2022b). Transposable elements in plants: Recent advancements, tools and prospects. Plant Mol Biol Report 40: 628–45.

Sahebi, M., Hanafi, M.M., van Wijnen, A.J., Rice, D., Rafii, M.Y., Azizi, P., ... and Noor, Y.M. (2018). Contribution of transposable elements in the plant's genome. Gene, 665, 155–66.

Saleh, A., Macia, A., and Muotri, A.R. (2019). Transposable elements, inflammation, and neurological disease. Frontiers in neurology, 10, 894.

Schulman, A.H. (2012). Hitching a ride: nonautonomous retrotransposons and parasitism as a lifestyle. Plant transposable elements: impact on genome structure and function, 71–88.

Schulman, A.H. (2013). Retrotransposon replication in plants. Current Opinion in Virology, 3(6), 604-614.

Seidl, M.F., and Thomma B.P.H.J. (2017) Transposable elements direct the coevolution between plants and microbes. Trends Genet 33: 842–51.

Serrato-Capuchina, A., and Matute, D.R. (2018) The role of transposable elements in speciation. Genes (Basel) 9: 254.

Smith, L. M. (2015). Mechanisms of transposable element evolution in plants and their effects on gene expression. Nuclear Functions in Plant Transcription, Signaling and Development, 133–64.

Sundaram, V., and Wysocka, J. (2020). Transposable elements as a potent source of diverse cis-regulatory sequences in mammalian genomes. Philosophical Transactions of the Royal Society B, 375(1795), 20190347.

Tsukahara, S., Kawabe, A., Kobayashi, A., Ito, T., Aizu, T., Shin-i, T., et al., (2012). Centromere-targeted *de novo* integrations of an LTR retrotransposon of arabidopsis lyrata. *Genes Dev.* 26, 705–713. doi: 10.1101/gad.183871.111.

Tollis, M. and Boissinot, S. (2012). The evolutionary dynamics of transposable elements in eukaryote genomes. Repetitive DNA, 7, 68–91.

Vangelisti, A., Simoni, S., Usai, G., et al. (2021). LTR-retrotransposon dynamics in common fig (Ficus carica L.) genome. BMC Plant Biol 21: 1–18.

Vicient, C.M., and Casacuberta, J.M. Impact of transposable elements on polyploid plant genomes. *Ann. Bot.* 2017, *120*, 195–07.

Voronova, A., Jansons, A., and Rucngis, D. (2011). Expression of retrotransposon-like sequences in Scots pine (Pinus sylvestris) in response to heat stress. Environ Exp Biol 9: 121–27.

Vuorinen, A.L., Kalendar, R., Fahima, T., Korpelainen, H., Nevo, E., and Schulman, A.H. (2018). Retrotransposon-based genetic diversity assessment in wild emmer wheat (*Triticum turgidum ssp. dicoccoides*). *Agronomy* 8, 107. doi: 10.3390/agronomy8070107.

Wang, L., He, Y., Qiu, H., et al. (2017). Mdoryco1-1, a bidirectionally transcriptional Ty1-copia retrotransposon from Malus$\times$ domestica. Sci Hortic (Amsterdam) 220: 283–90.

Wicker, T., Sabot, F., Hua-Van, A., et al. (2007). A unified classification system for eukaryotic transposable elements. Nat Rev Genet. 8: 973–82. https://doi.org/10.1038/nrg2165.

Woodrow, P., Pontecorvo, G., and Ciarmiello, L.F. (2012). Isolation of Ty1-copia retrotransposon in myrtle genome and development of S-SAP molecular marker. Molecular biology reports, 39, 3409–18.

Xu, Y. and Du, J. (2014). Young but not relatively old retrotransposons are preferentially located in gene-rich euchromatic regions in tomato (Solanum lycopersicum) plants. *The Plant Journal*, 80(4), 582–91.

Yadav, C.B. and Prasad, M. (2017). Transposable Elements in Setaria Genomes. The Foxtail Millet Genome, 23–35.

Yoth, M., Jensen, S., and Brasset, E. (2022) The Intricate Evolutionary Balance between Transposable Elements and Their Host: Who Will Kick at Goal and Convert the Next Try? Biology (Basel) 11: 710.

4

Genomic and Epigenomic Regulation of Retrotransposons

Saurabh Oswal,[1] *Medeline Joseph*[1] *and Priya Sundarrajan*[1*]

1. Introduction

Retrotransposons are transposable genetic elements that travel through reverse transcription of an intermediary RNA with the assistance of reverse transcriptase/ RNaseH to generate extrachromosomal DNA before being reinserted into the genome. Plants include a large number of retrotransposons, which are imperative in the evolution of plant genes and genomes. In many instances, and in a matter of millions of years, greater than 50% of nuclear DNA is made up of retrotransposons. The retrotransposons and retroviruses discovered in other eukaryotic species share structural and functional similarities with plant retrotransposons. In contrast to certain other eukaryotes, plants have significant peculiarities in the genomic structure of retrotransposons, including unique patterns of chromosomal dispersion, typically high copy numbers, and extremely diverse populations. By inserting into or near genes, retrotransposons can cause mutations (Kumar and Bennetzen, 2003). Transposable elements can modify gene expression upon insertion into or near gene loci, which might result in mutant phenotypes. Barbara McClintock's discovery of transposable elements was aided by this property (Grandbastien, 2015). The revelation, occurring years after its inception, that the distinct characteristic of wrinkled-seed peas, which formed the fundamental basis of Gregor Mendel's principles of genetics, is, in fact, the result of a DNA transposon embedding itself within a starch-branching enzyme, not only brought about a fresh perspective but also shed light on the matter for the entire scientific community (Bhattacharyya et al., 1990).

[1]   St. Xavier's College (Autonomous), Mumbai-400001 affiliated with the University of Mumbai.

*   Corresponding author: priya.s@xaviers.edu

The division of retrotransposons can be understood by considering two distinct types: LTR (long terminal repeat) retrotransposons and non-LTR retrotransposons. Within the LTR category, we can further categorize them into subgroups called Ty1 copia and Ty3 gypsy. These subgroups present variations in both sequence similarity and gene product sequencing order. Remarkably both Ty1 copia and Ty3 gypsy retrotransposons have a substantial presence across a variety of plant species with large genomes throughout the whole plant kingdom. Conversely. Non LTR retrotransposons consist of long interspersed repetitive elements (LINEs) and short interspersed repetitive elements (SINEs). Comparable to LRT type, LINEs, and SINEs also exhibit a considerable count in terms of copies present within the genome (Kumar and Bennetzen, 2003).

1.1 LTR Retrotransposons

There are long terminal repeats (LTRs) at both terminal ends of the LTR retrotransposons. These repeats adjoin the core domains, commonly referred to as the gag and pol domains, which encode the structural and enzymatic proteins required for the transcription and transposition of the retrotransposons (Grandbastien, 2015).

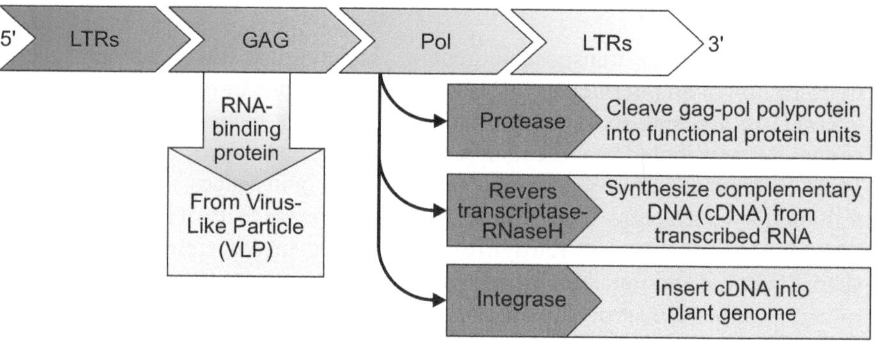

Fig. 1 LTR retrotransposon's domains and its functions

The gag domain encodes for structural proteins which ultimately mediates reverse transcription in VLP while the pol domain encodes for protease, the reverse transcriptase–RNaseH (RT/RH), and the integrase. The location and the functions of the domains are described in Figure 1. LTR retrotransposons is classified as copia-type and gypsy-type based on the coding domain sequences (PR -Protease, Int-Integrase, RT-RH—Reverse Transcriptase-RNaseH) as described in Fig. 2 (Grandbastien, 2015).

There are strong parallels between the LTR retrotransposons' replication methods and the intracellular phase of retroviral life cycles. The 5′ LTR generates an RNA template that interacts with the GAG protein, leading to the formation of

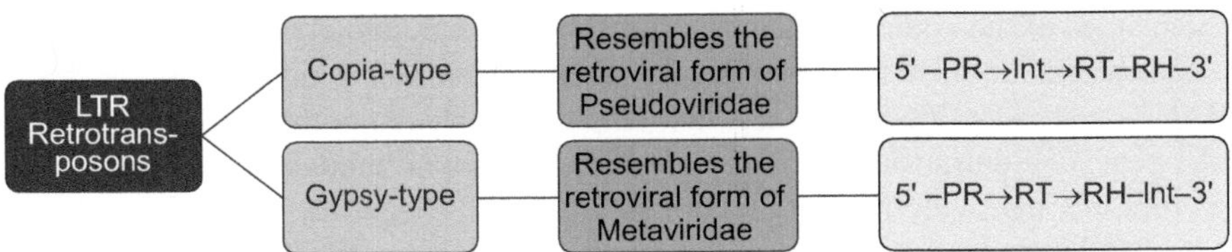

Fig. 2   Classification of LTR retrotransposons based on coding domain sequences

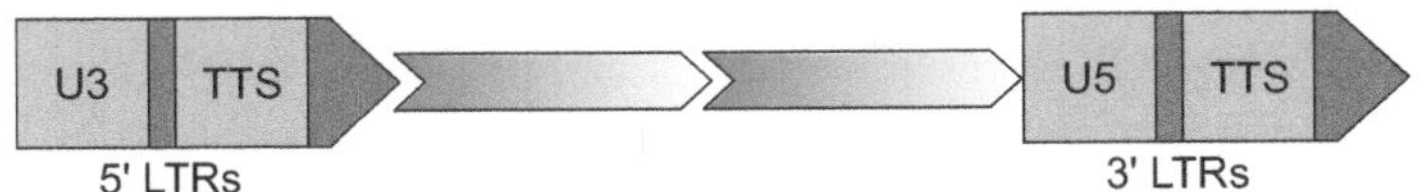

Fig. 3   Core components of LTRs

a virus-like particle (VLP) in the cytoplasm, resembling the core of a retroviral virion. The three core components, U3, R, and U5, are frequently identified in LTRs.

The U3 region is a significant component situated before the Transcription Start Site (TSS). It holds vital promoter and regulatory motifs essential for controlling transcriptional processes necessary in creating both RNA templates and mRNA/protein translations. Notably, it should be acknowledged that upstream of the Transcription Termination Site (TTS) lies the U5 region containing regulatory sequences alongside a functional promoter. Interestingly enough, these components have extraordinary potential to stimulate nearby regions' transcription activity hence affecting adjacent genes' overall expression levels. The R region, another core component of LTR, is situated between U3 and U5 this segment signifies the final sequence found within the LTR (Grandbastien, 2015).

1.2   Non-LTR Retrotransposons

Non-LTR retrotransposons do not have LTRs and instead resemble integrated mRNAs. One or two open reading frames (ORFs) with reverse transcriptase (RT) activity and a restriction enzyme-like (REL) domain are present in non-LTR retrotransposons. These retrotransposons have varying lengths and sequences in their 5′ and 3′ untranslated regions (UTRs). In particular, non-LTR elements often possess a distinctive sequence or structure in their 3′ UTR, which is detected by the reverse transcriptase ORF (Han, 2010).

Long terminal repeats are absent from non-LTR retrotransposons like LINEs, which instead feature a poly(A) tail that identifies the 3′ terminus of the element. LINEs typically span many kilobases and have two ORFs that cooperate to allow for independent retro-transposition. While ORF2 encodes the endonuclease and reverse transcriptase domains, ORF1 encodes a gag protein. A crucial enzyme for retro-transposition, reverse transcriptase exhibits conservation in seven domains typical of retroviral RNA-directed DNA polymerases. Due to overlapping open-reading frames, frameshifts occur frequently, resulting in untranslated regions (UTR) on both sides of the coding region. Many LINEs have cysteine-rich, zinc-finger-like sequences in both ORFs, which are potential nucleic acid-binding domains. Early LINEs were identified, such as Cin4, which was located in the *Zea mays* (maize) A1 gene's 30-untranslated region. The maize genome is thought to include 50–100 copies of Cin4, with the longest copy being 6.6 kb (Schmidt, 1999).

SINEs are another kind of non-LTR retrotransposons. SINEs have a composite structure and can have a length of up to several hundred basepairs. SINEs' 5′ region shares similarities with tRNA genes or, as was demonstrated for animal SINEs, 7SL RNA genes. An unrelated DNA sequence follows the tRNA region, and the 3′ region of many SINEs resembles the 3′ end of LINEs. SINEs are terminated by poly(A) tracts or A- or T-rich sequences. The tRNA-like portion of SINEs contains two motifs with strong evolutionary conservation. These regions, known as box A and box B, act as internal promoters for the transcription of the element by RNA polymerase III, much like tRNA genes do. SINEs are unable to transpose independently because they cannot encode their reverse transcriptase. However, they propagate through retrotransposition, much like LINEs, and produce brief target site duplications upon reintegration (Schmidt, 1999).

1.3 *Chromosomal Locations of Retrotransposons in Plants*

Retrotransposons such as Ty1-copia, Ty3-gypsy, LINE, and SINE are commonly found throughout plant chromosomes but exhibit variation within different families and chromosomal regions. For example, some Ty1-copia retrotransposons cluster in the centromeric areas of maize, rye, and wheat, while specific Ty3-gypsy retrotransposons are absent from the terminal heterochromatic section of *Allium cepa* chromosomes. It is noteworthy that retrotransposon sequences have not been detected in plant nucleoli. Most genes are located in euchromatic regions, where retrotransposons are frequently observed. In maize, these euchromatic retrotransposons primarily occupy intergenic areas based on sequence analyses of large genomic clones (Kumar and Bennetzen, 2003).

In situ hybridization has also been used to investigate the chromosomal distributions of LINEs and SINEs, the non-LTR retrotransposons. Although some components cluster in specific chromosomal sites, both types of elements exhibit dispersed chromosomal patterns of distribution. For instance, in the majority of the chromosomal arms of sugar beets, numerous LINEs are preferentially aggregated in the subtelomeric regions. Research conducted on the SINE element S1Bn from Brassica napus through in situ hybridization revealed several findings. The S1Bn element exhibited a scattered distribution pattern within the chromosomes, with hybridization signals detected in the centromeric regions. Furthermore, it displayed a tendency to co-localize with rDNA sites (Kumar and Bennetzen, 2003).

Many in situ and genome sequencing research suggested that the proportion of retrotransposons in euchromatic areas was higher than in heterochromatic regions. Retrotransposons, which were first discovered adjacent to genes, were frequently used in this research as the in-situ hybridization probe, and genic areas were the main focus of genome sequencing efforts, which tainted the results. According to recent findings from the sequencing of Arabidopsis centromeric regions, retrotransposons were shown to be particularly prevalent near centromeres and are typically organised in nested series, like those discovered in intergenic regions of the more complex maize genome (Kumar and Bennetzen, 2003). Recent studies suggest that the integration/nesting of LTR retrotransposons is not entirely a random event. Ty3 (gypsy) retrotransposons were found to have a comparatively higher nesting frequency than Ty1 (copia) retrotransposons (Jedlicka et. al, 2019). It was observed that the Ty3 gypsy retrotransposon family preferentially integrates into the same LTR retrotransposons in dicot species and a correlation was established between sequence composition, secondary structures, and chromatin landscape with the integration of nested LTR retrotransposons (Jedlicka et al., 2019).

2. Plant Retrotransposon Replication, Integration, and Mobilization in the Plant Genome

The presence of end terminal repeats surrounding the open reading frames (ORFs), plant retrotransposons are largely separated into the Long Terminal Repeats (LTRs) and Non-LTRs groups. Terminal repetitions that provide the information needed for transcription surround the LTR retrotransposons. They are located on each side of a potential retro transposition-related protein coding internal sequence. A peculiar aspect of the integration of such sequences is that it renders the LTRs identical during each integration event. In rare cases, the gene product produced by the two main ORFs, notably the Gag and Pol, may be merged into one (Sabot et al., 2006). GAG encodes the structural protein involved in nucleocapsid organization, while pol is in charge of reverse transcription and the subsequent integration of additional copies. Pol is a multiprotein complex that includes domains for the RT (reverse transcriptase) enzyme, which performs reverse transcription, the RNAseH enzyme, which breaks down RNA template strands, the IN (Integrase) enzyme, which integrates new retrotransposon copies in the host genome, and the AP (aspartic proteinase), which carries out post translational processing of Pol ORFs gene products. The organization of the pol ORFs further creates two subcategories *gypsy* and *copia* groups. LTR retrotransposons that partially or completely lack ORFs are divided into three different categories: terminal repeats in miniature (TRIMs) (Witte et al., 2001), Morgane (Sabot et al., 2006), and large retrotransposons derivatives (LARDs) with no ORFs (Kalendar et al., 2004). In the section that follows, the retroviral model is used to explore the life cycle of retrotransposons and explain the significance of each stage. "Transcription, translation, dimerization and packing, reverse transcription, and integration", are the stages that make up retrotransposition in the order they occur (Sabot and A H Schulman, 2006). Retrotransposon groups that lack ORFs are regarded as non-autonomous, and the process involved in overcoming the challenges associated with maintaining its copy number in the genome is described. Autonomous retrotransposons can express proteins crucial for various stages of retrotransposition. Individual retrotransposons can have varying competence concerning transcription or translation which is dependent on the presence of the requisite ORFs. Active elements can complement non-autonomous or partially inactive elements in *cis* if in same family or in *trans* from other families. The level of complementarity may be such that it reduces the ability of the active or autonomous element to proliferate, hence enhancing the parasitic property of the inactive or non-autonomous retrotransposons (Hu et al., 1997). One can predict the emergence of a new subfamily of non-autonomous retrotransposons due to complementation.

2.1 Plant Retrotransposon Replication

The Plant class 1 transposons or retrotransposons require an RNA intermediate and follow a "Copy and Paste", replication mechanism. The Plant LTR retrotransposons rely on *pol*II promoter mechanism for its expression. U3, R, and U5 regions further split the LTR region. The viral promoter and enhancer components are in the U3 region (Bennett et al., 2019). mRNA initiation site is found in the R region, which culminates into a polyadenylation termination site (Bennett et al., 2019). A putative function of the U5 region is to separate the tRNA primer binding site's R region which is utilized to start reverse transcription (Bennett et al., 2019). Transcription continues until the 3′ R region, positioned inside the 3′LTR region, before the 5′ R region, which is placed downstream of the TATA box, where expression starts. The entire process is driven in the 5′ LTR to 3′ LTR direction (Kumar and Bennetzen, 1999). The outcome of transcription is a bicistronic mRNA which encodes two genes *Gag* and *Pol*. The entire process of LTR retrotransposon transcription relies on host cellular transcription machinery and in that sense, it behaves in a parasitic manner. The outcome of this process is a polyadenylated mRNA which will eventually traverse out of the nuclear membrane into the cytoplasm. However, polyadenylation of the mRNA sequence relies on polyadenylation signal strength produced by the LTR region (Besansky, 1990; Suck and Traut, 2000).

2.2 Packaging of Plant Retrotransposons

The subsequent step after transcription is nucleocapsid formation. In general, the GAG protein contains three key domains which are also found in the GAG of plant LTR retrotransposon namely the capsid domain for polymerization, the nucleocapsid domain which houses zinc-fingers and residues involved in nucleic acid interactions, and the third domain in plant LTR retrotransposon shares some similarity with the matrix domain (Jääskeläinen et al., 1999). The matrix domain engages in interaction with the envelope proteins (Adamson and Jones, 2004). Subsequently, these protein entities form Virus-like particles (VLP) which facilitate the reverse transcription of specific mature mRNA. Packaging refers to the process through which the target mRNA is taken over inside the VLP. With regard to the GAG protein that builds the VLP, the packaging is a rather selective process for the involved RNA sequence (Sabot and Schulman, 2006). The specificity is mediated by packaging sequence (PSI) and the zinc-fingers or basic residues on the nucleocapsid domain that recognize secondary RNA structures (Evans et al., 2004). In human immunodeficiency virus (HIV) retroviruses the PSI sequence is located adjacent to the primer binding site (PBS) and preceding the start codon of GAG (Harrison et al., 1995). The PSI sequence's position in Plant LTR retrotransposons is unknown, nonetheless. According to (Sabot and Schulman, 2006) high RNA structural conservation along with family-specific motifs in interaction regions alludes to a common mechanism for packaging in retrotransposons and retroviruses.

2.3   Reverse Transcription and Integration of Plant Retrotransposon

The cDNA synthesis from the retrotransposon mRNA is a process that is significantly similar to the one found in retroviruses and yeasts (Joan Curcio et al., 2015). The cDNA synthesis takes place in the nucleocapsid. The final result is a double-stranded cDNA with two identical LTRs (Wilhelm and Wilhelm, 2001). An interesting phenomenon occurs called template switching when two RNA transcripts, whether related or unrelated, get packaged for reverse transcription inside the nucleocapsid. As a consequence of this event chimeric products are formed and give rise to complex elements like "LTR-internal sequence-LTR-internal sequence-LTR" (Vicient et al., 2005).

IN proteins mediate the integration of newly generated cDNA strands. IN binds to the LTR in a sequence-dependent manner preferentially near the motif and border regions. The IN proteins, once bound to the target genomic DNA make an asymmetric double stranded break which is approximately 2–16 bp long. The likelihood for an integration event to occur depends on whether the target site is already a retrotransposon or if it lies in the heterochromatin region. The endogenous DNA repair mechanism fixes the double strand break. The double stranded break repair gives rise to terminal site duplications (TSD). The non-autonomous retrotransposons follow two potential trans-parasitic routes for packaging and integration steps in their life cycle. It may share the same IN motif from an autonomous partner sequence or carry a common IN motif facilitating its interaction with IN protein from a wide range of sources. In an event where the selection pressure is focused on non-autonomous retrotransposons that will subsequently favor the propagation of such retrotransposons over autonomous retrotransposons, where the cellular machinery would limit the increase in copy number of autonomous retrotransposon (Sabot and Schulman, 2006).

3.   Influence of Abiotic and Biotic Stress on Epigenetic Regulation of Retrotransposons

In retrotransposons, the replicative mode of transcription is known to happen mainly during stressful conditions and tissue culturing. In Arabidopsis thaliana, ONSEN (which means 'hot spring' in Japanese) is a copia-type retrotransposon that was discovered to be transcriptionally active and also synthesize extrachromosomal

DNA copies under heat stress. The fact that ONSEN was transcriptionally active and migratory in the heat-stressed offspring shows that its regulation is not influenced by chromosomal location. Transgenerational transposition was shown to be transmitted from both sexes in reciprocal crosses with active ONSEN, demonstrating that the transposition is inhibited independently of gametophytic control (Matsunaga et al., 2012). Likewise, various plants show different levels of retrotransposon regulation during different stresses.

3.1 Maize

Maize was one of the few plants initially selected for studying transposable elements. Barbara McClinton discovered transposable elements in maize because of genomic stress. According to her theory, stressful conditions might cause a range of genetic changes that would then be subject to either natural or artificial selection. She and several other researchers theorized that the environment in tissue culture may potentially be a source of "genomic stress," which would result in the activation of transposable elements (McClintock, 1984).

Ac and Spm transposable elements were observed to be activated in the regenerated maize plants while in tissue culture-derived alfalfa, an erratic floral color variant was found. This behavior was believed to be controlled by a transposable element. Based on a DNA analysis of Ac elements, which undergo a cycle of activation and dormancy, it has been observed that when an Ac element is active, certain sites that are methylated during dormancy undergo demethylation (Peschke et al., 1987).

Bs1 and Cin1 are two similar LTR retrotransposons found in maize among many others. Bs1 stands out among other plant transposable elements because it is enclosed by long perfect direct repeats. The proviral versions of vertebrate retroviruses, however, have this structure and have been discovered to contain similar components in yeast and Drosophila. The initial documented instance of the activation of LTR retrotransposons due to environmental stress was observed in the transposition of the Bs1 element found in the descendants of maize parental plants that were exposed to infestation by the barley stripe mosaic virus (Johns et al., 1985). The maize Bs1 retrotransposon offers the strongest evidence for the theory that a retrotransposon may transduce a cellular gene which has been proved by EST (Expressed Sequence Tag) analysis (Elrouby and Bureau, 2010), however, it has never been demonstrated how Bs1 reacts transcriptionally to virus attacks.

Lu et al., (2011) through their research found that under drought conditions for three-leaf stage maize plants, in the roots four of the eight transposon and retrotransposon genes that were transcribed were up-regulated. It is speculated that the lack of water has an impact on the integrity of the maize genome because one mechanism that leads to genome instability is the transposition of transposons and retrotransposons. The indirect suggestion is that dryness may have an impact on these processes.

3.2 *Tobacco*

The first isolated tobacco transposable element was Tnt1 which also appeared to be the most fully developed mobile retrotransposons because from sequence comparison the Tnt1 ORF was found to encode a polypeptide with all the required components for independent transfer via reverse transcription of an intermediate RNA (Grandbastien et al., 1989), unlike Bs1 which is a defective element that lacks a functional pol ORF (Sanmiguel and Vitte, 2009).

Pouteau et al. (1991) speculate that except for the roots, healthy tobacco tissues do not express Tnt1. However, in newly isolated tobacco protoplasts, the first stage of tissue culture, Tnt1 is significantly expressed. They proposed that a fast change of the leaf cell programme, such as significant RNA degradation, happens during protoplast isolation, starting the process of rebooting the cell genome. This condition was called 'genomic shock' by McClintock (1984) while stating that tissue culture constitutes one of the stress situations for the cell as well as its genome.

Melayah et. al. (2001) conducted a study using the 'retrotransposon-anchored SSAP strategy', which allows the screening of several insertion sites for high copy number gene sequences. Their study looked into the stress-related amplification of the tobacco Tnt1 gene, which is depicted graphically in Figure 4. These findings provide evidence that transposition occurs after Tnt1 transcription and that fungus extracts effectively enhance Tnt1 mobility.

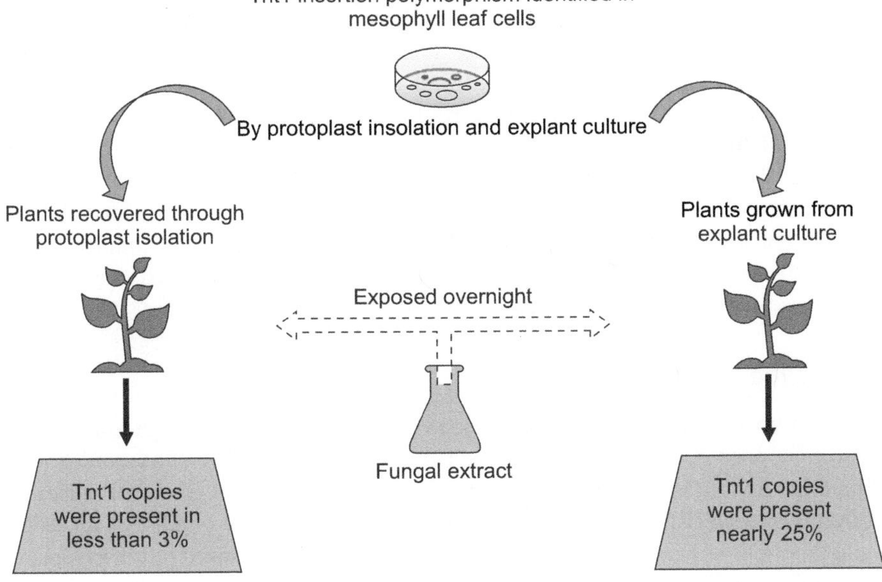

Fig. 4 Study on "stress-induced amplification of tobacco Tnt1 element".

In cultured cells, the transcription of Tto1 initiated from the long terminal repeat (LTR) region alone exhibited activity. However, the introduction of protoplast formation techniques enhanced the transcription process. Tto1 copy numbers increased approximately tenfold in established cell lines, as well as in transgenic plants and plants derived from tissue cultures. These results imply that activation of Tto1 primarily occurs during tissue culture. Unlike Tnt1 RNA, which was only detectable in protoplasts, Tto1 RNA (5.0 kb) was observed both in intact cells and protoplasts (Hirochika, 1993).

Takeda et al. (2001) from their study on transposition in tobacco demonstrated that stress from wounding can increase Tto1 expression in tobacco leaf tissue. Tto1 has cis-regulatory areas that react to damage and methyl jasmonate, according to research on transgenic tobacco plants carrying the Tto1-LTR: -glucuronidase fusion gene (LTR: GUS). A significant amount of Tto1 expression in cultured tissues of transgenic tobacco plants was shown to be driven by the Tto1 LTR promoter. It was discovered that the LTR's particular 13-bp repeated motif (TGGTAGGTGAGAT) functions as a critical cis-regulatory element, enabling responsiveness to tissue culture, wound signals, and methyl jasmonate. Furthermore, the promoter's many copies of this 13-bp motif allowed fungus elicitors to activate it. Okadaic acid and K252a were discovered to greatly and modestly increase the expression regulated by the 13-bp motif, indicating that phosphorylation and dephosphorylation of proteins are involved in the linked signaling pathways (Takeda et al., 1999).

In 2001, Takeda and coworkers investigated to see if tobacco contains the retrotransposon Tto1 that can be retro-transposed outside of tissue/cell culture conditions. They examined the DNA intermediates of Tto1 to understand this phenomenon. They effectively identified linear extrachromosomal Tto1 DNA molecules within a particular Gag-particle region using transgenic *Arabidopsis callus* with high levels of Tto1 RNA production in a ddm1 hypomethylation mutant context. Takeda et al. (2001) found Tto1 linear DNA molecules in tobacco leaf particle fractions obtained from callus cultures and methyl jasmonate-treated tobacco leaves but not in unaffected leaves using ligation-mediated PCR amplification. These findings provided proof that DNA intermediates are produced as a result of the transcriptional activation of Tto1 caused by defense-related stressors.

3.3 Rice

Using two molecular techniques, researchers carried out thorough analyses of retrotransposons in rice. One approach focused on the highly conserved primer binding site (PBS) sequence found in plant retrotransposons. The PBS acts as a binding site for tRNA, which is complementary to the 3' end of a tRNA molecule and functions as a primer for reverse transcription. They identified retrotransposons by looking at a collection of rice genomic DNA and using an oligonucleotide that matched the 14 nucleotides at the 3' end of the starting tRNA. Relying on this screening, an estimated 1000 retrotransposons were found in the rice genome, although the actual number may be higher considering the discovery of a retrotransposon with a complementary PBS for lysine tRNA. This led to the characterization of three families named Tos1-Tos3 (Hirochika et al., 1996).

The second strategy concentrated on the conserved amino acid sequences in the Ty1-copia group retrotransposons' reverse transcriptase domain. Using PCR techniques, comprehensive surveys were conducted, resulting in the identification of 17 distinct families named Tos4-Tos20. Additionally, the identification of retrotransposons in nucleic acid data repositories was successful using computer-aided sequence similarity scanning. The 5′ and 3′ regions of common plant genes had over twenty retrotransposon sequences. These findings demonstrate that retrotransposons are a significant form of transposable element in rice (Hirochika et al., 1996).

Although rice's retrotransposons normally are dormant during normal growth circumstances, researchers discovered that five of the 32 families specifically showed activity when grown in tissue culture. Among these families, Tos17 was identified as the most active retrotransposon, and further investigation focused on its behavior. The research indicated that the activity of Tos17 is primarily regulated at the transcriptional level. By studying the target locations of transposition, it was revealed that the activation of Tos17 significantly contributes to mutations induced by rice tissue culture. Leveraging the tissue culture-induced activation of Tos17, scientists developed a site-selected mutagenesis system. With the aid of this approach, mutant rice plants grown from tissue culture that have a Tos17 insertion in the gene of interest can be recognized. The technique uses the PCR method, which facilitates the identification of desirable mutants by employing one primer for the ends of Tos17 and another for the gene of interest (Hirochika et al., 1996).

In 2001, Komatsu et al., (2003) found the LINE-type retrotransposon Karma in rice plants, where it underwent continuous retrotransposition over several generations. They also found DNA hypomethylation triggering Karma transcription in rice cell culture. However, transcription was observed to be inadequate for retrotransposition since neither cultured cells nor the first generation of plants that were grown from them showed an increase in the copy number. Despite this observation, copy number expansion was seen in the following generations of regenerated plants as well, indicating that the posttranscriptional control of Karma retrotransposition is development dependent.

3.4 *Arabidopsis Thaliana*

Retrotransposons are frequently linked to restrictive chromatin changes in plants, which are controlled by RNA-directed DNA methylation. However, retrotransposons have developed mechanisms to evade this epigenetic silencing and overcome strict regulation. An interesting example is observed in Arabidopsis thaliana, where the retrotransposon ONSEN shows transcriptional activation under heat stress conditions. This activation occurs when the temperature exceeds a threshold of 37°C, but only after prolonged exposure to heat (Matsunaga et al., 2012).

Table 1   Activation of retrotransposons in various plants

	Plants	Activated chromosomes	Activating conditions		Consequences	References
1	*Zea mays* (Maize)	*Ac, Spm*	Unmethylated in tissue culture environment		Structurally alter chromosome	(Peschke et al., 1987)
		Bs1	Infected with barley strip mosaic virus		Transposition inherited by the progeny	(Johns et al., 1985) (Elrouby and Bureau, 2010)
		4-8 retrotransposons (not yet classified)	Water deficiency (Drought/Dehydration)		High copy number	(Lu et al., 2011)
2	*Nicotiana tabacum* (Tobacco)	*Tnt1*	(i)	Expressed in root apex of the plant and protoplast in tissue culture	Altering genomic size	(Pouteau et al., 1991)
			(ii)	Fungal infection/ elicitor		(Melayah et al., 2001)
		Tto1	(i)	Activated in cultured cells	Increased expression and lead to synthesis of DNA intermediate	(Hirochika, 1993)
			(ii)	Wounding		(Takeda et al., 2001)
			(iii)	Induced by methyl jasmonate		
3	*Oryza sativa* (Rice)	*Tos17*	Tissue culturing		Tissue culture-induced mutation	(Hirochika et al., 1996)
		Karma (non-LTR)	Hypomethylation in rice cell culture, also seen in growing cells		Copy number expansion	(Komatsu et al., 2003)
4	*Arabidopsis thaliana* (Thale cress)	*ONSEN*	(i)	Heat-induced DNA hypomethylation	Synthesis and incorporation of DNA intermediate	(Cavrak et al., 2014)
			(ii)	Heat shock factor (HSFA1)		

Cavrak et al. (2014) characterized the kinetics and mechanism of ONSEN activation during heat stress by measuring the ONSEN RNA level using qRT-PCR and also studying non-digested DNA samples produced during the stress. They found that the RNA levels peaked after 24 hours and remained stable for the final

30 hours of the stress treatment, but that the synthesis of linear ONSEN DNA intermediates peaked after 30 hours after being exposed to heat in tiny amounts during the first 12 hours. This suggested that the synthesis of linear ONSEN DNA intermediates occurs six hours after the activation of transcription in response to heat stress, indicating a requirement for a certain threshold of RNA and a specific duration for reverse transcription. They also discovered that ONSEN activation in response to heat stress was not only determined by decreased DNA methylation at the retrotransposon's promoter.

There are 21 heat shock factors in Arabidopsis, and HSFA1 has been recognized as the main factor influencing heat-responsive gene expression. Cavrak et al. (2014) also looked for RNA and DNA intermediates of ONSEN in HSFA1 mutants to see if HSFA1 promotes retrotransposon transcriptional activation. Moreover, they discovered that HSFA1b or d, but not HSFA1a, activated ONSEN to levels similar to those found in the wild type. In contrast, neither the triple mutant nor the plants in which HSFA1e is the sole functioning HSFA1 factor produced any ONSEN RNA or DNA intermediate in response to heat stress.

Conclusion

Earlier Retrotransposons including class two transposons were popularly considered, "Junk DNA". However, since the work published by Barbara McClintock, extensive research in comprehending the several processes of retrotransposons and their mobilization in the plant genome has provided a fundamental basis for the emergence of unique traits in plants e. g. the wrinkled phenotype in peas. The said feature of the pea plant was also part of the experiment conducted by Gregor Mendel to develop the Mendelian inheritance laws. Evolutionary genomics researchers have shown that plant genomes comprise more than 50% to 90% of retrotransposons. This knowledge has enabled researchers to investigate novel methods for increasing crop diversity as well as unravel mysteries associated with the evolution and adaptation of the plant genome. Retrotransposons are notoriously involved in creating several mutations and until recently the integration of LTR retrotransposons was considered a random process however according to (Jedlicka et al., 2019), in plant genomes LTR retrotransposons are not randomly nested. It has been noted that Ty3/gypsy families nest more frequently than Ty1/copia families. Only dicot species showed preferential insertions into the same LTR retrotransposon family, and they were more prevalent in the Ty3/gypsy than the Ty1/copia families (Jedlicka et al., 2019). Sequence arrangement, secondary structure (palindromes), and chromatin architecture are all connected to the integration of nested LTR retrotransposons (Jedlicka et al., 2019). Although nested LTR retrotransposons were more frequently found in the 3′ UTR of other LTR retrotransposons, LTRs were less frequently targeted (Jedlicka et al., 2019). The surge in data registered in retrotransposon database(s) has given rise to algorithms for deep learning and machine learning to identify retrotransposon sequences from large plant genomes. Fundamentally it appears that to construct a well-defined machine learning algorithm important features like retrotransposon length, ORFs, LTR length and motifs like the poly-A tails, AATAAA, and TATA box have been considered (Cavrak et al., 2014). Since retrotransposons take part actively in the expression of genes it is extremely crucial to study how the expression of retrotransposons is regulated by host mechanisms and environmental factors. Key epigenetic modulations initiated by changes in the environment target the activity of retrotransposons, and this information has benefited crop cultivators across the globe to maximize their crop yields.

References

Adamson, C.S., and Jones, I. M. (2004). The molecular basis of HIV capsid assembly—five years of progress. Reviews in medical virology, 14(2), 107–21.

Bennett, J.E., Dolin, R., and Blaser, M.J. (2019). Mandell, Douglas, and Bennett's principles and practice of infectious diseases E-book. Elsevier health sciences.

Besansky, N.J. (1990). A retrotransposable element from the mosquito Anopheles gambiae. Molecular and cellular biology, 10(3), 863–71.

Bhattacharyya, M.K., Smith, A.M., Ellis, T.H.N., Hedley, C., and Martin, C. (1990). The wrinkled-seed character of pea described by Mendel is caused by a transposon-like insertion in a gene encoding starch-branching enzyme. Cell, 60(1), 115–22. https://doi.org/10.1016/0092-8674(90)90721-P.

Cavrak, V.V., Lettner, N., Jamge, S., Kosarewicz, A., Bayer, L.M., and Mittelsten Scheid, O. (2014). How a retrotransposon exploits the plant's heat stress response for its activation. PLoS Genetics, 10(1). https://doi.org/10.1371/JOURNAL.PGEN.1004115.

Elrouby, N., and Bureau, T. E. (2010). Bs1, a New Chimeric Gene Formed by Retrotransposon-Mediated Exon Shuffling in Maize. Plant Physiology, 153(3), 1413. https://doi.org/10.1104/PP.110.157420.

Evans, M.J., Bacharach, E., and Goff, S.P. (2004). RNA sequences in the Moloney murine leukemia virus genome bound by the Gag precursor protein in the yeast three-hybrid system. *Journal of virology*, 78(14), 7677–84.

Grandbastien, M.A. (2015). LTR retrotransposons, handy hitchhikers of plant regulation and stress response. Biochimica et Biophysica Acta, 1849(4), 403–416. https://doi.org/10.1016/J.BBAGRM.2014.07.017.

Grandbastien, M.A., Spielmann, A., and Caboche, M. (1989). Tnt1, a mobile retroviral-like transposable element of tobacco isolated by plant cell genetics. *Nature* 1989 337:6205, 337(6205), 376–80. https://doi.org/10.1038/337376a0.

Han, J.S. (2010). Non-long terminal repeat (non-LTR) retrotransposons: mechanisms, recent developments, and unanswered questions. Mobile DNA, 1(1), 15. https://doi.org/10.1186/1759-8753-1-15.

Harrison, G.P., Hunter, E., and Lever, A.M. (1995). Secondary structure model of the Mason-Pfizer monkey virus 5'leader sequence: identification of a structural motif common to a variety of retroviruses. *Journal of Virology*, 69(4), 2175–86.

Hirochika, H. (1993). Activation of tobacco retrotransposons during tissue culture. The EMBO Journal, 12(6), 2521–28. https://doi.org/10.1002/J.1460-2075.1993.TB05907.X.

Hirochika, H., Sugimoto, K., Otsuki, Y., Tsugawa, H., and Kanda, M. (1996). Retrotransposons of rice involved in mutations induced by tissue culture. Proceedings of the National Academy of Sciences of the United States of America, 93(15), 7783–88. https://doi.org/10.1073/pnas.93.15.7783.

Hu, W.S., Bowman, E.H., Delviks, K.A., and Pathak, V.K. (1997). Homologous recombination occurs in a distinct retroviral subpopulation and exhibits high negative interference. *Journal of Virology*, 71(8), 6028–36.

Jääskeläinen, M., Mykkänen, A.H., Arna, T., Vicient, C.M., Suoniemi, A., Kalendar, R., ... and Schulman, A.H. (1999). Retrotransposon BARE-1: expression of encoded proteins and formation of virus-like particles in barley cells. *The Plant Journal*, 20(4), 413–22.

Jedlicka, P., Lexa, M., Vanat, I., Hobza, R., and Kejnovsky, E. (2019). Nested plant LTR retrotransposons target specific regions of other elements, while all LTR retrotransposons often target palindromes and nucleosome-occupied regions: in silico study. Mobile DNA, 10(1), 1–14.

Joan Curcio, M., Lutz, S., and Lesage, P. (2015). The Ty1 LTR-retrotransposon of budding yeast, Saccharomyces cerevisiae. Mobile DNA III, 925–64.

Johns, M.A., Mottinger, J., and Freeling, M. (1985). A low copy number, copia-like transposon in maize. The EMBO Journal, 4(5), 1093–1101. https://doi.org/10.1002/J.1460-2075.1985.TB03745.X.

Kalendar, R., Vicient, C.M., Peleg, O., Anamthawat-Jonsson, K., Bolshoy, A., and Schulman, A.H. (2004). Large retrotransposon derivatives: abundant, conserved but nonautonomous retroelements of barley and related genomes. Genetics, 166(3), 1437–50.

Komatsu, M., Shimamoto, K., and Kyozuka, J. (2003). Two-step regulation and continuous retrotransposition of the rice LINE-type retrotransposon Karma. The Plant Cell, 15(8), 1934–44. https://doi.org/10.1105/TPC.011809.

Kumar, A., and Bennetzen, J.L. (1999). Plant retrotransposons. Annual review of genetics, 33(1), 479–532.

Lu, H.F., Dong, H.T., Sun, C. bin., Qing, D.J., Li, N., Wu, Z.K., Wang, Z.Q., and Li, Y.Z. (2011). The panorama of physiological responses and gene expression of whole plant of maize inbred line YQ7-96 at the three-leaf stage under water deficit and re-watering. Theoretical and Applied Genetics, 123(6), 943–58. https://doi.org/10.1007/S00122-011-1638-0/METRICS.

Matsunaga, W., Kobayashi, A., Kato, A., and Ito, H. (2012). The effects of heat induction and the siRNA biogenesis pathway on the transgenerational transposition of ONSEN, a copia-like retrotransposon in Arabidopsis thaliana. Plant and Cell Physiology, 53(5), 824–33. https://doi.org/10.1093/PCP/PCR179.

McClintock, B. (1984). The significance of responses of the genome to challenge. *Science*, 226(4676), 792–801. https://doi.org/10.1126/SCIENCE.15739260/ASSET/A3643FE5-38B1-457C-A50B-DC2C01D1BBAE/ASSETS/SCIENCE.15739260.FP.PNG.

Melayah, D., Bonnivard, E., Chalhoub, B., Audeon, C., and Grandbastien, M.A. (2001). The mobility of the tobacco Tnt1 retrotransposon correlates with its transcriptional activation by fungal factors. The Plant Journal : For Cell and Molecular Biology, 28(2), 159–168. https://doi.org/10.1046/J.1365-313X.2001.01141.X

Peschke, V.M., Phillips, R.L., and Gengenbach, B.G. (1987). Discovery of transposable element activity among progeny of tissue culture–derived maize plants. *Science* (New York, N.Y.), 238(4828), 804–807. https://doi.org/10.1126/SCIENCE.238.4828.804.

Pouteau, S., Huttner, E., Grandbastien, M.A., and Caboche, M. (1991). Specific expression of the tobacco Tnt1 retrotransposon in protoplasts. *The EMBO Journal*, 10(7), 1911–1918. https://doi.org/10.1002/J.1460-2075.1991.TB07717.X.

Sabot, F., and Schulman, A.H. (2006). Parasitism and the retrotransposon life cycle in plants: a hitchhiker's guide to the genome. heredity, 97(6), 381–88.

Sabot, F., Sourdille, P., Chantret, N., and Bernard, M. (2006). Morgane, a new LTR retrotransposon group, and its subfamilies in wheats. Genetica, 128(1-3), 439–47.

Sanmiguel, P., and Vitte, C. (2009). The LTR-retrotransposons of maize. Handbook of Maize: Genetics and Genomics, 9780387778631, 307–27. https://doi.org/10.1007/978-0-387-77863-1_15/COVER.

Schmidt, T. (1999). LINEs, SINEs and repetitive DNA: non-LTR retrotransposons in plant genomes. Plant Molecular Biology, 40(6), 903–10. https://doi.org/10.1023/A:1006212929794.

Suck, G., and Traut, W. (2000). TROMB, a new retrotransposon of the gypsy–Ty3 group from the fly Megaselia scalaris. Gene, 255(1), 51–57.

Takeda, S., Sugimoto, K., Kakutani, T., and Hirochika, H. (2001). Linear DNA intermediates of the Tto1 retrotransposon in Gag particles accumulated in stressed tobacco and Arabidopsis thaliana. The Plant Journal : For Cell and Molecular Biology, 28(3), 307–317. https://doi.org/10.1046/J.1365-313X.2001.01151.X.

Takeda, S., Sugimoto, K., Otsuki, H., and Hirochika, H. (1999). A 13-bp cis-regulatory element in the LTR promoter of the tobacco retrotransposon Tto1 is involved in responsiveness to tissue culture, wounding, methyl jasmonate and fungal elicitors. *The Plant Journal*, 18(4), 383–393. https://doi.org/10.1046/J.1365-313X.1999.00460.X

Vicient, C.M., Kalendar, R., and Schulman, A.H. (2005). Variability, recombination, and mosaic evolution of the barley BARE-1 retrotransposon. *Journal of Molecular Evolution*, 61, 275–91.

Wilhelm, M., and Wilhelm, F.X. (2001). Reverse transcription of retroviruses and LTR retrotransposons. Cellular and Molecular Life Sciences CMLS, 58, 1246–62.

Witte, C.P., Le, Q.H., Bureau, T., and Kumar, A. (2001). Terminal-repeat retrotransposons in miniature (TRIM) are involved in restructuring plant genomes. Proceedings of the National Academy of Sciences, 98(24), 13778–83.

5

Computational Approaches for Retrotransposon Investigations

Abdul Rehman,[1] *Sadaf Batool,*[1] *Muhammad Shahid,*[1] *Shahroz Rahman,*[1] *Muhammad Waqas*[1] *and Farrukh Azeem*[1*]

1. Introduction

The eukaryotic genome restrains a large number of retrotransposons, which is an exceptionally infrequent group of transposable elements. Along with eukaryotes, these Class 1 transposons express their characteristics in retroviruses like HIV. The technique used by the retrotransposons to perform their particular function is "Copy-and-Paste"(Nadeem, et al., 2018). Retrotransposition begins with the conversion of transposable DNA into RNA. Then they generate a second copy and paste at different genomic transposons (location) through which DNA is again attained by reverse transcription. Retrotransposons can be autonomous or non-autonomous. It also creates repeated DNA sequences in the full-length genome as a result of this process. The association of these transposable elements with their hosts is long-term which can be a strong reason that retrotransposons widely spread in many higher organisms (Kidwell and Lisch, 1997). On the other hand, the simply generated transposons are linked for a short time and are dependent on other hosts for their survival. The retrotransposons can be either long interspersed nuclear elements (LINEs) or short interspersed nuclear elements (SINEs). In total 30% of the human genome is comprised of retrotransposons in which LINEs make up approximately 20 percent of the human genome (Cordaux and Batzer, 2009).

[1] Department of Bioinformatics and Biotechnology, Government College University, Faisalabad, Pakistan.

* Corresponding author: farrukh@gcuf.edu.pk

1.1 Retrotransposons and Types

Retrotransposons are classified into two types based on their mechanism of transportation and their DNA sequence configuration. The two major types of retrotransposons are:

1.1.1 LTRs (Long Terminal Repeats)

LTRs are identified by encoding both structural and enzymatic proteins and are direct repeat sequences that skirt the internal coding region. These retrotransposons are featured based on having hundreds of repeats at both ends. In eukaryotes, it is present in full length. LTRs are not capable of jumping but they are the only prehistoric genomic artefacts. The two retroviral genes gag, and pol are involved in reverse transcription along with its enzyme which transcript the retrotransposons into cDNA and integrate them into the genome. The third gene env goes with gag and pol to promote the translation to generate the viral proteins. To facilitate the sepsis of the target cells the env genes are regulated (Finnegan, 2012). This infected cell is originated from the virion which requires the envelope from the cell membrane.

Eventually, it is evaluated that the mechanism of LTR is generated from the yeast work. But, in general, it is seen that the LTR retrotransposons technique shares strong similarities with divergent hosts. The contraption of these retrotransposons is from the activator present at 5′ end of LTR which initiates transcription of the RNA with anatomically encoded RNA polymerase II. Further, for reverse transcription and homogenization steps, RNA is translated into the cytoplasm which synthesizes the proteins into virus-like particles (VLP). The RNA involved in this mechanism is a purine-rich sequence that has degraded genomic RNA to transfer the newly synthesized templates from one end to another end. Moving a step forward to process the reverse transcription ordinarily after the combination of two RNA molecules into one VLP which made the RNA into complete DNA copy by the reaction taking place in reverse transcription which pairs with a sequence near the binding site (5′ LTR) from the first primed tRNA. The strong stop DNA is then passed from 5′ LTR to 3′LTR which further takes on the reverse transcription towards integration (Wilhelm and Wilhelm, 2001). Succeeding from the first primer formation, the second priming is activated at the poly-purine site which is close to the 3′ LTR. The already formed primer from the poly-purine site moves forward to transfer another strand in the cDNA. The resulting site gives the double-stranded complementary DNA.

The diversity shown by LTR retrotransposons is of great significance as it manifested its appearance in different ancestors of different organisms. However, after analysis, it intimates that all the retrotransposons are LTRs and flanked encoded gag and poly genes. In the study it is observed that 50% of the retrotransposons contain single open reading frames by merging both the genes (gag and poly) and these logistics are possible to be found mostly in plant constituents (Feschotte et al., 2002). As per the coding information, accessed by the analysis, there is an abrupt structural change in the genome of the organism. The second class of non-autonomous LTR retrotransposons is identified in the plants as TRIMs, which were primarily found in potatoes. They lack internal coding as compared to other retrotransposons.

Variations in genome management are not limited to the coding information because some groups of meta-viruses contain non-coding DNA information between the 3′ LTR and pol gene. It is expected that advances in sequencing technology will lead to the identification of new data and current genetic elements in the eukaryotic genome.

1.1.2   Non-LTR retrotransposons

Non-long terminal repeats (non-LTR) are also present in the eukaryotic genome, especially in abundance in the human genome. Unlike LTRs, they contain one or two ORFs. Retrotransposon's domain is also present almost in non-LTRs as well and they are seen to be present as endonuclease domain. Both UTR region 5′ and 3′ of non-LTR retrotransposons show variations (Kazazian Jr, 2004). Because of the promoter activity and its abrupt alternation taking place in the 5′ untranslated region of non-LTR, it is seen that the promoter sequence between different elements of distinct species is not conserved. An enzyme i.e., reverse transcriptase ORF is involved in the recognition of the 3′ UTR region of non-LTRs having a particular structure. Probably, the 3′ UTR edge can also consist of short repeat sequences which are also known as SINEs, and poly-A. Due to inconstancy in the untranslated regions of non-LTRs retrotransposons, there is no specific exploration of the transcription process (its initiation and termination) (Han 2010).

Non-LTR retrotransposons go for a replication process involving transcription, exportation, and translation. Firstly, full-length activated elements are transcribed then the mRNA is exported out from the nucleus. After this, the retrotransposons proteins in an open reading frame are translated and aggregated to form RNP (ribonucleoprotein particle) passing through the cytoplasmic granule. By recognition and scanning framework shows that the first ORF is most likely to be translated in this process. Despite accommodating with retrotransposons ORF1 considerably has mutated coding region. The ritual cap in the 5′ UTR becomes dependent and regulates the translation like in the human genome L1. It is observed that the elements transcribed by pol genes can be translated with the help of IRES (internal ribosomal entry site). ORF2 translation depends on the initiation of the first open reading frame translation. When the RNP is imported into the nucleus via targeted prime reverse transcription, integration begins (Han and Boeke 2005). Recent and closer studies revealed that RNPs are localized in the granules and cytoplasmic sites. After the primer synthesis and its transcription, further steps are hypothesized in which the second scrape cleaves and synthesizes new strands. Some host-encoded factors are involved to complete the integration step. After this, the inhibition trampling of non-LTR retrotransposons takes place which clears out the foreign and germ cells from the transposable elements but experimentally this step is un-cleared yet.

1.1.3   SINEs and LINEs elements

SINEs (short interspersed nuclear elements) and LINEs (long interspersed nuclear elements) are Alu elements because of their role in the restriction site of enzymes. Both SINEs and LINEs are non-autonomous retrotransposons. Using RNA polymerase II and III these elements generate new genomic sites. LINEs are the class of non-LTR retrotransposons widely present in the eukaryotic genome. Like, in the human genome, L1 is present in rich amounts and comprises approximately 4,000 full-length of these elements. SINEs are the group of non-LTRs in which integration of one transposed site takes in any part of the genome (Biedler 2005). They are involved in the regulation of gene expression, and increment in the genetic variation in the desired part of the genome so that it permits the organism to have more adaptive ability.

1.2   Importance of Retrotransposons in Agricultural Sciences

Agricultural sciences is the sciences managing food and fiber creation and handling. They incorporate the innovations of soil development, crop development and collection, creature creation, and the handling of plant and creature items for human utilization and use. Until the 1930s, the advantages of rural examination got for the most part from work-saving developments, similar to the cotton gin (Wu et al., 2014). When the yield possibilities of the major financial harvests were expanded through horticultural examination, notwithstanding, crop creation per section of land expanded emphatically. Somewhere in the range of 1940 and 1980 in the US, for instance, per-section of land yields of corn significantly increased, those of wheat and soybeans multiplied, and ranch yield each hour of homestead work expanded just about 10 overlap as capital was filled in for work. New methods of food protection made it conceivable to move them over more noteworthy distances, thusly working with changes among areas of creation and utilization, with additional advantages to creation effectiveness.

Retrotransposons play a vital role and have great significance in agricultural sciences. In particular, LTR-retrotransposons (LTR-Rs) are the most bountiful TEs, particularly in plants possessing over 70% of many plant genomes (YALDIZ, Camlica et al., 2018). Until now, a few investigations have revealed transformative parts of the LTR-Rs for genealogy explicit genome development producing species hindrances by outrageous genome size varieties. But the genome development, hardly any examinations uncovered that LTR-Rs drive the rise of new qualities through retroduplication occasions. Be that as it may, the majority of the new qualities created by the LTR-Rs-intervened retroduplication were known as uncharacterized or pseudo qualities.

Starting from the first pepper genome project was finished in quite a while, have built two more all over again genomes of hot pepper (*Capsicum baccatum* and *C. chinense*, from this point forward Baccatum and Chinense) to make references addressing the sort Capsicum. We further developed characteristics of quality explanation and pseudomolecule chromosomes for the prior pepper genome (*C. annuum*, henceforth Annuum). The quality of gene annotations impacts subsequent functional and comparative analyses. However, most plant gene annotations remain as initial versions because the researchers who executed the plant genome project did not update the annotations further. At the point when we looked at the past and the refreshed form of comments, 9,000 qualities primarily containing speculative proteins and TE-related qualities were dispensed with, and 10,000 qualities connected with different capabilities were recently recognized in the refreshed quality model (Jaiswal et al., 2021).

2. Methods to Study Retrotransposons

Retrotransposons are studied using various mechanisms, toolkits, conventional and non-conventional methods as well as computational methods. The following evaluation of the methods of retrotransposons will give the contraption of the retroelements in the cell cycle and animal paradigm. The two are retroviruses and retroelements by manipulating the cell's enzymes along with its associated factors are involved in transcription. As a result, they undergo a similar mechanism appending the sequence removal by splicing the intron and as well as acceptor position (Goodier, 2016).

2.1 *Conventional Methods to Study Retrotransposons*

L1 retrotransposons (class of non-LTR) analysis is a compliant and influential kit to study the outcome and mechanism of retrotransposons. In cell culture, retro-transposition analysis is essentially done by two methods using a Neomycin Resistance Cassette.

2.1.1 *Traditional method*

In the initial analysis of retrotransposons, HeLa cells are scattered and developed in a basal medium i.e., DMEM at the concentration of 2×10^5 to 4×10^5 cells/encrusted in 6 well-bowled to almost 70% convergence. Immunoblotting is performed in 1ml of BRL (Opti-mem) along with 4µl lipofectamine reagent as well as 1µg of plasmid DNA. The cell cultures are augmented having 20% FCS beside 1ml of DMEM, this action is performed after the 5 hours with 7% of CO_2 and the temperature given to this medium is 37°C. The replacement of this culture is done after 16 hours . Then, for the appearance of the trajectory involved in hygromycin-resistance after two days trypsin is inserted and planted in tissue cells into 75 cm^2 vessels and 200 µg/ml hygromycin. Another neomycin component is puromycin when taken 5 µg/ml instead of hygromycin defiance.

The approximate time duration provided for the cell growth is 10 to fortnight beneath two neo-cassettes which are hygromycin and puromycin. Later on, they are inserted with trypsin enzyme and computed with an instrument named hemocytometer. These cultures are re-planted with recognizable densities in media augmented with 300–400 µg/ml of G418. The neo-selectable markers are expressed and this method is designated for the cells that undergo retro-transposition. After 10–14 days cells are stained and anchored with 0.4% Giemsa. The evaluation of the retro-transposition regulation depends on the number of tissue cells prior to and following the G418 draft (Philippe, et al., 2016).

The traditional assay is a slow process as it accumulates cautiously in genomes and cells of eukaryotes including higher plants and mammalian genes. In this process of L1 retro-transposition the m-neoI sheath contains the retrotransposons with alone troopers as cells because preference for the antidote G418 results in the demise of cells especially those that do not sustain the de-novo retro-transposition proceedings. Ostertag et al., (2001) swap out the neomycin genes with EFGP gene in γ-globin introns to dig up the agitation of retro-transposition more specifically in opposition to retro-transposition in negative cells. By using a flow catalogue of flashing cells, this variant permits for the easy evaluation of retro-transposition.

The γ-globin introns are also bounded by the acceptor sites and splice donors. In the 3′ UTR the mneoI cassette is lodged in currently active humans with the Moran's manufactured structural elements. The CMV motors expressed the full fusion of L1-mneoI transcription and ended up with a terminating signal that is SV40 polyadenylation. Some resistors encoded to the hygromycin which is the vector strength of the cells. Survival colonies propagated with the insolation of the DNA along with the insertion outcomes which are examined by some other techniques including PCR, Southern blotting, DNA sequence analysis, etc. In animal models, like humans and mice, the retro-transposition analysis gives an emerging area of revolution. In the transgenic process, gene discovery the elements involved in this traditional method make up the key role in transmitting the characters or retroelements in the next generation of eukaryotes including mammals as well as higher plants (Bennetzen, 2000).

2.1.2 *Transient analysis*

The transient assay involves the variations in methods and mechanisms of the retrotransposons at the cell or tissue level. The process starts by placing the two brackets of the co-expressing cells in a corresponding plate. One class has the feature of retro-transposition productivity to control the regulation of transfection. The second class is planted on the G418 medium in which the L1 construct is merged into the genome of the specified cell by the retro-transposition and globin introns in neomycin genes are separated by splicing which has to be designated by the cells. A brief elaboration of the protocol of the HeLa cells is discussed below:

In 6 well-plate culture bowl the HeLa cells are placed at the concentration of 2×10^5 per well. For the transfection of the cell, a Roche reagent of 3 µl (Fugene non-liposomal transfection) is utilized as well as plasmid, and pGreen Lantern reporter plasmid of 0.5 to 1µg is also involved for the transfection regulation. After that the cells are plumped for 3 days and then trypsin is inserted in them.

An instrument flow cytometry is exploited for computing the green flash. Wells without insertion of the trypsin (also known as test wells) are manifested with 400 micrograms per ml in G418 medium for about 12 days. Two aldehydes i.e., formaldehyde and glutaraldehyde in their specific concentration at 4°C for 30 mins are anchored in G418 positive bacterium medium.

Later on, they are pigmented with bromophenol blue and Giemsa with 0.1% and 0.4% respectively so that it helps to visualize all the procedures with great efficiency. In the corresponding plating of the cell in the culture dish with the above-discussed reagents, single colonies can propagate easily and can be extracted for other processes like Southern blotting, DNA isolation PCR, etc. On large proportions, percentages of the reagents and containers can also be more like greater quantities of the cell is essential for studying the retro-transposition on a large scale (Baldo et al., 2000).

Another approach for the transportation of retrotransposons uses the aid-reliant adenovirus for transfection. There are also some other variations involved in a component of the open reading frame which is ORF1 epitope tag, self-splicing of introns usage in retrotransposons. There are some other methods including no selectable marker or some antibiotic selected marker for further study and evaluations.

2.2   Non-Conventional Methods

The non-conventional methods play a major role in the study of the eukaryotic genomes, especially in mammals and higher plants. The mechanism performed by the retrotransposons is discussed below:

2.2.1   Translation of open reading frame

LINE-1 (Long Interspersed Element) or L1 is the lineage of the non-LTR retrotransposons in the genome of mammals. Because of the 5′ shorten, having inside managements and shelter draining mutations in open reading frames results in the retro-transposition flawed in a vast amount (approx. 99.8%). The calculations revealed that there are ~ 80–100 retro-transposition elements in humans and their regulation is sustained to influence genome evolution. The untranslated region and non-overlapping open reading frames (ORF1 and ORF2) are present in L1 elements. ORF2 has a significant role in the translation of retrotransposons by unconventional termination mechanisms.

Principally, the ORF2 can be translated in two ways which can be L1 transcription modification to post transcription or two cistrons L1 mRNA. Resulting of the vitro translation it is more likely that ORF2p is translated from full length cistrons L1 mRNA. Furthermore, there is a hypothesis that shows that IRES in an L1-inter ORF spacer is obligatory for this human ORF2p translation mechanism. It is hard to find out the notion of ORF2p both in vitro and in vivo medium. In the retro-transposition analysis in cell culture, it is seen that the starting codon

AUG initiates the translation of the retrotransposons. On the other hand, ORF2p translation and retro-transposition regulation decrease due to the inducement of stable fold or termination codon having the blockage of the conversion of the ribosomes being scanned.

Then, we have real evidence that both the instinctive and synthetic codons in L1 are also involved in the initiation of translation of ORFs in the retrotransposons. Deletion of the sequences and analysis of the concluded constructs for the retro-transposition gives the quantified genetic cell analysis. It has a great effect on the efficiency of the retro-transposition as when there is partial deletion of the sequences it results in the reduction of the approx. 50 wild-type levels of the retrotransposons. After the separation from the stop codon corollary mutant lessened up to some control level. To summarize this, we can say that the size of the ORF2 inter-spacer makes an impact on the regulation of retro-transposition and the translation of ORF2.

Both the components of the open reading frames are encoded distinctively. After the fusion of the required proteins in the ORF1 and ORF2, the mutant obtained for the retro-transposition is imperfect. This fusion does not affect the HeLa cells in the culture during translation and it can be seen that these fusion proteins are not defective. All of this available data revealed that in the retro-transposition of these reading frames the presence of the stop codon is of great significance. An investigation of the requirement of the ORF2 AUG codon is noted on the low level of the retro-transposition. This codon is also known as the dependent manner.

All this work on the genome model of humans shows a resemblance with the mechanism performed by other viruses, retroelements, and retrotransposons. A virus RHDV goes under the translation process with the starting codon-independent re-initiation method. It is now clear from the above discussion that the translation of the human L1 ORF2 is done by a non-conventional mechanism. In other mammalian cells like rats, mice, and rodents they have human L1 retrotransposons and they are more likely to be unknown from a translation process which is common to all the mammalian genomes (Ostertag and Kazazian, 2001).

2.2.2   Laboratory method

(a) **Cell culture state:**   Using a DMEM medium with a high level of glucose amplification the HeLa cells were plated. The reagents used included 2 mM L-Glutamine, 10% FBS (fetal bovine serum), 0.1 mg/mL streptomycin, and 100 U/mL penicillin . Same as the case with CHO cells within the same amount of the above mentioned, reagents were cultured. However, there is an addition of 0.1 mM of non-essential amino acids. Both of these cells were planted in a humid incubator of 7% CO_2 at 37°C (Ostertag and Kazazian, 2001).

(b) **Composition of DNA:**   Purification on midi prep columns was done for the plasmid DNA. Using the agarose-ethidium bromide gel the electrophoresis from the transfection of the DNA was performed.

(c) **Cell-cultured retro-transposition analysis:**   This step performed in a laboratory to study the retrotransposons unconventionally is almost similar to that of the transient analysis done in the culture dish. Specifications, trypsinization, instrumentation, reagents, and their quantities are the same as in that short-termed process. Although the stain used for fixing the cells is Brilliant Blue and Crystal Voilet in 0.1% at room temperature.

(d) Trans-accomplishment analysis: It is also the same step as in the transient analysis. However, in the G418 medium (400 µg/ml) the HeLa cells were deprived after almost 72 hours. This medium is also used for fixing the cells to check out the regulation of the trans-complementation analysis.

(e) RNA preparation: In the working culture flask, the transfection reagent was used as per requirement. Cells are washed, incubated, and then transferred to a clean conical flask. Then the same samples were incubated for 5 min at room temperature. Then, further chemicals and reagents were added to the sample in specific concentrations at a particular temperature. Then, the RNA pebble was washed after centrifugation and then again suspended in the ethanol. The outcoming RNA preparation was incubated using the spectrophotometer.

(f) Realtime RT-PCR: After the ribonuclease preservation analysis and rescue of integrants of the G418 colonies the PCR is performed to complete the analysis. Using the leukemia virus and primer (Oligo dT). cDNAs present as an outcome used for the semi-quantification PCR in a thermal cycler. Some dilutions in specific concentrations were used for the amplification of two different base pairs of GFP fragments. The amount of mRNA in the sample was normalized and mRNA content was quantified from the transfected L1 elements (Schulman et al., 2004).

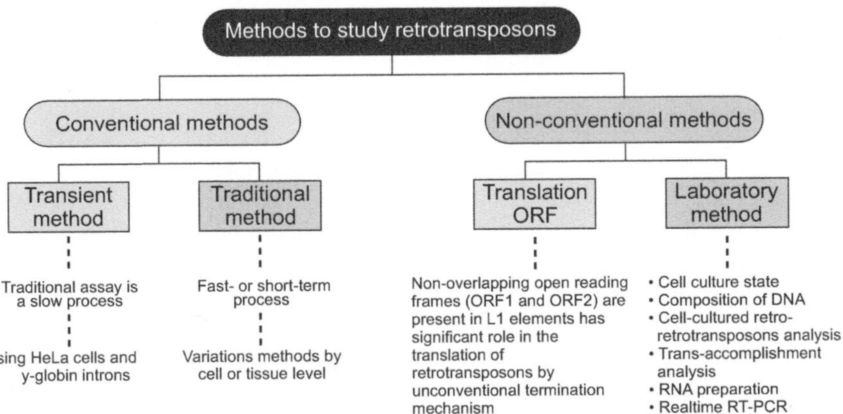

Fig. 1   Workflow for the Methods to study retrotransposons bearing in mind the Conventional and Non-Conventional Methods

3. Computational Approaches to Study Retrotransposons

Retro or DNA transposable elements are the versatile sequences found in almost all eukaryotic genomes. They engrossed a large area of genomes. They are involved in genomic evolution, mutations, and movement of genes by influencing gene variation, gene expression, etc. This era is the age of evolution and automated approaches to study for better results, less time consumption, and detail of our desired study. In the past, there was the problem of limited sequencing and computational capacity to analyze these transposable elements on a wide range. So here, we will discuss some computational approaches to study the retrotransposons based on structure, homology, de-novo based, comparative genomics, and machine learning.

3.1 Structure Based

Based on structure, a program named LTR_STRUC is used for the sequencing and scanning of the nucleotide files of LTRs to analyze them and obtain results from their best hits. It also identifies the transposable elements completely and observes their conserved structure. The retrotransposons in long terminal repeats can be fetched out at every terminal of these retroelements. The algorithms and methodologies under this approach identify the presence of duplications or replication transposition, poly-A tails for LINEs, and short conserved regions (motif) like PBS, PPT, and TSDs for LTR-retroelements.

The structure-based approaches do not require repetition of every transposable element in the genome as well as a collection of the known TEs. So, by using these methods retroelements can find out with some duplicated numbers. Some limitations of the structure-based method are that they are less convenient while probing for the degenerated transposable elements or those with a lack of conserved structural features like TIRs and LTRs. Moreover, it is seen that the algorithm under this approach only well-elaborated the strong structural mark ups of these elements to study the retrotransposons.

3.2 Homology Based

In the homology-based approaches known TEs are utilized as reference sequences to identify and classify RTs based on similarity. The identification activity is elementary if there is availability of any TE collection or recurrent database of species studied before. There can be two ways to get the library of transposable elements like LTRs, non-LTRs, and TIRs that is by library assembly by de novo approach or subsisting databases. Any sequence alignment tool like BLAST or Repeat-Masker can search out the perfect hits of transposable elements with such a high similarity instead of any threshold. It also helps in achieving the library creation to perform homology to identify the retrotransposons easily.

However, there are a few more suites of tools which besides the fact that they struggle while fetching the most possible or close sequences of known or degraded sequences are probably the one of the most reliable and accurate for identifying these elements. Hidden Markov Models (HMM) are also used by some homology-based approaches as they show close relation with genomes but while fetching out some distant or novel sequences they may resist to some extent for similarity among the sequences. This approach also needs categorical lineage having reference retrotransposons elements domains. Because of the high variability of RTs in sequences and being most conserved, these genes are also analyzed by Phylogenetic analysis having RT domain.

The limitations of some tools related to this approach are that they are less computational and automated. The struggle during the utilization of this approach is that it gives only specific properties of particular species having TEs, difficulty in creating the library of the known TEs sequences, a vast number of these elements at the nucleotide level.

3.3 De Novo Identification Based

By using or taking benefit from the TEs which are habitually repetitive this computational approach requires a similar sequence at the multiple sites and positions within the sequence. It takes out the prior knowledge for its functioning like mated TEs which are significant in the process of retro-transposition having basic and structural features. This approach works in a way that once it is identified, the resultant sequences are then gathered, filtered, and featured. De novo method is applied in two ways:

- **Self-comparison:** The process in which the genome and its parts are aligned itself. The filtering of aligned sequences makes this way case sensitive.
- **Spacing or Counting:** The process in which the k-mers of the nucleotides or RT elements are counted exactly or approximately.

As this approach is one of the expensive computational approaches, it gives the best results while finding out the high similarity between the novel TEs and their sequences. It is known as de novo because it does not demand any additional data about the query sequence. The limitation of this approach is that it is not effective in fetching out the low-copy sequences. The most used tools in this approach include PILER and RECON (Edgar and Myers 2005; Flutre et al., 2011).

3.4 Comparative Genomics Based

In this method, the undo regions of the TEs are identified by comparing the whole genome sequences. This approach was presented by two scientists Caspi and Pachter when they found the extreme changes between the closely related genomes using the multiple sequence alignment (MSA). The differences created between the genomes are then analyzed and classified to study retrotransposons along with their domains. The advantage of this approach is that there are already closely related genomes present and it also has significance in identifying the new members and families of transposable elements. The drawback of this method is that it cannot find out common ancestral genes and shows vast divergence between novel and closely related species.

3.5 Machine Learning Based

The machine learning approach is an emerging automated technique to learn the algorithms in the research area. The two categories of ML include:

- **Supervised learning:** In which each sample passes through classification and regression using the readily available dataset. It is also involved in training several algorithms when it has to evaluate out large amount of data.
- **Unsupervised learning:** In which ancestral pattern of data is learnt for clustering and PCA (principal component analysis).

The basic requirement of this approach is the accurate representation of data. Recent techniques of machine learning for retro-transposable elements include neural networks, graphical models, SVM, and HMM. These techniques are capable of handling the randomness and uncertainty of data as well as they are master in the generalization of vast amounts of biological data. For the classification of LTRs retrotransposons by using random forest but for identification conventional methods are used. HMM technique uses software "RED" so that it identifies the repeats in the sequence data. Despite using the RT domain this technique uses the LTR retrotransposons for alignment and constructing a phylogenetic tree. Several searches described how much ML is beneficial for studying biological problems but a few of them used this approach.

3.6 Miscellaneous Approaches

There are many miscellaneous approaches for studying RTs having both short-read sequencing data and long-read sequencing data. Like in humans, these

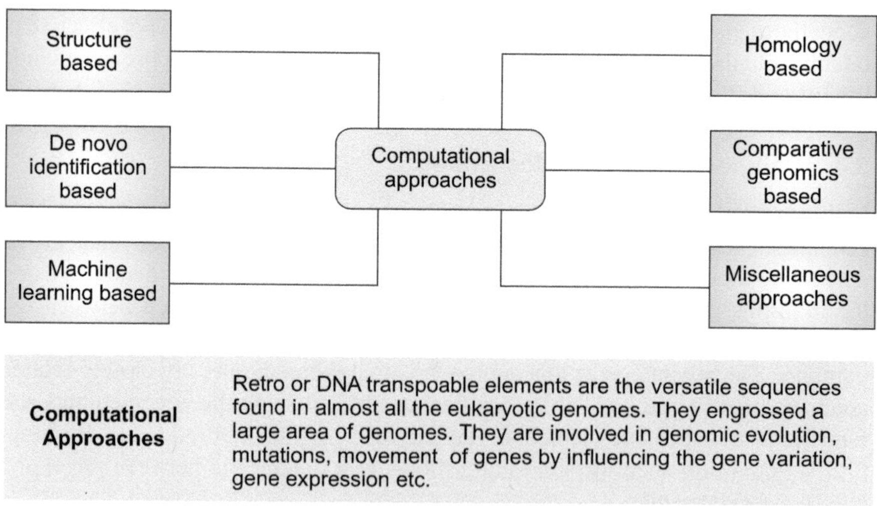

Fig. 2 Mind map for the Computational approaches to study retrotransposons all the approaches that can be used for finding the retrotransposons.

retrotransposons can be identified using high-throughput sequencing data analysis. Over time new, computational, and automated approaches are making it more reliable for identification and analysis. These methods include the Retroseq (Keane, Wong et al., 2013), Illumina, etc. Transposon DB is the collection of the 10 databases (ConTEdb, DPTEdb, MnTEdb, PMITEdb, SPTEDdb, RiTE, Soyetedb, TrepDB, RepBase, mipsREdat-PGSB) which is the largest one to have the transposons sequences available (Riehl et al., 2022).

4. Software and Online Resources for Retrotransposons

Transposable elements like RTs are demonstrated in several software, online tools, and resources so that it is easier for the researcher to find out or get the results desire. The software and online resources used to study retrotransposons are as follows:

Table 1   Information for software and online resources used to study retrotransposons

S. No.	Tools	URLs	Description
1	RepEnrich	https://github.com/nskvir/RepEnrich	RepEnrich uses high-throughput sequencing data to calculate the enrichment of repeated elements
2	RepBase	https://www.girinst.org/repbase/	The most widely utilized database of repetitive DNA elements is called Repbase
3	Bowtie	https://bio.tools/bowtie	A short read aligner with high memory efficiency is called Bowtie
4	Bowtie2	https://github.com/BenLangmead/bowtie2	For quickly and effectively matching sequencing reads to lengthy reference sequences, use Bowtie 2
5	TEtools	https://github.com/l-modolo/TEtools	TE tools for galaxy-based TE RNASeq and small RNASeq analysis
6	FeatureCount	https://bioinformaticshome.com/tools/rna-seq/descriptions/FeatureCounts.html#gsc.tab=0	gDNA-seq and RNA-seq data may be quantified as counts using the tool featureCounts. Additionally, it works well with single-cell RNA-seq (scRNA-seq) data
7	TEtoolkit	https://pypi.org/project/TEToolkit/	Tools for differential enrichment analysis of transposable elements and other highly repetitive regions

Contd.

Table 1   *Contd.*

S. No.	Tools	URLs	Description
8	TEtranscripts Unique	https://pypi.org/project/TEtranscripts/	Tools for estimating differential enrichment of Transposable Elements and other highly repetitive regions
9	SQUIRE	http://www.squire-statement.org/	Standards for quality improvement Reporting excellence is referred to as SQUIRE. The SQUIRE standards offer a structure for disclosing brand-new information on how to enhance healthcare
10	BWA	https://github.com/lh3/bwa	For short-read alignment, use the Burrow-Wheeler Aligner (see minimap2 for long-read alignment)
11	Repeat Masker tools	https://www.repeatmasker.org/	A programme called RepeatMasker checks DNA sequences for low complexity and interspersed repeats
12	Novoalign	http://www.novocraft.com	A potent tool made for the mapping of short reads from the Illumina, Ion Torrent, and 454 NGS technologies onto a reference genome
13	STAR	https://github.com/alexdobin/STAR	Spliced transcripts alignment to a reference, RNA-seq aligner, Alexander Dobin, 2009–2022
14	Alu-detect tool	https://github.com/compbio-UofT/alu-detect	Alu insertions may be found using the software programme alu-detect on Illumina HTS reads
15	Tangram	https://github.com/jiantao/Tangram	Detecting structural variation quickly toolbox
16	MELT	https://melt.igs.umaryland.edu/	A Java-based software programme called the Mobile Element Locator Tool (MELT) is used to find, annotate, and genotype non-reference Mobile Element Insertions (MEIs) in Illumina DNA paired-end whole genome sequencing (WGS) data

Conclusion

Retrotransposons is that class of transposons or TEs which is based on the copy-paste mechanism. Both the LTRs and non-LTRs have the RT domain which acts as an activator for these terminal repeats to classify, filter, and then generalize all the data in a better way. Different methods are involved in the synthesis and mechanism of the retrotransposons which aided the researcher in taking a dig into these elements, their importance, and analysis in eukaryotes. It plays a productive role in agriculture leading to different crops. Different techniques like structural based, homology based, de novo identification, and comparative genomics are applied on a wide range. In this era, computational methods, software, and tools are used at a high level for the analysis of these transposable elements. But still, more approaches, tools are on head for study.

References

Baldo, M.A., Adachi C., and Forrest, S.R. (2000). "Transient analysis of organic electrophosphorescence. II. Transient analysis of triplet-triplet annihilation." Physical Review B 62(16): 10967.

Bennetzen, J.L. (2000). "Transposable element contributions to plant gene and genome evolution." Plant molecular biology 42(1): 251–69.

Biedler, J.J.K. (2005). Non-LTR retrotransposons in mosquitoes: Diversity, evolution, and analysis of potentially active elements, Virginia Polytechnic Institute and State University.

Cordaux, R., and Batzer, M.A. (2009). "The impact of retrotransposons on human genome evolution." Nature reviews genetics 10(10): 691–703.

Edgar, R.C., and Myers, E.W. (2005). "PILER: identification and classification of genomic repeats." Bioinformatics 21(suppl_1): i152–i158.

Feschotte, C., Jiang, N., and Wessler, S.R. (2002). "Plant transposable elements: where genetics meets genomics." Nature Reviews Genetics 3(5): 329–41.

Finnegan, D.J. (2012). "Retrotransposons." Current Biology 22(11): R432–R437.

Flutre, T., Duprat, E., Feuillet, C., and Quesneville, H (2011). "Considering transposable element diversification in de novo annotation approaches." PloS one 6(1): e16526.

Goodier, J.L. (2016). "Restricting retrotransposons: a review." Mobile DNA 7(1): 1–30.

Han, J.S. (2010). "Non-long terminal repeat (non-LTR) retrotransposons: mechanisms, recent developments, and unanswered questions." Mobile Dna 1(1): 1–12.

Han, J.S., and Boeke, J.D. (2005). "LINE-1 retrotransposons: modulators of quantity and quality of mammalian gene expression?" Bioessays 27(8): 775–84.

Jaiswal, V., Gahlaut, V., Kumar, N., and Ramchiary, N. (2021). Genetics, genomics and breeding of chili pepper Capsicum frutescens L. and other Capsicum species. Advances in plant breeding strategies: Vegetable crops, Springer: 59–86.

Kazazian Jr, H.H. (2004). "Mobile elements: drivers of genome evolution." *Science* 303(5664): 1626–32.

Keane, T.M., Wong, K., and Adams, D.J. (2013). "RetroSeq: transposable element discovery from next-generation sequencing data." Bioinformatics 29(3): 389–90.

Kidwell, M.G. and Lisch, D. (1997). "Transposable elements as sources of variation in animals and plants." Proceedings of the National Academy of Sciences 94(15): 7704–11.

Nadeem, M.A., Nawaz, M.A., Shahid, M.Q., Doğan, Y., Comertpay, G., Yıldız, M., Hatipoğlu, R., Ahmad, F., Alsaleh, A., and Labhane, N. (2018). "DNA molecular markers in plant breeding: current status and recent advancements in genomic selection and genome editing." Biotechnology & Biotechnological Equipment 32(2): 261–85.

Ostertag, E.M. and Kazazian Jr, H.H. (2001). "Biology of mamalian L1 retrotransposons." Annual review of genetics 35: 501.

Philippe, C., Vargas-Landin, B.B., Doucet, A.J., van Essen, D., Vera-Otarola, J., Kuciak, M., Corbin, A., Nigumann, P., and Cristofari, G. (2016). "Activation of individual L1 retrotransposon instances is restricted to cell-type dependent permissive loci." Elife 5: e13926.

Riehl, K., Riccio, C., Miska, E.A., and Hemberg, M. (2022). "TransposonUltimate: software for transposon classification, annotation and detection." Nucleic Acids Research 50(11): e64.

Schulman, A.H., Flavell, A.J., and Ellis, T. (2004). The application of LTR retrotransposons as molecular markers in plants. Mobile genetic elements, Springer: 145–73.

Wilhelm, M., and Wilhelm, F.-X. (2001). "Reverse transcription of retroviruses and LTR retrotransposons." Cellular and Molecular Life Sciences CMLS 58(9): 1246–62.

Wu, G., Fanzo, J., Miller, D.D., Pingali, P., Post, M., Steiner, J.L. and Thalacker-Mercer, A.E. (2014). "Production and supply of high-quality food protein for human consumption: sustainability, challenges, and innovations." Annals of the New York Academy of Sciences 1321(1): 1–19.

YALDIZ, G., Camlica, m., Nadeem, M.A. Nawaz, M.A., and BALOCH, T.S (2018). "Genetic diversity assessment in Nicotiana tabacum L. with iPBS-retrotransposons." Turkish journal of agriculture and forestry 42(3): 154–64.

Retrotransposons in Plant Genomics
(*Zea mays*)

Priyanka Devi,[1] *Preedhi Kapoor,*[2] *Prasann Kumar*[1*]
and Joginder Singh[3*]

1. Introduction

The Class I transposable elements (TEs) were initially found in mammals and yeast. In contrast to class II transposable elements, which are transposed in a DNA-mediated (hence, cut-and-paste) manner, retrotransposons insert into the genome following the reverse transposition mode through an RNA intermediate and reverse transcriptase (hence, copy-and-paste), which can lead to an increase in the number of retrotransposon copies and the generation of stable insertion mutations in the host genome. Numerous papers have been published concerning retrotransposons since the first study on maize retrotransposons in plants (Shepherd et al., 1984). These reports demonstrate the widespread distribution of retrotransposons in plant genomes (Feschotte et al., 2002). According to conventional wisdom, retrotransposons constitute the vast majority of plant-dwelling transposable elements. Retrotransposons, for instance, take up 50% of the maize genome and 90% of the wheat genome, respectively (Gregory 2005; Wicker and Keller, 2007). New research suggests that different retrotransposon

[1] Department of Agronomy, School of Agriculture, Lovely Professional University, Phagwara, 144411 (Punjab), India.
[2] Department of Biochemistry, School of Bioengineering and Biosciences, Lovely Professional University, Phagwara, 144411(Punjab), India.
[3] Department of Microbiology, School of Bioengineering and Biosciences, Lovely Professional University, Phagwara, 144411(Punjab), India.
* Corresponding author: prasann0659@gmail.com[1]; joginder.15005@lpu.co.in[3]

subfamilies have different transposition activity and distribution patterns, which are typically linked to significant alterations in genome architecture and ultimately have a stronger influence on genetic and epigenetic modifications (Piegu et al., 2006; Vitte et al., 2007). Thus, we learn that retrotransposons play a crucial role in agricultural genetic improvement research and are the primary driver behind plant development. Although retrotransposons were first discovered in animal and yeast genomes, mounting evidence suggests that they are also abundant in plant genomes and may make up a significant portion of certain genomes, particularly those with enormous genome sizes. They play a crucial role in the structural development of plant genomes, particularly those of higher plants, being one of the most fluid of genomic components, fluctuating considerably in copy quantity over very short evolutionary timescales. Both the population structure (how many copies of each retrotransposon type are present) and transpositional activity of these mobile elements have changed over the development of higher plant genomes (how frequently do different transposons replicate, and what percentage of transposons are active).

Maize has been researched more than any other plant throughout genetic research's past. In addition to its importance in agriculture and the economy as a crop used for food, feed, and fuel, maize also possesses distinctive biological properties that make it an excellent research model for studying genetic diversity and the evolution of genomes (Coe et al., 2001; Bennetzen and Hake 2009). There is a lot of variety in the maize gene pool from both wild and domesticated ancestors. Because of the use of well-established breeding strategies, inbred lines, mutant collections, easily-followed phenotypes, and large, distinctive chromosomes, this first plant genetic map, evidence of meiotic recombination linked with the recombination of genetic traits (Chandler et al., 2000), but instead assistance for the chromosomal principle of inherited wealth were all made feasible. Epigenetic mechanisms (in the form of paramutation) as well as transposable elements (TEs) (later shown to be universal to the majority, if not all, of eukaryotic along with prokaryotic organisms) both played a role in the evolution of the human genome were discovered via further study utilizing maize as a genetics model (Bennetzen and Hake 2009). The organization of the maize genome as well as its evolutionary history have been illuminated by the accumulating cytogenetic and genetic data and, more recently, by the massive sequencing information generated from genome project attempts in grasses. Extensive resequencing through next-generation sequencing technologies including state-of-the-art comparative genomic hybridization (CGH) techniques (Lai et al., 2010) has led to the accumulation of structural and sequence data for many maize genotypes. Additionally, the growth of the grass family, Poaceae has contributed a significant number of full genome sequences, which serve as an important resource for comparative genomics (Wang et al., 2011). Four non-maize grass species have a rather full physical assembly at this time: Plants such as rice (*Oryza sativa* L.), sorghum (*Sorghum bicolor*), purple false broom (*Brachypodium distachyon*), as well as foxtail millet (*Setaria italica*) (Vogel et al., 2010).

There is a huge amount of data on how genomes are structured and organized because of genomic technology. Marker-based genetic maps, "chromosome paintings" generated using fluorescent in situ hybridization, and completely genomic DNA sequences are all examples of this type of information. However, the difficulties of interpreting genomic data in an evolutionary context remain largely unchanged from those encountered by Stebbins (1) and the other authors of the evolutionary synthesis, even though genomic techniques have altered the volume and quality of data. Two major obstacles need to be overcome: deducing the mechanics of evolution and building a complete picture of evolutionary change. In this work, we will use maize (Zea mays) as a model system to examine the processes that have contributed to the development of plant nuclear genomes. The near-term completion of the Arabidopsis (*Arabidopsis thaliana*) genome, and the subsequent completion of the rice (*Oryza sativa* L.) genome, will undoubtedly shed light on many of the mysteries surrounding the evolution of plant genomes, making it somewhat premature to discuss the subject at this time. Keep in mind that the reason why scientists are sequencing plants like Arabidopsis and rice is because their genomes are so unusually compact and well-organized (Paterson et al., 2009). Understanding the evolution of plant nuclear genomes like the maize genome, for which whole DNA sequences will not be easily available, would be an enormous task even once these genomes are sequenced. In this lecture, we will explore nucleotide replacement in the context of phenotypic variation. By revealing the evolutionary forces at work on genomes, patterns of genetic diversity shed light on the population genetic processes that affect various parts of the genome. While the study primarily addresses maize, we do make some species-level generalizations as well (AN et al., 2009).

2. Categorization of Retroelements with Control of Retrotransposition

Retrotransposons are one of the two major groups of transposable elements found in animal and plant genomes during the past decade and are classified according to their mode of replication. They are unique from other transposons (class II elements that use DNA in movement, including Ac and En/Spm) because they may transpose via an RNA intermediate (also termed class I elements). There are two primary kinds of plant retrotransposons, each with its own distinct structure and transposition cycle:

First, long terminal repeat (LTR) retrotransposons (Ty1-copia group, also known as Pseudoviridae and gypsy group - Metaviridae), which are retrotransposons with LTRs and encode products with structural similarities to the retroviral gag- and pol-encoded proteins.

The second kind of retrotransposon is the non-LTR variety, which includes the LINE and SINE elements (Leigh et al., 2003). These retrotransposons similarly encode proteins that resemble gag and pol but don't have the long terminal repeats. Both LTR retrotransposons and vertebrate retroviruses, which together make up the retrovirus-related transposon family in animals, arise from a reverse transcription of an RNA template by creating a DNA daughter copy, and their replication cycle requires an intermediary cytoplasmic step. Since retrotransposons may replicate by transposing their genetic material into the genomes of their hosts, they pose a significant threat due to this process. Retrotransposition is strictly regulated to protect the health of the host organism, which is essential to the life of the retrotransposon itself. There are both host factors and element-encoded functions involved in this regulation. Transcription is a key regulatory process because it controls both RNA template generation and the synthesis of mRNA, which is necessary for protein synthesis. The 5′LTR of LTR retrotransposons often contains cis-regulatory sequences that regulate transcription, such as the U3 region upstream of the transcription start site or the 3′LTR, which contains terminator and polyadenylation signals, are involved in transcriptional regulation. Evidence from genome sequencing projects to date indicates that LTRs of cereal retrotransposons can range in size from a few hundred base pairs (Tos17, 28) to more than five thousand base pairs (Sukkula, 79, Grande, 58) (Garcia et al., 2003). They all share a common 5′TG…CA3′ structure thanks to the short inverted repetitions at their ends. The 5′ and 3′LTRs include reverse transcriptase priming sites that are used to generate the (−) and (+) strands, respectively. Primers bind to a particular region of the gene called the template strand, which is distinguished from the template strand by an internal domain that dictates one or more open readings (Queen et al., 2004).

Life cycles that occur inside cells and the primary encoded proteins (gag, which codes proteins for virus particles, as well as pol, which encodes the enzymatic activities for replication) (Malik et al., 2000) are shared by LTR retrotransposons and closely related infectious and endogenous retroviruses (Wilhelm and Wilhelm 2001), However, retrotransposons do not have a domain called env, which codes for an envelope glycoprotein that is essential for infectiousness. Retroviruses are responsible for encoding glycoproteins that are found on the viral envelope. These glycoproteins attach to the cellular membrane that stimulates the release of viral core particles from infected cells. Not only that, but they also serve as infection mediators by identifying target cells' receptors. Certain gypsy-like retrotransposons in plant genomes (Athila, Calypso, Bagy-2, and Rigy-2) and a small number of copia-like elements (SIRE-1) have recently been found to contain the presence of an env domain, which codes for a hypothetical envelope protein and is often regarded as an indicator of an infectious retroelement. Interestingly, neither D. melanogaster nor any other invertebrate or vertebrate has been reported to have env-like sequences in their copia-like retrotransposons (Eickbush and Malik, 2002). Given that env-like sequences have been found. It's possible that the env gene was picked up by both the copia as well as gy psy retrotransposon families separately. On the contrary, retroviral offshoots of copia and gypsy retrotransposons infiltrated plant genomes and later lost their env gene. To yet, it has been impossible to determine which process began first. Nonetheless, plant retroviruses provide credence to the idea that plants may engage in the apparent horizontal spread of plant viruses (Wright and Voytas, 2002). Transposable elements make up a sizable group of genomic entities on the cusp between genes and viruses. They follow the same path as the cell and organism in which they dwell, much like cellular genes. However, similar to viruses, they embed the information necessary to replicate themselves (Du and Messing, 2006). A family of transposable elements, while constituting an integral part of a normal genome, yet occupies a distinct evolutionary niche distinct from that of the genic elements that make up that genome. This is likely why every genome has so many different transposable element families, each with its unique transposition process. There has been a recent push to standardize the way eukaryotic transposable elements are categorized and named This was found to be the case (Wicker et al., 2007). That way, retrotransposons might be labelled as "RLG" or "RLC" according to the "class, order, superfamily" nomenclature. In other words, they belong to the retroelement class (R), the long terminal repeat (LTR) order (L), and either the Copia or Gypsy superfamilies (C and G) (G). Take "RLC Opie" and "RLG Huck" as two examples. For the sake of brevity, however, we shall avoid prefacing most references to element families with this qualifier. Additionally, we shall not use the "LTR-" prefix when referring to retrotransposons. We shall refer to retrotransposon as an example exclusively to refer to LTR-retrotransposons, even though the words "retroelement" and "retrotransposon" are sometimes used interchangeably.

3. Distribution

In plants, genomes have been shown to include examples of every known retrotransposon family (Schnable et al., 2009). Their pervasiveness in plant life indicates they must have a long evolutionary history (V.L. Chandler et al., 2000). The LTR-retrotransposons appear to be the most common kind of transposable element in plants (Birchler and Han et al., 2009). Many different types of angiosperms have been reported to have LTR retrotransposons from both the Ty1-copia and gypsy families (flowering plants). So far, the Ty1-copia group has been the most well studied family of plant LTR retrotransposons. A large proportion of the genome in at least some plant species is made up of Ty1-copia group retrotransposons (Woodhouse et al., 2010). Many plant genomes have been revealed to have extremely diverse populations of Ty1- copia group retrotransposons (Bruggmam et al., 2006). Dicot plant species have undergone both intraspecific and interspecific phylogenetic analyses of these transposons, and the results imply that Ty1-copia group retrotransposons existed and split into diverse subgroups prior to the emergence of contemporary plant orders. Genetic information from these clusters has been passed down through the generations vertically to their offspring species, with horizontal gene transfer across clusters being extremely uncommon or nonexistent. Like in dicots, Ty1-copia group retrotransposons were detected in monocots, and some of these retrotransposons are reported to be interspecific (Wei et al., 2007). All current research provides more evidence that Ty1-copia group retrotransposons and other transposon groups play an important role in plant-genome evolution through their amplification as well as subsequent spread (Morgante et al., 2007). Although gypsy-like components were first reported in arthropods and a few taxonomically skewed plants, they were later discovered to be ubiquitous in grains and other plants. Using the cloned RT domains, Vershinin et al., 2002 identified four families of gypsy-like components.

It has been estimated that the four gypsy-like families obtained comprise around three percent of the barley genome based on the number of hybridizing representatives to gridded and blotted BAC clones. Are there alive retrotransposons? Only a small number of elements, predominantly LTR retrotransposons, have been revealed to have transcriptional activity in the previous decade. The vast majority of these were discovered following the insertion of a foreign DNA sequence into or near a host gene. The majority of mobile elements have been characterized using PCR amplification of genomic DNA or cDNA, it has now been proven that they transpose. Similar to what has been seen with Green alga has stony components. Analysis of comparable LTR sequences in newly transposed copies of Volvox carterii has provided evidence for retrotransposition activity. Common retrotransposons in tobacco (Tnt1A and Tto1), barley (BARE-1), rice (Tos17), and maize have all been shown to contain specific transcripts that initiate the LTR of the element. With the use of constructs in which the LTR of an element was used to regulate the expression of reporter genes, we were able to, additional confirmation of LTR transcriptional capacity was obtained for Tnt1A, BARE-1, and Tto1.

4. General Systems Characteristics of the Corn (Maize) Genome

The genetically diploid maize genome consists of 10 chromosomes and is estimated to be between 2.3 and 2.7 GB in size (Zhou et al., 2009). The majority of the maize genome is made up of no genic, repetitive portion, similar to other big genomes in plant species, and is broken up by islands of distinct DNA with low copy numbers that only carry a single gene or a small collection of genes. The terms "transposable elements" (TEs), "ribosomal DNA" (rDNA), and "high-copy short-tandem repeats" are all abbreviations for the same thing (HCSTRs) are all examples of repetitive elements that can be found primarily in telomeres, centromeres, and heterochromatin knobs, and they all play an important role in the species' wide variety of variation. The DNA sequences known as transposable elements (TEs) may replicate independently and move to new locations throughout the genome. Depending on regardless of whether the intermediary of transposition is RNA or DNA, they are categorized as Class I (retrotransposons) or Class II (DNA transposons), respectively (Feschotte et al., 2009). Retrotransposons are mobile genetic elements that replicate in place by reverse transcription, which results in a gain of one element due to each copy inserting itself into a different location in the genome. Cut-and-paste transposition is the norm for the vast majority of Class II transposable items. It is hypothesized that members of one family of plant DNA transposons that are engaged in replicative transposition have a DNA replication mechanism in the form of a rolling circle (Kapitonov and Jurka, 2001). Both class I and class II transposons can be either autonomous or non-autonomous meaning they contain everything needed for transposition, or no autonomous, meaning their transposition requires the existence of the corresponding autonomous element (Feschotte et al., 2009).

Extended repetitions towards the end (LTRs) the majority of mobile genetic elements are retrotransposons, which are mobile genetic elements similar to retroviruses and may be identified by their possession of lengthy terminal repeats frequent class of TEs in plants. Target site duplication (TSD) borders, widespread CpG and CHG methylation, substantial clustering, and frequent nesting inside other LTR retrotransposons are all characteristics shared by the vast majority of LTR retrotransposons (Hollick and Springer, 2009). The maize genome contains about a million LTR retrotransposon segments, or over 75 percent of the nuclear genome has been sequenced. The precise number of elements is unknown; however, estimations have ranged from 150,000 to 250,000 elements due to the nested structure of retrotransposition in maize and the dispersion of the material (Baucom et al., 2009). As of now, 441 families of LTR retrotransposons have been identified; the vast majority of them are present in the genome at extremely low copy counts (less than 10 copies each), making up just a tiny fraction of the total genome size. However, nearly 80% of all retroelements belong to just two families (Copia and Gypsy). Copia-like elements are often found in generic areas, whereas Gypsy-like elements are disproportionately prevalent in pericentromeric and other heterochromatic regions (Schnable et al., 2009). Finally, non-LTR retroelements with less clarity include LINES (Long Interspersed Nuclear Elements) and their shorter cousins, SINES (Short Interspersed Nuclear Elements). The transposition of SINEs requires the existence of the autonomous LINE, and SINEs contain an internal RNA polymerase III promoter sequence and a single homopolymer terminus (Baucom et al., 2009). The terminal sialic acid (TSD) and homopolymer (often poly A) at one end are characteristic features of LINEs, which are normally formed after insertion (T). In maize, LINES, and SINES make up just around 1% of the genome, making them uncommon. In comparison to retrotransposons, DNA transposons (which are part of Class II elements) make up a lesser chunk of the maize genome (about 8.6%). Class II maize transposons were initially described by Barbara McClintock (Zhou et al., 2009) while she was investigating the Ac/Ds elements' role in causing chromosomal duplications, inversions, and translocations. Tc1-Mariner, hAT, Mutator, PIF Harbinger, and their MITES (Miniature Inverted Transposable Elements) nonautonomous descendants have now been discovered and categorized as new families of DNA transposons (Wicker et al., 2007). Terminal inverted repeats (TIRs) and short terminal simple duplications (TSDs) define these Class II components. Nonautonomous elements are often shorter, have less conserved internal sequences, and/or carry captured DNA fragments, while the transposase and/or additional genes necessary for autonomous elements' transposition are encoded by those elements (Feschotte et al., 2002). It has been shown that class II elements, such as retrotransposons, are not uniformly dispersed across the genome, and except for the CACTA family, most show an insertional bias towards genic regions (Schnable et al., 2009).

Methylation may play a role in the regulation and silencing of class I and II transposable elements since activation is associated with the de-methylation of TIRs (Jia et al., 2009). Knobs, centromeric repeats, telomeres, and ribosomal DNA (rDNA) are only a few examples of genomic regions that contain high-copy tandem repeats. Sequences that help build the kinetochore and bind microtubules to chromosomes during mitosis and meiosis can be found in or near centromeres repeats and retrotransposons. The CentC unit, which is 156 base pairs long, is found in the thousands in maize centromeres (Birchler and Han 2009). The rapid evolution of centromeres and the lack of sequence similarity between species means that repeats inside centromeres are likely to have different functions in different organisms. Nonetheless, all maize chromosomes contain identical repetitions (Dawe et al., 2009). The number of copies in most centromeres is unknown because of the challenges in sequencing extremely long sections full of tandem repeats. Chromosomes 2 and 5, which are assumed to be the shortest, are the only ones with fully completed centromeres (Wolfgruber et al., 2009). Estimates put the CentC content of assembly of the maize ZmB73v1 reference genome has reached 54% completion. An examination of DNA fibers that have been stretched indicates that the total length of CentC arrays can range anywhere from less than 100 kb to several megabases. Four different types of retrotransposons may be found inside the CentC tandem repeat sequences. These retrotransposons are referred to as CRM1, CRM2, CRM3/CentA, and CRM4 (Schnable et al., 2009). Not only that but also Cent1 and Cent2 repeat sequences, there is a third repeat sequence, Cent4, located on or near chromosome 4's major constriction (Page et al., 2001). Cytologically, heterochromatic knobs, which appear as black, spherical structures, are made up of derivatives of two repeats that are megabase-sized tandem repetitions of two repeats (180 and 350 bp long), which together account for 0.6% to 6% of the whole genome.

They are located in over twenty distinct pachytene chromosomal regions (Ananiev et al., 2005), and structural variations between different types of maize imply substantial intraspecific variability in the maize genome, as we shall see. Retro elements can also be installed into knobs (Rafalski and Ananiev, 2009). The rDNA regions have thousands of tandem repeats, each of which codes for a different rRNA. An estimated 1,600–23,000 9–kb tandem copies of genes encoding the 45S RNA precursor for the 18S, 5.8S, and 28S ribosomal RNAs are present on the short arm of chromosome 6 in the maize genome. The nucleolus organizer region (NOR) is comprised of this configuration, and it is another early cytogenetic characteristic in the maize genome. A non-transcribed spacer separates this repetition of ribosomal genes from one another. There is a second, smaller cluster of genes for 5S ribosomal RNA on the long arm of chromosome 2 (Buckler et al., 2006). These Genes are structured as tandem repetitions of 342 base pairs (Li and Arumuganathan, 2001) Telomeres, which were first named by Muller in 1938, but defined by McClintock in maize several years earlier, include tandem-repeated telomeric and subtelomeric sequences to protect the DNA from the irreversible damage that can be caused by the frequent rearrangements that occur naturally at the ends of DNA molecules. Telomeres were first named by Muller in 1938. There are thousands of satellites with a basic sequence, and there are a few megatracts with trinucleotide repeats (such as AGT as well as AGC) scattered across the maize genome.

5.   Variation within the Maize Species

The distribution of heterochromatin knobs and C bands was found to be significantly different amongst different lines in early cytogenetic investigations. Some maize and teosintes also have extra chromosomes, known as B chromosomes. DNA content variations have been shown to correlate favourably with cytogenetic variations (Tito et al., 1991). Copy counts involving tandemly repeated sequences like those seen in ribosomal DNA and knob repeats, differed by as much as two to threefold among North American genotypes, as determined by Southern hybridization and the work of Rivin et al. The typical nucleus content of Zea mays is 5.5 pg/2n (Biradar and Rayburn, 1993), however there are reports of intraspecific variability of up to 38.8%. Recent sequencing data has shown that the amount of natural genetic variation present in the maize genome varies considerably between lines (Ching et al., 2002). One substitution per one hundred bases is the average rate of single nucleotide polymorphism between two inbred lines of maize. Compared to other animals, this amount of interspecies polymorphism is notable; the average rate of polymorphism is also higher than what is shown between humans and chimpanzees, which is 10 times higher than what is found between humans (Buckler et al., 2006).

Increases in DNA content per nucleus appear to have no discernible influence on the phenotype of maize plants. LTR retrotransposons have made the most recent contributions to the size of the maize genome, and their abundance and distribution have been demonstrated to vary widely among haplotypes (Kato et al., 2004). Tandem repeats at centromeres, knobs, and ribosomal DNA loci have been reported to exhibit copy number variation (Springer et al., 2009). Recently, copy number variation (CNV) and presence-absence variation have received a lot of attention (PAV). Now as genome-wide hybridization techniques have advanced and next-generation sequencing has revealed more about the sequences of various maize lines, it is becoming increasingly clear that copy number variations (CNVs) and point allele variations (PAVs) play a substantial influence in the heterosis of the maize genome. Comparing the genomes of the inbred lines B73 and Mo17 using CGH allowed Springer et al., 2009 to establish the degree to which these two inbred lines varied structurally from one another (CGH). In B73, several megabase-size regions were found, but none of these were found in Mo17. Using PCR analysis on 22 more lines, they were able to locate a region on chromosome 6 that was either present or absent in the lines as a single haplotype block. This region was two megabases in size. The comparative genomic hybridization (CGH) screening of 13 North American maize inbreeds relative to the reference inbred B73 using an expression array revealed 2,109 possible CNVs. The authors evaluated a selection of 15 CNV loci through PCR and confirmed that 12 loci (80%) were actual insertion/ deletion events. It was discovered that just two of the CNV regions were hundreds of kilobases in size, with the remainder of the CNVs being less than 10 kb in size.

Swanson-Wagner et al., 2010 used array-based CGH to compare the content and copy number variation of 32,500 genes between 19 unique maize inbred lines and 14 teosinte accessions, all concerning the B73 reference genome. They found that there was a significant amount of variation across the board. It was found that almost ten percent of the targets were unique, with 3,410 genes having a lower copy number or being absent in B73, 479 genes having a larger copy number, and 3,480 genes having a bigger copy number than in B73. B73 only had one copy of most down genes, hence they were classified as PAV. Several genes were more prevalent in certain lines while being missing or at lower levels in others. They found it fascinating that most of these polymorphisms existed long before maize diverged from teosinte.

6.   Maize Genome Evolves and Diverges

Multiple major mechanisms have contributed to the evolution of the maize genome and the generation of intraspecific genome diversity, including: (1) whole genome duplications (polyploidization) but rather segmental duplications, (2) DNA transposition as well as retrotransposition, (3) transposon encapsulate as well as translocation of genes or gene segments, (4) assortment but also gene conversion events, and (5) single base mutations or expansion/contraction of simple sequence repeats (SSRs). In addition to the genetic variation introduced into maize via interpopulational gene transfer and introgression with closely related species (teosinte) (Beló et al., 2010), the following processes are also responsible for generating new variants.

7.   Include Polyploidization and Duplication

Important steps in the evolution of the maize genome have been facilitated by polyploidization (Messing et al., 2009), as it has in the genomes of many other cereals and plants. Extensive chromosomal duplications in maize were originally detected in linkage and comparative genomic analyses, supporting the existence of whole-genome as well as segmental duplications via broad crossings. Recent comparative sequencing data supports the assumption that two polyploidization events have occurred in the maize genome. The first event occurred at 70–80 Mya and involved entire genome duplication and subsequent gene loss in a cereal ancestor. There is just one copy of the genes that were duplicated as a result of this phenomenon, rice and sorghum are now coincident. Given that 99 percent of these genes are shared between species, we may infer that gene loss happened prior to the evolution of cereals (Paterson et al., 2004). Through a process known as biased fractionation (Woodhouse et al., 2010), genes have been selectively lost from one homolog. After maize and sorghum diverged from a common ancestor, a second polyploidization event occurred between 5 and 12 Mya. At some time between 4.8 and 11.9 Myr (Swigonov et al., 2004), two maize progenitors mated, giving rise to a tetraploid. Subsequently, the genome underwent widespread deletion and rearrangement of chromosomal rearrangements and up to half-duplicated genes, bringing it back to a diploid state. The number of high-confidence protein-coding genes in sorghum (27,640) is lower than estimates for Brachypodium (25,532), rice (29,717), and sorghum (27,640) (Devos 2010). Maize B73 projected under high stringency is 32,540. Due to gaps in the existing physical construction, this estimate is likely too low. Although stable tetraploid maize varieties have been recorded (Birchler and Han, 2009), the redundancy induced by early polyploidization and the subsequent easing of selection restrictions is the most significant consequence of polyploidization on contemporary maize.

8.   The Growth of the Maize Genome Through Retrotransposition

In the Andropogoneae family, close cousins maize and sorghum have corn has chromosomes much as wheat, however, its genome is around three times larger (800 Mbp). The above-described secondary polyploidization only partially explains this variation. In the last 10 Myr, LTR retrotransposition has greatly increased the intergenic spacing and total size of the maize genome. The fraction of long terminal repeat retrotransposons (LTR retrotransposons) in a grass genome increases with its size but stays the same. Retrotransposon content ranges from 54.5% in sorghum to 75.9% in maize (Devos, 2010), with Brachypodium and rice's tiny genomes having values of 23.3% and 25.8%, respectively. One of the earliest discoveries observed as sequence data accumulated was the high quantity and nonrandom distribution of LTR retroelements in maize (Wei et al., 2009). Mutation analysis allows for the date of these elements due to their identical lengthy terminal repeats now of transposition (Liu et al., 2007). Early investigations of the region's nested retroelements suggested a significant retrotransposition event had happened during the past 3 Myr (Haberer et al., 2005). Liu et al., 2007, who found two amplification peaks in gene-free regions, studied LTR retrotransposon insertion kinetics: one approximately 1.5–2 Mya and another within the last 500,000 years. One single peak of amplification was discovered in gene-containing areas, and it occurred during the last million years. LTR retrotransposition is conservative since it uses an RNA intermediary and discards the original element, yet this self-serving DNA is responsible for a significant amount of expansion in the maize genome.

9.   How Genomes Genomes are Diminished

Important factors that may limit the propagating results of LTR retrotransposition include uneven and illegitimate recombination. The internal sequence and one LTR sequence are thought to be lost due to intrachromosomal (intrastrand) homology disequilibrium associated with direct duplications of over fifty base pairs in length (bp) (in this example, between neighbouring LTRs). It is hypothesised that a "solo" LTR is formed by this procedure. Uneven crossing between homologous LTR sequences on opposite chromosomes can have more dramatic consequences, such as a mutually exclusive deletion or duplication event, inversions, and mutually exclusive translocations (Vitte and Bennetzen, 2006). Many internal deletions and shortened LTR retrotransposons are thought to have originated through illegitimate recombination, which may occur between homologies of shorter lengths than unequal homologous recombination (Ma et al., 2004). Recombination of this type is thought to result in DNA loss through a process of mismatches between strands of DNA, also known as NHJ or slip-strand mispairing (Garfinkel, 2005). It is hypothesised that all three of these processes work together to limit the expansion of the plant genome, which is caused by ploidy increases and the amplification of repetitive DNA (Vitte and Bennetzen, 2006).

10. Genealogical Coinherence and Transposable Elements

Changes in the amount of heterochromatic DNA outside of coding regions have long been hypothesized to account for intraspecific genome variation (Jiang et al., 2011). Since it was initially reported by Fu and Dooner (2002), however, instances of violation of gene microcolinearity have been discovered in numerous settings. The BAC contigs flanking the bz locus in two North American inbred lines were sequenced. The contigs were 230 kb and 110 kb long for the B73 and McC lines, respectively. Significant differences were found between the two lines in both the number and location of retrotransposons, which are elements that can move genes around. Even more surprisingly, four of the ten genes found in a cluster in the McC sequence were not present in B73. Moreover, only 25% to 84% of sequences were shared amongst different maize lines, as determined by further sequencing analysis of the bz locus. It has been discovered that there may be many sites on each chromosome where comparable polymorphisms for the presence/absence of genic sequences occur (Brunner et al., 2005).

Helitrons have been connected to the intraspecific violation of genetic collinearity in maize. In 2005, Lai et al., were the first to report the role of Helitrons in causing genome variation in maize. Their work involved a comparative bioinformatics analysis of the bz region in McC and B73, which revealed the presence/absence of two Helitrons, HelA, and HelB, which together accounted for all of the allelic variation at this locus. However, Helitron elements do not produce TSDs and are not flanked by terminal inverted repeats like the other Class II TEs. They preferentially insert in AT dinucleotides and feature a hairpin-forming sequence of 18–25 bp near the 3′ end (Brunner et al., 2005). An expanded analysis of the genetic differences between the inbred lines B73 and Mo17 (Morgante et al., 2005) found evidence that Helitron sequences were responsible for the majority of the differential insertions in the genome between the two lines. Recently, a Helitron grabbed the maize CYP72A27-Zm gene and transferred it into an Opie-2 retrotransposon, representing a complete cytochrome P450 monooxygenase (P450) gene (Gupta et al., 2005). Unlike the vast majority of other Helitron genes that have been discovered, which appear to be truncated versions of their progenitor genes. There are several instances of full Helitron elements in the DNA sequence. More than 20,000 Helitron fragments and 1,930 complete Helitrons were found in a single research (Yang and Bennetzen, 2009). Researchers in another investigation found 2,791 Helitron elements that did not behave autonomously (Du et al., 2009). Complete sequencing of a single maize inbred line gives biassed information on the breadth and variety of gene capture, transposition, and amplification by Helitrons, although most of the elements found so far are nonautonomous Helitrons including chimeric regions from different genes.

Helitron is not the only member of the Mutator superfamily with chimaera-making potential; the family's non-autonomous elements, Pack-MULES, can also grab portions of a nuclear gene(s) (Jiang et al., 2004). Moreover, molecular evidence has shown that these novel chimaeras are capable of transcription and translation, indicating that a novel protein-coding sequence may be generated and developed using this strategy of gene fragment capture inside non-autonomous elements. Many recent studies have shown that Pack-MULEs preferentially grab GC-rich genomic sequences and exhibited biased insertion into the 5 end of coding areas (Hanada et al., 2009). This behaviour is analogous to that which was observed with Helitron elements during transposition and amplification, which can cause variations in intraspecific synteny. Because pack-MULE elements tend to insert themselves at the beginning of the transcribed region, the endogenous genes into which they are inserted can have their expression profoundly altered because of the presence of these elements. These elements have sequences that are related to short RNAs, and these sequences can affect the expression profile of the genic sequences that have been recorded.

11. The Role of Retrotransposons in Plant Genome Evolution

Retrotransposons are well known to constitute a sizable fraction of the plant genome, but just what biological significance do they have and what function do they play there? Fortunately, thanks in large part to the information provided by the genome sequencing project and the fact that our understanding of this topic is constantly improving thanks to the use of cutting-edge genetic and biological technology, we have a much clearer picture of the situation. First, retrotransposons have a significant role in regulating the structure and function of genes, as well as the appearance of new genes. When retrotransposons insert themselves into genes or the sequence, they are referred to as genomic integration immediately next to them; they can trigger intricate processes that regulate metabolism. Retrotransposon insertion is linked to the sequencing of one-sixth of all genes in rice. The distance that separates the spot where retrotransposons were inserted and the place where transcription began determines how the gene is translated after retrotransposon insertion. Changes in the splicing model occur in the post-transcriptional phase when retrotransposons enter into translation-sensitive regions of genes. The LTR structure of retrotransposons, which includes promoter and terminator sequences, can also be used to alter gene function. Gaining new functionalities (neofunctionalization) or splitting the original function (sub-functionalization) among duplicated genes in the highly polyploidy plant genome may be driven by the insertion of a new promoter or terminator sequences sitting inside the elements in the internal gene structure. Some retrotransposons have a specific preference for a group of genes whose incomplete sequence is abundantly recorded in cDNA databases.

This morphological feature collectively showed that retrotransposons play a vital role in controlling how genes look and work (Krom et al., 2008). Using genetic mapping, scientists have recently been able to clone SUN, a crucial gene responsible for the tomato has elongated fruit morphology. Fruit shape-related loci were initially shown to have arisen from the long terminal repeat retrotransposon rider was responsible for an unusual 24.7 kb gene duplication. Increased expression of SUN loci after retro-transposition into a novel genomic context led to the full realization of an elongated fruit form that had been in progress for some time (Xiao et al., 2008). Transgenic plants of the plant species Arabidopsis underwent RNA interference (RNAi) after SINE retrotransposon transcription, which altered the pattern and timing of the transgenic gene of interest (Pouch-Pelissier et al., 2008). It is clear from the studies that retrotransposons play a significant role in the emergence of novel gene structures or functions and in significantly amplifying phenotypic variability across plants. Second, retrotransposons' "copy and paste" replication mechanism, which may increase retrotransposon copy number and potentially extend genome size through the process of recombination, plays a critical role in defining genome size and organization. It was speculated that the small genome sizes of Arabidopsis and other plants resulted from their low levels of retrotransposition activities, whereas maize's massive genome size was attributable in part to the high frequency with which retrotransposons were transposed (Bennetzen et al., 1998). In recent years, it has become clear that the quantity of long terminal repeat (LTR) retrotransposons (RTRs) correlates with genomic flow patterns across the rice genus (Ammiraju et al., 2007). More than 90,000 retrotransposon copies from three LTR-retrotransposon families have accumulated in the genome of Oryza australiensis, a wild relative of the Asian cultivated rice O. sativa, over the past three million years, causing a rapid two-fold increase in genome size (Piegu et al., 2006). Genome-wide analysis of the Copia retrotransposon distribution and evolution in Triticeae, rice, and Arabidopsis revealed the conservation of ancient evolutionary lineages and the distinct dynamics of independent Ty1-copia families (Wicker and Keller, 2007). Comparing the distribution of TEs in Arabidopsis and *Brassica oleracea* revealed that while virtually all TE lineages were shared across the two species, the number of TE elements in each lineage was usually larger in B. oleracea, and retrotransposons filled considerable parts in B. oleracea. These results backed the idea that retrotransposons contributed to the fast growth of the B. oleracea genome after the two species diverged from a common ancestor (Zhang and Wessler, 2004).

12. Families of Major Retrotransposons in Maize

Historically, genetic definitions were used to classify families of transposable elements. Elements were regarded to be related if trans-acting components of one element might hasten the transposition of another. As the cost of sequencing has decreased rapidly, structural data has become abundant, whereas functional data has become scarce. Accordingly, sequence-based definitions of "families" have been and are being developed to identify families without functional evidence. It is unclear whether these two categorization strategies overlap. How sequence-sensitive is this family's transposition machinery? Successful insertion of a nascent element into the genome requires binding to the termini of the retrotransposon and inserting them into the host DNA. Even short, the fact that yeast Ty1IN can integrate DNA segments comprising as little as 12 bp of each Ty1 LTR terminus with a 50% efficiency demonstrates that lengths of terminal similarity may be sufficient to bind relatively distantly related components into functional families (Eichinger and Boeke, 1990). Tabular data 2 displays possible meta-family groupings based on the termini of 14 widely reported retrotransposon families in maize. In continuation with this theme, Table 3 displays the sequence similarity of meta-family members near to internal borders of LTRs.

12.1 An Opie-Ji

A member of the Copia superfamily, Opie-Ji is the most common maize family. Typically, there is a 50%+ difference between Opie and Ji LTRs. Thus, they make up different families, at least according to one definition. However, having similar LTR termini suggests that Opie, Ji, Ruda, and perhaps even Giepum all belong to the same genetic lineage (Table 2). This family's LTRs include several copies of a short sequence called the "maize palindromic unit" (MPU). The consensus sequence for this MPU is CACCGGACANTGTCCGGTG. Therefore, the MPU is a very common palindrome since it is duplicated many thousands of times across the maize genome. These regions probably provide a regulatory function. During the early phases of microspore development, there is a dramatic increase in the transcription of Ji (also called "PREM-2") (Turcich et al., 1996).

12.2 Cinful-Zeon

To put it simply, Cinful is a member of the Gypsy royal dynasty. They both seem broken; Cinful lacks gag and Zeon pol. LTRs from Cinful and Zeon show enough sequence similarity, especially at their ends (Table 2), to imply they use the same gene products for transposition.

12.3 Huck

Huck belongs to the retrotransposon family Gypsy and is particularly big and GC-rich. A POL-encoding subfamily and a POL-lacking subfamily are two of many possible groups. This family may be more abundant than the Opie-Ji family in the maize genome, according to certain calculations (Meyers et al., 2001). The Huck element makes up at least 10% of the maize genome, and this is the case even if lower estimates of the number of Huck insertions in the maize genome are right. With an average GC content of 62%, Huck suggests that this family's amplification has added a few percentage points to the GC content of the maize genome.

12.4 Prem1

Prem1 belongs to the Gypsy superfamily of elements and has particularly big long terminal repeats (LTRs). Many members of this family have undergone extensive deletion and sequence divergence, suggesting that they were inserted a very long time ago. However, entire insertions with their termini intact have been identified. Xilon (Fu and Dooner, 2002), and Tekay all appear to belong to a larger Prem1 family based on their shared sequence features, particularly at the LTR termini. Early to mid-stage uninucleate microspores, as well as root and cob tissues, show significant levels of Prem1 expression.

12.5 Huge

Grande, a member of the Gypsy superfamily, has the biggest internal domain of all the high-copy maize retrotransposons (Table 1). Multiple clusters of 50-150 base tandem repeats, which appear to have stable secondary structures, are also present in these elements (Monfort et al., 1995). This quality is shared by a tobacco component with no name and the maize protein Cinful-Zeon (Sanz-Alferez et al., 2003). These tandem repetitions likely serve some structural role, such as a GAG binding site or a genomic RNA dimerization site, as they are conserved across quite different retrotransposon families.

13. Characteristics of Retrotransposons, which are Virological in Origin

Similarities between retrotransposon and retrovirus behaviour have been observed. This is to be anticipated, given their tight familial ties. However, the nature and scope of their connection remain mysterious.

13.1 Movement of Genes between Cells

Researchers have shown that retrotransposons, like their retroviral relatives, may hijack host genes and insert them into their genome. One logical way for this to have happened is reverse transcription in a VLP with several RNAs, given RT's preference for swapping templates. Table 1 shows that many, if not all, retrotransposons belonging to the Gypsy superfamily contain DNA sequences of unknown origin, some of which may have been taken from fully functioning maize genes. However, maize Bs1 retrotransposon has the most convincing evidence

Table 1 Distribution and proportion of features in the annotated genomes of maize

S.No.	Sequence completed	Maize: B73 2160 Mb	
		Copies	% Genome
1	Class I (retroelements)	1,139,990	75.6
2	LTR Ty1/copia	404,000	23.7
3	LTR Ty3/gypsy	477,000	46.4
4	LTR other	222,000	4.5
5	Non-LTR (LINE)	35,000	1
6	Class II (DNA Transposon)	1,990	0
7	hAT	31,800	1.1
8	Mutator	N/A	N/A
9	Tc1/mariner	N/A	N/A
10	MULE	12,900	1
11	MITE (Stowaway)	14,000	0.1
12	Other Class II	N/A	2.2
13	Protein-encoding genes	32,540	6

of retrotransposon transduction of a cellular gene. Despite possessing a spliced proton ATPase gene (654 bp), Bs1 is a flawed element since it lacks a functioning pol ORF. The Bs1 gene has been shown to contain pieces of at least two different cellular genes, according to a later investigation (Elrouby and Bureau, 2001). There is debate over what role these transduction occurrences have in function. There is still the possibility that such host gene trafficking may confer a novel resistance to host suppression of transposition or act as a stepping stone on the path to the acquisition of a functional env, allowing the element to "become viral" as well as exit the host cell. The existence of plant retroviruses is controversial, while the existence of plant retrotransposons is not. These gene fragments may serve just as "stuffers," increasing the size and content of a non-mobile element so that it is more competitive for inclusion into VLPs of a certain retrotransposon family.

13.2 Proving the Value of Plant Retroviruses

Related to retroviruses, retrotransposons play an important role in genome maintenance. In theory, retrotransposons are similar to retroviruses, with the exception that they may jump from one cell to another inside the same organism, so inserting themselves into the genome of a new host. The env gene, named for the "envelope protein" it produces, is positioned downstream of the pol gene in retroviruses and confers this capacity. As opposed to pol and even gag to a lesser extent, env's sequence has not been properly maintained. The gypsy retrotransposon in Drosophila was the first non-vertebrate retrovirus to be discovered. There is currently no hard proof that plants may be infected by retroviruses, but there is a lot of circumstantial evidence to support the idea that such viruses do exist. Several

Table 2 Different LTR termini and ends of the internal domain of several common maize retrotransposon families

S. No.	Family	Type	5′ End	3′ End	Primer binding site 5′ end	PolyPurine Tract 3′ end
1.	Opie	RLC	TGAAAGGGAAATGTGCCCTT	CCCCCCTCTAGGTGCTCTCA	ATTGGTATCRGAGCCGTTCT	ATCACCAAAAAGGGGGAGAT
2.	Giepum	RLC	TGAAAGCATCTAGGSCCCTG	TGGGCATCGTGATCCTTTCA	ATTGGTATCAAAGCCTTGTT	ATTACCAAAAAGGGGGAGAT
3.	Cinful	RLC	TGTTGGGACCATGCTTCGTC	TTGAGAACAAGTCCCCAACA	TTGGCGCCCACCTCCGGTGA	TAGCACCGCGAAGGGGCTAC
4.	Zeon	RLC	TGTTGGGGGCCTTCGGCTTC	TTGAGAACAAGTCCCCAACA	TTGGCGCCCACCTCCGGTGA	TTAATATTGCGAGGGGCTAC
5.	Diguus	RLC	TGTAACACCCTGAATTTTGG	CTTCAAAACCGGGTGTGACA	TAAGTGGTATCAAAGTCGTG	CCTGTTAAGGGGGTTAGATT
6.	Huck	RLC	TGTCGGGGACCATAATTAGG	CCKGTCTCGAAACGCCGACA	GTTGGCGCGCCAGGTAGGGG	CTTCGAGGCTCGGGGGCTAC
7.	Fourf	RLC	TGTTGGATCTTTTATGGGCT	AAATGCCTATATTTCTAACA	ATCCAAAAACCTAATGTTAG	CCCTAGAGTTTGGTGGGGAT
8.	Machiavelli	RLC	TGTTAGGATTTATGGGCTTG	ATTTGTCTAATTATTCAACA	ATCCAAAAACCTTATTGTAG	TTTGAGATCTGGTGGGGAT

plant Gypsy superfamily retrotransposons have had ORFs discovered downstream of pol (Wright and Voytas, 2002), and one example has been found in a Copia superfamily retrotransposon. The absence of compelling proof is disappointing. Potentially less efficient than in animals, a retroviral lifestyle may be impeded by the structure of plant cells. In light of the lack of evidence for their role in the production of particles of infectious plant retroviruses, these downstream ORFs probably have a different, unrelated purpose.

13.3   Capturing and Taming Retrotransposons as Hosts

When a new copy of a retrotransposon is introduced into the genome, it can be modified by several mechanisms, such as inconsistent recombination, the buildup of indels, and DNA methylation. Recombination and the accumulation of small deletions are thought to eventually lead to the elimination of the inserted sequence, while methylation is thought to cause transcriptional inactivation of the retrotransposon and the gradual decay of its sequence through the accumulation of substitutions (particularly transitions, as we discuss below). In some circumstances, however, the inserted copy may provide a selective advantage to the host plant, resulting in the "domestication" of the homologous element by the plant genome.

14.   DNA Methylation and Heterochromatinization

RNA-directed DNA methylation specifically targets retrotransposons, presumably silencing them in the process (Hamilton et al., 2002, Huettel et al., 2007, Matzke et al., 2007). The histone changes that prevent transcription and turn off transposition, hallmarks of the heterochromatic state, have been linked to cytosine methylation (Lippman et al., 2004). Transposable elements are insertional mutations that can produce chromosomal rearrangements; hence, it has been hypothesized that this methylation process alters and rearranges chromatin to render retrotransposons inactive, so reducing their impact on the "host" genome. Although 5′-CG-3′ and 5′-CNG-3′ sites are common places for cytosine 5-methylation in maize, genic sequences are not. Because of its prevalence, this modification of repetitive DNA has been used to generate maize gene-dense libraries (Palmer et al., 2003; Emberton et al., 2005). Upon sequencing these libraries, researchers found that DNA methylation in maize mostly affects transposable elements. Nonetheless, the results of this study showed that not all retrotransposon sequences are methylated, suggesting that some of them can avoid this sort of regulation. It is also worth noting that retrotransposon sequences have been discovered in EST databases (Meyers et al., 2001), however, some of these ESTs probably come from transcripts driven by flanking promoters rather than retrotransposon promoters. DNA methylation has been associated with an increased C-to-T transition rate. Compared to the 1:1 rate normally reported for genic sequences, including introns, retrotransposon sequences from maize and four additional angiosperm species showed a transition to transversion ratio of 1.5:1 (Vitte and Bennetzen, 2006). Due to 5-methylation, the rate of spontaneous deamination of cytosine is considerably increased, and uracil N-glycosylase-based repair pathways are unable to reverse the alteration. Thus, in comparison to unmethylated DNA, retrotransposons should accrue a shortfall of cytosines at CG and CNG locations and suffer fast mutations.

14.1 Inequitable Recombination

The relevance of uneven recombination in determining retrotransposon structure was highlighted by the discovery of altered retrotransposon copies resulting from this mechanism in various plant genomes. Below, we will go over the many rearrangements that can occur. They provide evidence that uneven recombination is rather uncommon in maize.

14.1.1 Alteration using migrating branches during recombination

Gene conversion has been blamed in certain cases for the lack of TSD. When retrotransposons undergo unequal recombination, the resulting exchange does not always result in a rearrangement of their structure. Many instances of potential recombination "abort," failing to result in the swapping of sequences at the site of initiation. In either case, a "conversion tract" is generated due to branch movement at the start of recombination. Two families of rice retrotransposons (Vitte and Panaud, 2003) exhibit this property. In a similar situation, recently, 2,145 identical bases of a Grande element were discovered in maize (Ma et al., 2005). In all likelihood, the flanking sequence of one LTR was replaced by the element's internal sequence when branch migration progressed beyond that LTR after an unequal recombination event between LTRs. If the repair is interrupted anywhere along the conversion tract, more complicated rearrangements might result. This finding suggests that uneven recombination may account for a more extensive set of rearrangements than previously thought (Ma et al., 2005).

14.2 Erosion of the Sequence Caused by Multiple Deletions

Deletions in retrotransposon sequences have been found often in Arabidopsis (Devos et al., 2002) and cereals like maize, suggesting that these elements actively participate in genome evolution. As in other organisms, such as yeast (Asami et al., 2002) or insects, deletion is a common method for eliminating retrotransposon sequences in plants (Petrov et al., 2000). Numerous plant genomes have had retrotransposon deletions meticulously investigated; these deletions range in size from 1 to 3766 base pairs (bp), with a typical of about 10-300 bp (Devos et al., 2002; Vitte and Bennetzen, 2006; Ma et al., 2004). A maize multi-locus investigation (Vitte and Bennetzen 2006) discovered deletions averaging between 1 bp and 116 bp, and deletions of up to 2.5 kb in retrotransposons in the maize adh1 region. Most of the deleted sequences have direct duplications surrounding them, indicating that they are leftovers from unintended recombination (Devos et al., 2002, Vitte and Bennetzen, 2006). In most cases, the size of these duplications is rather small, ranging from 2 bp to 53 bp on average; however, maize likely displays more variety (mean size of 11.3 bp, compared to the mean size of 5 bp in other species) than other organisms (Vitte and Bennetzen, 2006).

14.2.1   *"Retrotransposon domestication"*

It is possible that infrequent contributions to plant fitness can be attributed to whole or partial retrotransposon sequences. The element is tamed ("domesticated") by the host genome, a process generally referred to as "molecular domestication". There have been several reports of retrotransposons being domesticated in plants, particularly in maize. Many "wild-type" maize genes, for instance, have retrotransposons in their promoters. In addition to introducing new regulatory elements to a gene, retrotransposons can also introduce new retrotransposons. There is reason to believe that retrotransposons play a role in centromere function in maize and other cereals since reports of a retrotransposon maintained at particular centromere locations in distantly related grass species suggest they do.

15. Influence of Retrotransposons on the Development of the Maize Genome

15.1   *The Maize Genome is Primarily Comprised of Retrotransposons*

The 1C genome size of maize is around 2,700 Mbp. It has been estimated that at least 58-66% of the genome is repetitive based on genome-wide sequencing analysis of random BACs (100 BACs representing 14 Mb) or BAC ends (475,000 sequences representing 307 Mb) or an unprocessed sheared library (200,000 sequences representing 132 Mb) (Whitelaw et al., 2003). These estimates, however, are likely low since they do not take into account the fact that researchers are still sequencing BAC genes in maize. More than 80% of this percentage is made up of retrotransposons, making them the most common kind of repeat in the human genome (Messing et al., 2004). That retrotransposons predominate in the maize genome is a well-established fact. Table 1 lists the highly repeated retrotransposon families. However, there are other families with intermediate copy numbers (such as Fourf, Kake, and Victim) and low copy numbers (less than 10 copies for Reina). The five most common families in maize are there are five major maize gene families, including Opie-Ji, Cinful-Zeon, Huck, Prem1, and Grande, which are responsible for anywhere between 36% and 62% of the maize nuclear genome.

16. Where in the Genome Do Retrotransposons Live?

The distribution of high-copy retrotransposons between genes is distinctive, as Ac/Ds, En/Spm, and Mu DNA transposons are known to preferentially insert into

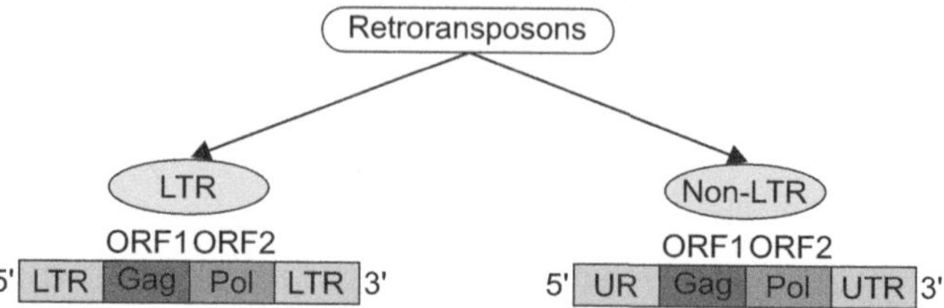

Fig. 1   LTR retrotransposons consist of gag and pol genes flanked by LTRs whereas Non-LTR retrotransposons encode both gag and pol genes and are not surrounded by LTRs.

genes and low-copy DNA, whereas MITEs and LINEs are frequently found in introns (Haberer et al., 2005). Cytogenetic analyses reveal spatial variation among retrotransposon families. Knob areas are enriched for Grande, Cinful-Zeon, and Opie-Ji, whereas centromeres regions are enriched for Huck (Mroczek and Dawe, 2003). More so than the Gyma family, Opie-Ji is much more prevalent in gene-containing than gene-free areas (Liu et al., 2007).

17.   The Era of Retrotransposons

Sequence divergence between a retrotransposon's two LTRs can be used to approximate the insertion date of the element. The sequences of its LTRs are assumed to have been identified upon insertion, and that nucleotide substitution has occurred at a consistent rate during the intervening period. The LTR divergence may be used to estimate when an insertion occurred given an approximation of the substitution rate. The retrotransposon's two LTRs can be easily estimated, but the substitution rate is more difficult to estimate. The rate of 1.3×108 substitutions per site per year is commonly cited for maize retrotransposons. It is based on the discovery that the rate of synonymous site substitution in the adh1 and adh2 genes is ten times slower than the pace at which retrotransposon sequences in rice evolve. The study was conducted by Ma and Bennetzen (2004), all "datable" maize retrotransposons appear to have been introduced during the last 4 My, with an average insertion period of around 1 My (Ma et al., 2005; Du, Swigonova et al., 2006; Vitte and Bennetzen 2006; Liu et al., 2007). In no way does this indicate that retrotransposons have only been active during the last 4 My. Older components generally lack at least one LTR, making it hard to determine their insertion date. Sites in rice were found where the insertion date of highly truncated copies could be calculated, and these copies were found to be very ancient (up to 10 My) (Ma et al., 2004; Vitte et al., 2007).

18.   Resulting Effect on the Size of the Maize Genome

The role of retrotransposons regarding the evolution of the maize genome and the genetic variation between size among angiosperms has been extensively studied by comparing orthologous sections from these plants. Genes are highly conserved, but

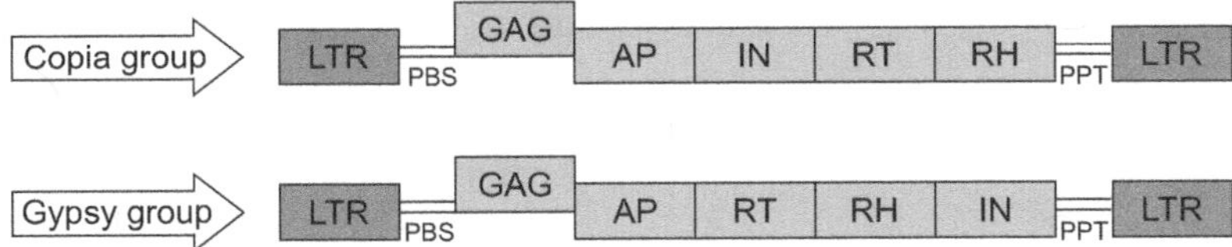

Fig. 2   Classification of Long Terminal Repeat Retrotransposons into two major groups

the non-coding regions between them, which are dominated by retrotransposons, are not, as shown by genes lrs1, lg2, lg3, and lrs1 were compared to adh1 (Ilic et al., 2003). Although maize and sorghum diverged from a common ancestor only 12 million years ago, the maize genome is nearly three times larger than the sorghum genome.

There are significant blocks of retrotransposons in maize that are absent in sorghum's orthologous locus, suggesting that retrotransposons are primarily responsible for the disparity in genome size between maize and sorghum. On the other hand, retrotransposons are slowly removed from the genome by the mechanisms discussed above and shown in Fig. 3. In a study comparing five angiosperms, including maize, researchers found that differences in size, content, and structure all originate from the same molecular mechanisms, albeit with varying relative efficiency across various lineages (Vitte and Bennetzen 2006). The timing and scope of these events will soon be more clearly defined, because of the increasing amount of sequencing data for maize (Langham et al., 2004).

19.   Effect on the Organization of the Maize Genome

Since retrotransposons are so abundant in the maize genome, they significantly impact how it is organized. Detailed sequencing study of maize BACs showed that retrotransposons are substantially intact and mostly ordered into nested structures, which runs counter to the initial assumptions that repeating DNA between genes would be a disorganized jumble. Gene density variation in maize has been linked to retrotransposon insertions, as shown by the analysis of duplicated regions of the genome (Swigonova et al., 2005). Uneven (ectopic) recombination can occur at retrotransposons because of their large copy number and random distribution across the genome. Depending on the relative positions and orientations of the recombining elements, ectopic recombination can result in a net loss or duplication of nuclear DNA between elements, inversions, and reciprocal translocations (Garfinkel 2005). The extent of the corresponding deletions in maize is currently unclear, however, it has been documented that Genomic deletions and the production of elements with mismatched terminal sequence differences (TSDs) are both possible outcomes of intra-strand inter-element recombination. However, retrotransposon-mediated inversions and translocations of greater complexity have been seen in other eukaryotes, their presence in maize has to be demonstrated (Garcia-Martinez and Martinez-Izquierdo, 2003).

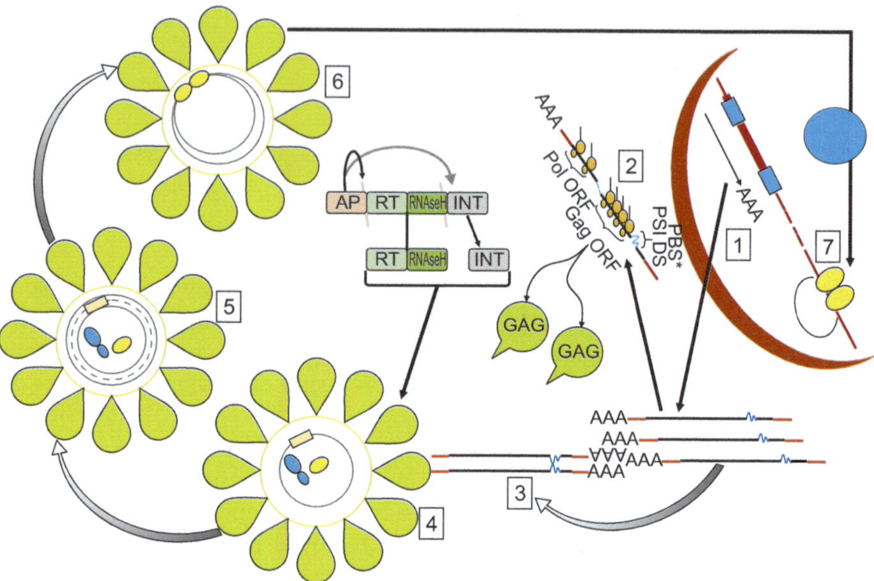

Fig. 3　LTR retrotransposon theoretical life cycle. 1) mRNA transcription begins from the 5′R to the 3′R. 2) Translation and protein synthesis in GAG and POL; internal cleavage of POL by AP in AP, RT-RNAseH, and IN. 3) RNA 'kissing-loop' dimerization via dimerization initiation signal (DIS) recognition. 4) RNA is packed and reverse transcription begins. When the GAGs assemble into the virus-like particles (VLPs), the RT-RNaseH dual protein takes over to carry out the reverse transcription. As a result, the packaged RNA can be used to construct the first strand of the cDNA. 5) The breakdown of the RNA matrix and formation of the second strand of cDNA. 6) Completing double-stranded cDNA synthesis and joining the IN to the long terminal repeats. 7) A double-stranded break is created, and the newly synthesized copy is integrated into a different region of the genome. (PSI-packaging signal sequence, PBS- Primer binding site).

20.　Effect on the Activity of a Gene in Maize

Insertion of retrotransposons within or close to genes can cause mutation. Specifically, the expression of many genes can be changed in some manner because of retrotransposon insertions inside or close to those genes. This is often accomplished by the production of aberrant transcripts (Kashkush et al., 2003) or by the modification of patterns of DNA methylation (Lippman et al., 2004; Huettel et al., 2007). In maize, it has been proven that retrotransposon insertions can lead to the inactivation of genes, as well as alterations in the size and quantity of transcripts, and lower levels of protein. In addition, it has been shown that maize is capable of tissue-specific alternative splicing. In addition, the expression of a paralogous gene (Fie2) nested within a retrotransposon was distinct from that of its non-nested paralog (Du et al., 2006). This difference in expression suggests that regulatory components of the retrotransposon bordering the Fie2 gene affected the gene's expression. Collectively, these data provide credence to the hypothesis that retrotransposons can induce substantial sequence rearrangements, leading to genomic variation. Due to its ancient polyploidy, maize's genome may have been more open to functional diversification via retrotransposons than other plant species. In other words, due to gene redundancy, a novel expression pattern would not be initially negatively chosen; this would make it possible for additional sequence mutations to lead to an advantageous shift in expression (Doust et al., 2009).

21.   Relative Variations Haplotypes in Maize

Full-genome sequencing will be performed on the B73 cultivar. While some research on additional maize inbreeds besides McC, work on the local genome structure has been done (Fu et al., 2010) and BSSS53, is still rather rare (Song et al., 2001). Initial comparative cytogenetic investigations revealed extensive chromosomal diversity across several maize races, whereas estimates of maize genome size (2C) range from 4.9 pg to 6.1 pg. These results imply sequence variance among maize races and probably among maize inbreeds.

Non-collinearity at the bz locus was found to be rather large when comparing the sequences of two inbreeds, McC and B73 (Fu and Dooner 2002). They can only align based on the genes they had in common, as the flanking retrotransposon blocks had varied compositions and sizes. Finally, when comparing identical results have been obtained in the z1C zein genomic portions of B73 and BSSS53, in addition to three randomly chosen parts of B73 and Mo17 (Song and Messing, 2003, Brunner et al., 2005). Comparing the bz genomic region of eight distinct maize cultivars allowed for the subsequent validation of a large number of retrotransposon insertions that exhibited polymorphism. There was a wide range of similarity between any two haplotypes, ranging from 25% up to 84%. Chimeric haplotypes have been found, which refers to periods of intense transposition and subsequent recombination reshuffling retrotransposon blocks as the source of the structural variation seen in today's maize cultivars (Wang and Dooner, 2006). Orthologous insertion date estimations demonstrate that non-shared retrotransposons are significantly more recent than shared ones, with the bulk of non-shared elements younger than 0.5 million years (Brunner et al., 2005). Although maize was domesticated from its wild parent (teosinte) about 9,000 years ago, almost all of the non-shared insertions probably occurred much earlier. Therefore, it may be deduced that the wild progenitor of maize already possessed the structural diversity seen in contemporary maize germplasm. This discovery is consistent with the assumption that maize still has 75% of the allelic diversity in teosinte, even though sequence diversity at domestication genes was decreased throughout the domestication process.

22.   Using Retrotransposons as a Tool for Investigating the History of Plant life

Due to their unique features and roles within the plant genome, retrotransposons have been used extensively in research on the genetic evolution of plants. The creation of an excellent illustration of this is the use of molecular markers based on retrotransposon heterogeneity or insertional polymorphism. Several different types of molecular markers based on amplification of retrotransposons have been created: sequence-specific amplification polymorphisms (SSAP), retrotransposon-based insertion polymorphisms (RBIP), inter-retrotransposon amplified polymorphisms (IRAP), along with retrotransposon-microsatellite amplified polymorphisms (REMAP). Marker systems for phylogenetic and diversity studies have been useful for a wide variety of crops, including peas, apples, and rapeseed (Allnutt et al., 2008; Antonius-Klemola et al., 2006). Retrotransposon-based molecular markers might be utilized in developing genetic maps and identifying genes, shedding insights into plant evolution in the process. Because of their rapid mutation rate, limited species-specific limitation, and ability to express normally in heterogeneous environments, retrotransposons are increasingly commonly utilized in the production of changed materials, gene tagging, and functional study. Tos17, a retrotransposon discovered through research on rice tissue culture, is particularly fond of inserting itself into genes and nearby genes. The technique of random insertion mutation in plants is now widely used as a tool for gene cloning and the investigation of the functions of rice genes. The model plant legume Medicago truncatula has undergone substantial insertional mutagenesis using the Tnt1 retrotransposon, which has considerably benefited gene cloning efforts in recent years.

Conclusions

To a first approximation, retrotransposons make up the complete maize genome. The retrotransposons that make up the majority of the genome are thus responsible for its overall structure and content. Although numerous, the great majority of these are so well isolated from the genic region of the genome that they were undiscovered until researchers began studying large-insert clones. Most of their traits go counter to what is expected from genetics. They multiply throughout the genome, but each copy only lasts a few million years before it is gone forever. They live in a selective grey region, following their ecology, rather than aiding the host organism's survival. While their eventual demise appears to be related to that of their host, a small percentage of them may be truly viruses, able to infect the genomes of other creatures. There should be more research done. We know a lot more about the structure and function of animal retroviruses than we do about plant retrotransposons, which have only been sketched broadly. This is why the release of a maize genome draught sequence is so important; it will yield a plethora of information. However, much more research into the biochemistry of plant retrotransposons is required to offer a full picture of these pervasive elements of the maize genome.

Conflicts: None

Reference

Allnutt, T.R., Roper, K., and Henry, C. (2008). Development and application of SINE multilocus and quantitative genetic markers to study oilseed rape (Brassica napus L.) crops. *Journal of agricultural and food chemistry*, 56(2), 426–32.

Ammiraju, J.S., Zuccolo, A., Yu, Y., Song, X., Piegu, B., Chevalier, F., and Wing, R.A. (2007). Evolutionary dynamics of an ancient retrotransposon family provides insights into evolution of genome size in the genus Oryza. *The Plant Journal*, 52(2), 342–51.

Ananiev, E.V., Chamberlin, M.A., Klaiber, J., and Svitashev, S. (2005). Microsatellite megatracts in the maize (Zea mays L.) genome. Genome, 48(6), 1061–69.

Antonius-Klemola, K., Kalendar, R., and Schulman, A.H. (2006). TRIM retrotransposons occur in apple and are polymorphic between varieties but not sports. Theoretical and Applied Genetics, 112(6), 999–1008.

Asami, Y., Jia, D.W., Tatebayashi, K., Yamagata, K., Tanokura, M., and Ikeda, H. (2002). Effect of the DNA topoisomerase II inhibitor VP-16 on illegitimate recombination in yeast chromosomes. Gene, 291(1-2), 251–57.

Baucom, R.S., Estill, J.C., Chaparro, C., Upshaw, N., Jogi, A., Deragon, J.M., ... and Bennetzen, J.L. (2009). Exceptional diversity, non-random distribution, and rapid evolution of retroelements in the B73 maize genome. PLoS genetics, 5(11), e1000732.

Beló, A., Beatty, M.K., Hondred, D., Fengler, K.A., Li, B., and Rafalski, A. (2010). Allelic genome structural variations in maize detected by array comparative genome hybridization. Theoretical and Applied Genetics, 120(2), 355–67.

Bennetzen, J.L. (2009). Maize genome structure and evolution. In Handbook of maize (pp. 179–99). Springer, New York, NY.

Bennetzen, J.L., and Hake, S.C. (Eds.). (2009). Handbook of maize: genetics and genomics. Springer Science & Business Media.

Bennetzen, J.L., SanMiguel, P., Chen, M., Tikhonov, A., Francki, M., and Avramova, Z. (1998). Grass genomes. Proceedings of the National Academy of Sciences, 95(5), 1975–1978.

Biradar, D.P., and Rayburn, A. (1993). Heterosis and nuclear DNA content in maize. Heredity, 71(3), 300–304.

Birchler, J.A., and Han, F. (2009). Maize centromeres: structure, function, epigenetics. Annual review of genetics, 43(1), 287–303.

Bruggmann, R., Bharti, A.K. and Gundlach, H. (2006) "Uneven chromosome contraction and expansion in the maize genome," Genome Research, vol. 16, no. 10, pp. 1241–51.

Brunner, S., Fengler, K., Morgante, M., Tingey, S., and Rafalski, A. (2005). Evolution of DNA sequence nonhomologies among maize inbreds. The Plant Cell, 17(2), 343–60.

Brunner, S., Pea, G., and Rafalski, A. (2005). Origins, genetic organization and transcription of a family of non-autonomous helitron elements in maize. *The Plant Journal*, 43(6), 799–810.

Buckler, E.S., Gaut, B.S., and McMullen, M.D. (2006). Molecular and functional diversity of maize. Current opinion in plant biology, 9(2), 172–76.

Coe E, Davis G, McMullen M, Musket T and Polacco M. (1995) UMC maize RFLP and genetic working map. Maize Genet Cooper Newsletter 69: 247–256.

Chandler, V.L., Eggleston, W.B., and Dorweiler, J.E. (2000). Paramutation in maize. Plant Gene Silencing, 1–25.

Ching, A.D.A., Caldwell, K.S., Jung, M., Dolan, M., S. Howie Smith, O., Tingey, S., and Rafalski, A.J. (2002). SNP frequency, haplotype structure and linkage disequilibrium in elite maize inbred lines. BMC genetics, 3(1), 1–14.

Dawe, R.K. (2009). Maize centromeres and knobs (neocentromeres). In Handbook of Maize (pp. 239–50). Springer, New York, NY.

Devos, K.M. (2010). Grass genome organization and evolution. Current opinion in plant biology, 13(2), 139–45.

Devos, K.M., Brown, J.K., and Bennetzen, J.L. (2002). Genome size reduction through illegitimate recombination counteracts genome expansion in Arabidopsis. Genome research, 12(7), 1075–79.

Doust, A.N., Kellogg, E.A., Devos, K.M., and Bennetzen, J.L. (2009). Foxtail millet: a sequence-driven grass model system. Plant physiology, 149(1), 137–41.

Du, C., Fefelova, N., Caronna, J., He, L., and Dooner, H.K. (2009). The polychromatic Helitron landscape of the maize genome. Proceedings of the National Academy of Sciences, 106(47), 19916–21.

Du, C., Swigoňová, Z., and Messing, J. (2006). Retrotranspositions in orthologous regions of closely related grass species. BMC evolutionary biology, 6(1), 1–12.

Eichinger, D.J., and Boeke, J.D. (1990). A specific terminal structure is required for Ty1 transposition. Genes & development, 4(3), 324–30.

Eickbush, T.H., and Malik, H. S. (2002). Mobile DNA II, Vol. 49.

Elrouby, N., and Bureau, T.E. (2001). A Novel Hybrid Open Reading Frame Formed by Multiple Cellular Gene Transductions by a Plant Long Terminal Repeat Retroelement 210. *Journal of Biological Chemistry*, 276(45), 41963–68.

Emberton, J., Ma, J., Yuan, Y., SanMiguel, P., and Bennetzen, J.L. (2005). Gene enrichment in maize with hypomethylated partial restriction (HMPR) libraries. Genome Research, 15(10), 1441–46.

Feschotte, C., Jiang, N., and Wessler, S.R. (2002). Plant transposable elements: where genetics meets genomics. Nature Reviews Genetics, 3(5), 329–41.

Feschotte, C., Keswani, U., Ranganathan, N., Guibotsy, M.L. and Levine, D. (2009). Exploring repetitive DNA landscapes using REPCLASS, a tool that automates the classification of transposable elements in eukaryotic genomes. Genome biology and evolution, 1, 205–220.

Fu, H., and Dooner, H.K. (2002). Intraspecific violation of genetic colinearity and its implications in maize. Proceedings of the National Academy of Sciences, 99(14), 9573–78.

García-Martínez, J., and Martínez-Izquierdo, J.A. (2003). Study on the evolution of the Grande retrotransposon in the Zea genus. Molecular biology and evolution, 20(5), 831–41.

Garcia-Perez, J.L., Doucet, A.J., Bucheton, A., Moran, J.V. and Gilbert, N. (2007). Distinct mechanisms for trans-mediated mobilization of cellular RNAs by the LINE-1 reverse transcriptase. Genome research, 17(5), 602–611.

Garfinkel, D.J. (2005). Genome evolution mediated by Ty elements in Saccharomyces. Cytogenetic and genome research, 110(1–4), 63–69.

Gregory, T.R. (2005). The evolution of the genome Elsevier Academic Press. San Diego CA USA.

Gupta, S., Gallavotti, A., Stryker, G.A., Schmidt, R.J., and Lal, S.K. (2005). A novel class of Helitron-related transposable elements in maize contain portions of multiple pseudogenes. Plant molecular biology, 57(1), 115–27.

Haberer, G., Young, S., Bharti, A. K., Gundlach, H., Raymond, C., Fuks, G., and Messing, J. (2005). Structure and architecture of the maize genome. *Plant physiology*, 139(4), 1612–24.

Haberer, G., Young, S., Bharti, A. K., Gundlach, H., Raymond, C., Fuks, G., and Messing, J. (2005). Structure and architecture of the maize genome. *Plant physiology*, 139(4), 1612–24.

Hamilton, A., Voinnet, O., Chappell, L., and Baulcombe, D. (2002). Two classes of short interfering RNA in RNA silencing. *The EMBO journal*, 21(17), 4671–79.

Hanada, K., Vallejo, V., Nobuta, K., Slotkin, R.K., Lisch, D., Meyers, B.C., and Jiang, N. (2009). The functional role of pack-MULEs in rice inferred from purifying selection and expression profile. The Plant Cell, 21(1), 25–38.

Hollick, J.B., and Springer, N. (2009). Epigenetic phenomena and epigenomics in maize. In Epigenomics (pp. 119–47). Springer, Dordrecht.

Huettel, B., Kanno, T., Daxinger, L., Bucher, E., van der Winden, J., Matzke, A.J., and Matzke, M. (2007). RNA-directed DNA methylation mediated by DRD1 and Pol IVb: A versatile pathway for transcriptional gene silencing in plants. Biochimica et Biophysica Acta (BBA)-Gene Structure and Expression, 1769(5–6), 358–74.

Ilic, K., SanMiguel, P.J., and Bennetzen, J.L. (2003). A complex history of rearrangement in an orthologous region of the maize, sorghum, and rice genomes. Proceedings of the National Academy of Sciences, 100(21), 12265–70.

Jia, Y., Lisch, D.R., Ohtsu, K., Scanlon, M.J., Nettleton, D., and Schnable, P.S. (2009). Loss of RNA–dependent RNA polymerase 2 (RDR2) function causes widespread and unexpected changes in the expression of transposons, genes, and 24-nt small RNAs. PLoS genetics, 5(11), e1000737.

Jiang, N., Bao, Z., Zhang, X., Eddy, S.R., and Wessler, S.R. (2004). Pack-MULE transposable elements mediate gene evolution in plants. Nature, 431(7008), 569–73.

Jiang, N., Ferguson, A.A., Slotkin, R.K., and Lisch, D. (2011). Pack-Mutator–like transposable elements (Pack-MULEs) induce directional modification of genes through biased insertion and DNA acquisition. Proceedings of the National Academy of Sciences, 108(4), 1537–42.

Kapitonov, V.V., and Jurka, J. (2001). Rolling-circle transposons in eukaryotes. Proceedings of the National Academy of Sciences, 98(15), 8714–19.

Kashkush, K., Feldman, M., and Levy, A.A. (2003). Transcriptional activation of retrotransposons alters the expression of adjacent genes in wheat. Nature genetics, 33(1), 102–106.

Kato, A., Lamb, J.C., and Birchler, J.A. (2004). Chromosome painting using repetitive DNA sequences as probes for somatic chromosome identification in maize. Proceedings of the National Academy of Sciences, 101(37), 13554–59.

Krom, N., Recla, J., and Ramakrishna, W. (2008). Analysis of genes associated with retrotransposons in the rice genome. Genetica, 134(3), 297–10.

Lai, J., Li, R., Xu, X., Jin, W., Xu, M., Zhao, H., and Wang, J. (2010). Genome-wide patterns of genetic variation among elite maize inbred lines. Nature genetics, 42(11), 1027–30.

Lai, J., Li, Y., Messing, J., and Dooner, H.K. (2005). Gene movement by Helitron transposons contributes to the haplotype variability of maize. Proceedings of the National Academy of Sciences, 102(25), 9068–73.

Langham, R.J., Walsh, J., Dunn, M., Ko, C., Goff, S.A., and Freeling, M. (2004). Genomic duplication, fractionation and the origin of regulatory novelty. *Genetics*, 166(2), 935–45.

Leigh, F., Kalendar, R., Lea, V., Lee, D., Donini, P., and Schulman, A.H. (2003). Comparison of the utility of barley retrotransposon families for genetic analysis by molecular marker techniques. Molecular Genetics and Genomics, 269(4), 464–74.

Li, L., and Arumuganathan, K. (2001). Physical mapping of 45S and 5S rDNA on maize metaphase and sorted chromosomes by FISH. Hereditas, 134(2), 141–45.

Lippman, Z., Gendrel, A.V., Black, M., Vaughn, M.W., Dedhia, N., Richard McCombie, W., ... and Martienssen, R. (2004). Role of transposable elements in heterochromatin and epigenetic control. Nature, 430(6998), 471–76.

Liu, R., Vitte, C., Ma, J., Mahama, A.A., Dhliwayo, T., Lee, M., and Bennetzen, J.L. (2007). A GeneTrek analysis of the maize genome. Proceedings of the National Academy of Sciences, 104(28), 11844–49.

Ma, J., and Bennetzen, J.L. (2004). Rapid recent growth and divergence of rice nuclear genomes. Proceedings of the National Academy of Sciences, 101(34), 12404–10.

Ma, J., Devos, K.M., and Bennetzen, J.L. (2004). Analyses of LTR-retrotransposon structures reveal recent and rapid genomic DNA loss in rice. Genome research, 14(5), 860–69.

Ma, J., SanMiguel, P., Lai, J., Messing, J., and Bennetzen, J.L. (2005). DNA rearrangement in orthologous orp regions of the maize, rice and sorghum genomes. *Genetics*, 170(3), 1209–20.

Malik, H.S., Henikoff, S. and Eickbush, T.H. (2000). Poised for contagion: evolutionary origins of the infectious abilities of invertebrate retroviruses. Genome research, 10(9), 1307–1318.

Matzke, M., Kanno, T., Huettel, B., Daxinger, L., and Matzke, A.J. (2007). Targets of RNA-directed DNA methylation. Current opinion in plant biology, 10(5), 512–19.

Messing, J. (2009). Synergy of two reference genomes for the grass family. Plant physiology, 149(1), 117–124.

Messing, J., Bharti, A.K., Karlowski, W.M., Gundlach, H., Kim, H.R., Yu, Y., ... and Wing, R.A. (2004). Sequence composition and genome organization of maize. Proceedings of the National Academy of Sciences, 101(40), 14349–54.

Meyers, B.C., Tingey, S.V., and Morgante, M. (2001). Abundance, distribution, and transcriptional activity of repetitive elements in the maize genome. Genome Research, 11(10), 1660–76.

Monfort, A., Vicient, C.M., Raz, R., Puigdomènech, P., and Martínez-Izquierdo, J.A. (1995). Molecular analysis of a putative transposable retroelement from the Zea genus with internal clusters of tandem repeats. DNA Research, 2(6), 255–61.

Morgante, M., Brunner, S., Pea, G., Fengler, K., Zuccolo, A., and Rafalski, A. (2005). Gene duplication and exon shuffling by helitron-like transposons generate intraspecies diversity in maize. *Nature genetics*, 37(9), 997–1002.

Morgante, M., De Paoli, E., and Radovic, S. (2007). Transposable elements and the plant pan-genomes. Current opinion in plant biology, 10(2), 149–55.

Mroczek, R.J., and Dawe, R.K. (2003). Distribution of retroelements in centromeres and neocentromeres of maize. Genetics, 165(2), 809–19.

Page, B.T., Wanous, M.K., and Birchler, J.A. (2001). Characterization of a maize chromosome 4 centromeric sequence: evidence for an evolutionary relationship with the B chromosome centromere. Genetics, 159(1), 291–302.

Palmer, L.E., Rabinowicz, P.D., O'Shaughnessy, A.L., Balija, V.S., Nascimento, L.U., Dike, S., ... and McCombie, W.R. (2003). Maize genome sequencing by methylation filtration. Science, 302(5653), 2115–17.

Paterson, A.H., Bowers, J.E., Bruggmann, R., Dubchak, I., Grimwood, J., Gundlach, H., ... and Rokhsar, D.S. (2009). The Sorghum bicolor genome and the diversification of grasses. Nature, 457(7229), 551–556.

Paterson, A.H., Bowers, J.E., and Chapman, B. (2004). Ancient polyploidization predating divergence of the cereals, and its consequences for comparative genomics. Proceedings of the National Academy of Sciences, 101(26), 9903–9908.

Petrov, D.A., Sangster, T.A., Johnston, J.S., Hartl, D.L., and Shaw, K.L. (2000). Evidence for DNA loss as a determinant of genome size. *Science*, 287(5455), 1060–62.

Piegu, B., Guyot, R., Picault, N., Roulin, A., Saniyal, A., Kim, H., ... and Panaud, O. (2006). Doubling genome size without polyploidization: dynamics of retrotransposition-driven genomic expansions in Oryza australiensis, a wild relative of rice. Genome research, 16(10), 1262–1269.

Pouch-Pélissier, M.N., Pélissier, T., Elmayan, T., Vaucheret, H., Boko, D., Jantsch, M.F., and Deragon, J.M. (2008). SINE RNA induces severe developmental defects in Arabidopsis thaliana and interacts with HYL1 (DRB1), a key member of the DCL1 complex. PLoS genetics, 4(6), e1000096.

Queen, R.A., Gribbon, B.M., James, C., Jack, P., and Flavell, A.J. (2004). Retrotransposon-based molecular markers for linkage and genetic diversity analysis in wheat. Molecular Genetics and Genomics, 271(1), 91–97.

Rafalski, A., and Ananiev, E. (2009). Genetic diversity, linkage disequilibrium and association mapping. In Handbook of Maize (pp. 201–219). Springer, New York, NY.

Sanz-Alferez, S., SanMiguel, P., Jin, Y.K., Springer, P.S., and Bennetzen, J.L. (2003). Structure and evolution of the Cinful retrotransposon family of maize. *Genome*, 46(5), 745–52.

Schnable, P.S., Ware, D., Fulton, R.S., Stein, J.C., Wei, F., Pasternak, S., ... and Presting, G.G. (2009). The B73 maize genome: complexity, diversity, and dynamics. *Science*, 326(5956), 1112–15.

Shepherd, N.S., Schwarz-Sommer, Z., Vel Spalve, J.B., Gupta, M., Wienand, U. and Saedler, H. (1984). Similarity of the Cin1 repetitive family of Zea mays to eukaryotic transposable elements. Nature, 307(5947), 185–187.

Song, R., and Messing, J. (2003). Gene expression of a gene family in maize based on noncollinear haplotypes. Proceedings of the National Academy of Sciences, 100(15), 9055–60.

Song, R., Llaca, V., Linton, E., and Messing, J. (2001). Sequence, regulation, and evolution of the maize 22-kD α zein gene family. Genome research, 11(11), 1817–25.

Springer, N.M., Ying, K., Fu, Y., Ji, T., Yeh, C.T., Jia, Y., ... and Schnable, P.S. (2009). Maize inbreds exhibit high levels of copy number variation (CNV) and presence/absence variation (PAV) in genome content. PLoS genetics, 5(11), e1000734.

Swanson-Wagner, R.A., Eichten, S.R., Kumari, S., Tiffin, P., Stein, J.C., Ware, D., and Springer, N.M. (2010). Pervasive gene content variation and copy number variation in maize and its undomesticated progenitor. *Genome research*, 20(12), 1689–99.

Swigoňová, Z., Lai, J., Ma, J., Ramakrishna, W., Llaca, V., Bennetzen, J.L., and Messing, J. (2004). Close split of sorghum and maize genome progenitors. *Genome research*, 14(10a), 1916–23.

Tito, C.M., Poggio, L., and Naranjo, C.A. (1991). *Cytogenetic studies in the genus Zea. Theoretical and Applied Genetics*, 83(1), 58–64.

Turcich, M.P., Bokhari-Riza, A., Hamilton, D.A., He, C., Messier, W., Stewart, C.B., and Mascarenhas, J.P. (1996). PREM-2, a copia-type retroelement in maize is expressed preferentially in early microspores. *Sexual Plant Reproduction*, 9(2), 65–74.

Vershinin, A.V., Druka, A., Alkhimova, A.G., Kleinhofs, A., and Heslop-Harrison, J.S. (2002). LINEs and gypsy-like retrotransposons in Hordeum species. *Plant molecular biology*, 49(1), 1–14.

Vitte, C., and Bennetzen, J.L. (2006). Analysis of retrotransposon structural diversity uncovers properties and propensities in angiosperm genome evolution. Proceedings of the National Academy of Sciences, 103(47), 17638–43.

Vitte, C., and Panaud, O. (2003). Formation of solo-LTRs through unequal homologous recombination counterbalances amplifications of LTR retrotransposons in rice Oryza sativa L. Molecular Biology and Evolution, 20(4), 528–40.

Vitte, C., Panaud, O., and Quesneville, H. (2007). LTR retrotransposons in rice (Oryza sativa, L.): recent burst amplifications followed by rapid DNA loss. BMC genomics, 8(1), 1–15.

Vogel, H., Badapanda, C. and Vilcinskas, A. (2011). Identification of immunity-related genes in the burying beetle Nicrophorus vespilloides by suppression subtractive hybridization. Insect molecular biology, 20(6), 787–800.

Wang, X., Tang, H., and Paterson, A.H. (2011). Seventy million years of concerted evolution of a homoeologous chromosome pair, in parallel, in major Poaceae lineages. The Plant Cell, 23(1), 27–37.

Wei, F., Coe, E.D., Nelson, W., Bharti, A.K., Engler, F., Butler, E., and Wing, R.A. (2007). Physical and genetic structure of the maize genome reflects its complex evolutionary history. PLoS genetics, 3(7), e123.

Whitelaw, C.A., Barbazuk, W.B., Pertea, G., Chan, A.P., Cheung, F., Lee, Y., ... and Quackenbush, J. (2003). Enrichment of gene-coding sequences in maize by genome filtration. *Science*, 302(5653), 2118–20.

Wicker, T., and Keller, B. (2007). Genome-wide comparative analysis of copia retrotransposons in triticeae, rice, and arabidopsis reveals conserved ancient evolutionary lineages and distinct dynamics of individual copia families. *Genome research*, 17(7), 1072–81.

Wicker, T., Sabot, F., Hua-Van, A., Bennetzen, J.L., Capy, P., Chalhoub, B., ... and Schulman, A.H. (2007). A unified classification system for eukaryotic transposable elements. Nature Reviews Genetics, 8(12), 973–82.

Wilhelm, M., and Wilhelm, F.X. (2001). Reverse transcription of retroviruses and LTR retrotransposons. *Cellular and Molecular Life Sciences CMLS*, 58(9), 1246–62.

Wolfgruber, T.K., Sharma, A., Schneider, K.L., Albert, P.S., Koo, D.H., Shi, J., ... and Presting, G.G. (2009). Maize centromere structure and evolution: sequence analysis of centromeres 2 and 5 reveals dynamic loci shaped primarily by retrotransposons. PLoS genetics, 5(11), e1000743.

Woodhouse, M.R., Schnable, J.C., Pedersen, B.S., Lyons, E., Lisch, D., Subramaniam, S., and Freeling, M. (2010). Following tetraploidy in maize, a short deletion mechanism removed genes preferentially from one of the two homeologs. PLoS biology, 8(6), e1000409.

Wright, D.A., and Voytas, D.F. (2002). Athila4 of Arabidopsis and Calypso of soybean define a lineage of endogenous plant retroviruses. Genome research, 12(1), 122–31.

Xiao, H., Jiang, N., Schaffner, E., Stockinger, E.J., and Van Der Knaap, E. (2008). A retrotransposon-mediated gene duplication underlies morphological variation of tomato fruit. *Science*, 319(5869), 1527–30.

Zhang, X., and Wessler, S.R. (2004). Genome-wide comparative analysis of the transposable elements in the related species Arabidopsis thaliana and Brassica oleracea. Proceedings of the National Academy of Sciences, 101(15), 5589–94.

Zhou, S., Wei, F., Nguyen, J., Bechner, M., Potamousis, K., Goldstein, S., ... and Schwartz, D.C. (2009). A single molecule scaffold for the maize genome. PLoS genetics, 5(11), e1000711.

7

Retrotransposons in *Triticum aestivum*

Hajira Imran,[1] *Iqra Siddique*[1] *and Alvina Gul*[1*]

1. Introduction

1.1 Overview of Retrotransposons in Triticum aestivum (Common Wheat)

Triticum aestivum, commonly called bread wheat, is one of the major cereal crops grown and consumed. Food security has become a significant concern as the world population has increased. Wheat is considered a staple in almost all parts of the world. To meet the needs of the people, about 50% increased demand for wheat production will be required to feed the world by 2050 (Ghonaim et al., 2021). *T. aestivum* has six sets of chromosomes. This type of wheat has been created by combining the two species *T. urartu* and *T. tauschii,* as well as an unknown B-genome. The A and B genomes were combined 400,000–800,000 years ago, while the addition of the D genome happened about 10,000 years ago. These two diploid species have similar genes, although their intergenic regions differ significantly due to their distinctive evolutionary histories. Now that high-quality genome assemblies are available, wheat is a valuable model for investigating the origins and diversification of transposable elements in polyploids (Grauda et al., 2016).

In recent years, breakthroughs in sequencing and assembly technologies have allowed for the development of complete whole genome assemblies for bread wheat and several of its relatives. This discovery helped scientists decode the intricate wheat genome. In 2017, experts from the private sector and academic institutions collaborated to create an assembly algorithm that could handle repetitive DNA

[1] Department of Plant Biotechnology, Atta-ur-Rahman School of Applied Biosciences, National University of Sciences and Technology, Islamabad, Pakistan.

* Corresponding author: alvina_gul@yahoo.com

stretches. This progress made it easier to sequence and assemble all other known wheat relatives. The release of the sequences of *A. tauschii* (the donor of the D genome) and *T. urartu* (the donor of the A genome) ushered in the age of wheat genomics, which was followed by the whole genome draughts of bread wheat and durum wheat (Portis et al., 2019). Because of these advancements, the relationship and influence of transposable elements in the wheat genome, polymorphism, and organization can be evaluated (Balážová et al., 2021).

1.2 Identification of Different Retrotransposon Families in Wheat

Retrotransposons are categorized into several subtypes based on their structure and transposition process. Some of the most common types of retrotransposons are:

1. Long Terminal Repeat retrotransposons (LTR): Long terminal repeats (LTRs) are found at the ends of the transposable element, which makes this the most common type of retrotransposon. LTR retrotransposons can be divided into two more groups: Gypsy and Copia (Galindo-González et al., 2017).
2. Non-LTR retrotransposons: These retrotransposons do not have LTRs and instead rely on RNA intermediates for reverse transcription and integration into the genome. LINEs (Long Interspersed Nuclear Elements) and SINEs (Short Interspersed Nuclear Elements) are subtypes (Wicker et al., 2022).
3. Telomere retrotransposons are a type of retrotransposon associated with the telomeres, the end regions of chromosomes. It is believed that they play a part in maintaining and regulating the length of the telomeres.

1.2.1 LTR retrotransposons

They can move around the genome and insert themselves into new locations. *In T. aestivum,* LTR retrotransposons have been found to be more active than non-LTR retrotransposons. This is likely due to their ability to replicate themselves more efficiently than non-LTR retrotransposons. The activity of LTR retrotransposons has been linked to variations in gene expression, regulation, and changes in the structure of chromosomes. The activity of retrotransposons can also affect the genome organization of species. By inserting themselves into different locations within the genome, they can create new alleles that may be beneficial or detrimental to the organism. This can lead to changes in the phenotype of an organism over time, as well as changes in its evolutionary trajectory. In conclusion, retrotransposon activity is an essential factor in the evolution of *T. aestivum.* Understanding how these elements move around the genome and affect gene expression and regulation can give us insight into how this species has evolved (Grandbastien, 2015).

Sequencing sRNA using High-throughput sequencing techniques in wheat plants revealed that siRNA pools were drastically diminished, and miRNAs react to changes in ploidy differently from TE-derived siRNAs. This results in siRNA instability, which lowers the CpG methylation of LTR retrotransposons and may be a factor in genomic unrest early in the speciation process (Ramakrishnan et al., 2021).

1.2.2   Ty1-copia retrotransposons in wheat

Ty1-copia retrotransposons are members of the LTR retrotransposons. This family of retrotransposons is distinguished by the ability to reverse transcribe RNA copies into DNA and integrate DNA copies into new genomic sites. It is believed that Ty1-copia retrotransposons make up 5% of the wheat genome and play a significant part in determining the structure of the wheat genome and its historical development. These retrotransposons have been implicated in forming new genes and genome size evolution. A critical feature of Ty1-copia retrotransposons is their ability to affect gene expression by inserting into or near genes, leading to the activation or silencing of these genes. This can result in significant changes in plant phenotypes, including developing new traits or improving existing traits (Liu et al., 2022).

1.2.3   Ty3-gypsy retrotransposons in wheat

Ty3-gypsy retrotransposons are another type of transposable element found in wheat. They are also known as LTR retrotransposons because they have long terminal repeats (LTRs) at their ends.Ty3-gypsy retrotransposons are abundant and diverse in wheat, with several families and subfamilies identified. They have been estimated to comprise around 70% of the wheat genome, although this varies depending on the wheat variety and the methods used to measure their abundance. It is thought that Ty3-gypsy retrotransposons have had a key influence in altering the wheat genome's structure and function throughout its evolutionary history. They can insert into and disrupt genes, leading to genetic diversity within and between wheat populations. They may also contribute to the evolution of new traits in wheat, such as disease-resistance (Salina et al., 2011). In addition, these retrotransposons can be helpful to genetic markers for wheat breeding and genetic diversity studies. They can also be used to develop molecular markers for wheat varieties, which can aid in identifying and selecting desirable traits in breeding programs. Overall, the study of Ty3-gypsy retrotransposons in wheat is an active area of research, with ongoing efforts to better understand their roles in genome evolution and their potential applications in wheat breeding and genetic improvement (Aguilar and Prieto, 2020).

1.2.4   CR1 retrotransposons in wheat

CR1 retrotransposons are a type of retrotransposon, which means they use an RNA intermediate to copy themselves and insert into new locations in the genome. They are characterized by a conserved region (CR) that encodes the reverse transcriptase responsible for copying the retrotransposon RNA into DNA. CR1 retrotransposons are a type of transposable element found in many plant genomes, including wheat. In wheat, CR1 retrotransposons have been found to make up a significant portion of the genome. One study estimated that CR1 retrotransposons account for approximately 17% of the wheat genome, with over 66,000 individual copies identified (Kojima, 2019).

1.2.5　*Comparison of Ty1-copia, Ty3-gypsy, and CR1 retrotransposons in wheat*

Ty3/Gypsy elements and Ty1/Copia elements, two of plants' most common mobile genetic pieces, have distinct distribution patterns. Aspects of the Ty3/Gypsy family are overrepresented in euchromatic sub-telomeric areas, whereas components of the Ty1/Copia family are overrepresented in heterochromatic pericentromeric regions (Jedlicka et al., 2019). Ty3/Gypsy elements are more diverse than Ty1/Copia elements and play essential roles in host epigenetic response. Both families were initially discovered in the fruit fly, Drosophila, but are now known to be present in large copy numbers in higher plants. The primary distinction between members of these superfamilies is the location of the polymerase-function-coding gene inside the polyprotein (POL) region. The pol gene in Ty1/Copia elements is structured into the protease (PR), integrase (INT), reverse transcriptase (RT), and ribonuclease H (RNase H) domains (PR-INT-RT-RNase H). Each Ty3/Gypsy element comprises four subunits: the PR, RT, RNase H, and the INT domain (Sant et al., 2000).

2.　Role of Retrotransposons in Wheat Genome Structure and Function

2.1　Involvement in Genome Rearrangements

One way of how retrotransposons influence the genome is through the creation of genome rearrangements. This can occur when retrotransposons insert themselves into the genome, causing changes in DNA sequence and resulting in new genes or disrupting existing ones. In some cases, these rearrangements can result in unique traits or adaptations in a population (Bariah et al., 2020a). Retrotransposons also significantly impact genome size, as they contribute to the accumulation of repetitive DNA sequences in the genome. This increase in genome size can have consequences for genome function, resulting in changes in gene expression, altered chromosome structure, and increased genome instability (Bariah et al., 2018).

2.2　Regulation of Gene Expression

Retrotransposons can significantly impact gene expression and regulation, which has been well-studied in several organisms, including *T. aestivum* (common wheat). One way retrotransposons can regulate gene expression is through their proximity to genes. For example, retrotransposons can insert themselves into the regulatory regions of genes, such as promoter regions or enhancer regions, altering their expression. This can result in the activation or repression of gene expression, leading to changes in the organism's phenotype (Zhang et al., 2017). Moreover, retrotransposons can control gene expression by contributing to the formation of new genes. This can occur when retrotransposons integrate into a gene, creating a new hybrid gene with novel functions. This process can lead to the evolution of unique traits in a population and the diversification of gene function within a species. Additionally, retrotransposons can act as epigenetic regulators, modulating gene expression through changes in chromatin structure. For example, some retrotransposons are associated with histone modifications that can result in the regulation of gene expression (Papolu et al., 2022).

2.3 Contribution to Genome Size and Diversity

It has become abundantly evident that the attack of retrotransposons is the primary component leading to the differential amplification of the Poaceae genome compared to other genomes. Throughout the past three million years, the size of the maize genome has increased due to retrotransposon insertions. The occurrence of repetitions at several levels—species, genera, families, and specifics—strongly implies that distinct retrotransposons have repeatedly invaded the genome. What causes the differential amplification of the genomes found within the same family? Nonetheless, it appears to depend on the kind of intruding retroelement. For instance, certain types of retrotransposons, such as the long terminal repeat variety, transfer more frequently into gene-deficient regions (Nadeem, 2021).

On the other hand, the micro inverted-repeat transposable elements type is partial to gene-dense areas. Due to the selection pressure, the replication for retroelements that favor gene-poor areas would be more significant than that of elements that prefer gene-rich regions. This is because gene-poor sites contain a smaller number of genes. Most noncoding regulatory regions (NTRs) in more giant genomes are made up of elements that tend to transpose into gene-poor areas (Senerchia et al., 2013).

3. Applications of Retrotransposons in Wheat Genetics and Genomics

The study of retrotransposons has essential relevance in wheat genetics and genomics because of the substantial role retrotransposons play in the development and control of the wheat genome. Some of the main applications of retrotransposons in these fields are:

3.1 Understanding the Wheat Genome

The study of retrotransposons is helping to increase our understanding of the complexity of the wheat genome and how it has evolved. For example, the presence and distribution of retrotransposons in the wheat genome can provide insight into the origin and evolution of different wheat species and cultivars. Such as polymorphism within the wheat genome can be detected using retrotransposons (Dubrovna et al., 2020).

3.2 Genome Improvement

Retrotransposons can be used to engineer the wheat genome to enhance desirable traits and to improve the crop's overall performance. For example, retrotransposons as mutagenic agents can be used to create new variations in the wheat genome, which can then be used in breeding programs to develop new cultivars with better yield, disease resistance, and other desirable traits (Yagci and Agar, 2022).

3.3 Study of Gene Expression

Retrotransposons can significantly impact gene expression, and their study can help better understand the regulation of gene expression in wheat. For example, studying retrotransposons can help identify genes regulated by retrotransposons and understand the mechanisms by which retrotransposons regulate gene expression.

3.4 Use as Molecular Markers

Retrotransposons are abundant in plant genomes and are highly variable, making them ideal candidates for developing molecular markers. These markers can be used to track the inheritance of retrotransposons and identify genetic variations associated with important agronomic traits such as yield, resistance to abiotic and biotic stresses, etc. In wheat, molecular markers based on retrotransposons have been used in various applications, including genome mapping, diversity analysis, and association mapping. For example, retrotransposon-based tags have been used to construct high-density linkage maps of the wheat genome, providing valuable information for understanding the wheat genome's structure and organization (Balážová et al., 2021).

Retrotransposon-based markers have also been used to study the genetic diversity of Triticum aestivum and other wheat species. This information can identify wheat accessions well adapted to specific environments and develop new varieties with improved agronomic traits. Because of the diversity and mobility of the many groups of TEs, they have become more valuable as molecular markers in recent years. Retrotransposon approaches have been directly compared to AFLP methods, and the results of these comparisons show that the retrotransposon markers are more informative in various crops. Changes in restriction sites and indels across the genome cause single-nucleotide polymorphism (SNP), the basis for polymorphism in AFLP markers. On the other hand, retrotransposon markers use integration events to generate an observable variation (Roncallo et al., 2019).

In contrast to Class II TEs, retrotransposons cannot remove themselves from the sites where they have been inserted. Because this integration is unidirectional, it is much easier to reconstruct pedigrees and phylogenies. This is possible because the ancestral state is readily apparent; it is the location where there is no content. On the other hand, it is only possible to infer directionality for some of the different genetic variations upon which markers are created. In this way, phylogeny can be investigated using retrotransposon insertions; for example, SINE elements have been utilized to trace human ancestry back to Africa, to determine the connection between whales and even-toed ungulates, and to infer the evolutionary relationships between different species of wild rice (Akakpo et al., 2020).

Most retrotransposon marker techniques use two fundamental qualities that retrotransposons have: their transpositional activity causes massive insertions, and their DNA contains conserved domains that may be used to create PCR primers (Zou et al., 2019). To develop fingerprints, several alternative approaches concentrate on the minute deletions and insertions that can be discovered inside otherwise stable TE domains. Most methods are anonymous, producing fingerprints from multiple retrotransposon insertion sites in the genome by using PCR primed on conserved motifs in the element and on some widespread and conserved motifs in the surrounding DNA. These fingerprints can be used to identify the retrotransposon insertion sites. Primers for LTR retrotransposons are often created from the LTRs close to the insertion site. Primers for LTR retrotransposons are typically designed from LTR sub-domains that are conserved within retrotransposon families but differ from family to family. Although regions internal to the LTR that contain conserved segments can be used for this purpose, in general, the LTRs are selected to reduce the target range that needs to be amplified and to assay insertion site polymorphism rather than events that occur internally to the attribute. The LTRs are shorter than the target size and must be enhanced (Kalendar et al., 2021).

Eukaryotic genomes include many retrotransposable elements and other similar elements. These elements insert themselves into new genomic sites using reverse transcription of an RNA intermediate. During the activation of recombinational events during the meiotic prophase, changes can be seen in the copy number of repeat elements and internal rearrangements on both homologous chromosomes. During transposition, which occurs during a species' continual evolution, LTR retrotransposons can randomly insert themselves into the genome. The XA study of development, species, and genomic differentiation might benefit from the plethora of knowledge this can give (ALTINTAŞ et al., 2021).

Applications of DNA markers based on retrotransposons have emerged as an essential component in studying genetic variability and diversity. Creating genomic maps and determining whether people or lines possess particular genetic polymorphism variants are only two examples of the use of these technologies. To understand the genetic diversity of agricultural plants, molecular genetic marker systems generated from LTR retrotransposons have been utilized. When identifying the effects of environmental stress on retrotransposon activation, the marker systems based on retrotransposons are successful. In addition, detecting TE expression, which may reveal host-TE interactions in novel and exciting ways by showing polymorphisms and the variety of the transposon transcriptional landscape, is an important step. In addition, LTR retrotransposons are linked to essential genes involved in prospective applications of genome assembly, genome variation, gene tagging, and functional analysis of genes. There is evidence that LTR retrotransposons play an essential role as markers in molecular breeding (Konovalov et al., 2010).

LTR retrotransposons were inserted into the genome of pepper (C. annuum) around 6 million years ago. These retrotransposons demonstrated chromosomal insertional preferences, which may be beneficial for constructing species-specific retrotransposon-based markers. There is a possibility that a system that combines active LTR retrotransposons with Inter-Retrotransposon Amplified Polymorphism (IRAP) markers would be an appropriate method for determining the genetic fidelity of tissue-culture-generated plants in sugarcane and for improving the management of germplasm in *Xanthosoma* and *Colocasia*. The IRAP marker system in LTR retrotransposon insertions of the flax genome revealed a high level of plant adaptability in a radioactive environment. It looked helpful in detecting retrotransposon polymorphisms (Grauda et al., 2016). The IRAP and REMAP markers found in the cassava genome both produced significant levels of polymorphism. They may be used to investigate the genetic variation of cassava cultivars and their connections with one another (Shingote et al., 2019). Examining the similarities and differences between two long terminal repeat (LTR) retrotransposons, BARE-1 and Jeli, may provide a possible source of polymorphic Sequence-Specific Amplification Polymorphism (SSAP) markers that may be used to assess the genetic diversity of diploid wheat. The LTR retrotransposon-based SSAP markers in cashew, and myrtle genomes displayed much larger polymorphic markers (Sadeqi et al., 2019). The genetic maps generated with several retrotransposon-based tags, such as iPBS (inter-Priming Binding Site) and REMAP (Retrotransposon-Microsatellite Amplified Polymorphism), exhibited regions of different marker densities. This indicates that the distribution of retrotransposons in lentils is not random and is widespread throughout the lentil genome. Identifying intraspecific variability by creating markers called retrotransposon-based insertion polymorphism (RBIP) obtained from sweet potatoes is possible. These markers can also be utilized as crucial Single sequence repeats (SSR) primers to generate linkage maps of diverse plant species, assisting in breeding and germplasm research. During the cultivar development for the Asian pear, it was discovered that the RBIP marker was replicated many times. It may offer a complete picture of the intricate relationships among Pyrus species and their evolutionary history. Similar to the above, a genome-wide analysis of RBIP markers in the Melilotus genome revealed considerable polymorphism information content (PIC), indicating that these markers are highly informative and can potentially be used to implement genetic improvement in the *Melilotus* genus (DEMİREL, 2020; Kalendar and Schulman, 2014).

In addition, the RBIP markers utilized for DNA profiling Japanese, Chinese, and European pear cultivars demonstrated that retrotransposons have transferred throughout the evolution of Asian pears or reflect the genetic link between Asian and European pears (Haliloğlu et al., 2023). Hence, appropriate combinations of retrotransposon insertions may prove advantageous for developing cultivar-specific DNA markers. Comparisons were made between the polymorphism markers created from several different retrotransposon families and the efficiency of the dominant (IRAP) and codominant (RBIP) marker systems for determining the level of genetic variation present in the various potato types. Potatoes have distinct DNA profiles for the retrotransposons Ty1/Copia and Ty3/Gypsy, both of which are active in the genome and may contribute to the structure of the potato genome. Analysis of high-throughput RBIP data suggested that it would give substantial support for the hypothesis of distinct domestications for *Pisum sativum* species. This, in turn, provides a comprehensive knowledge of the variety and development of Pisum. In a similar vein, the several LTR retrotransposon-based markers that have been developed from peas, broad beans, and Norway spruce have the potential to be beneficial in illuminating polymorphisms that are connected with the respective retrotransposons within the Pisum genus. As a result of the non-random distribution of abundant LTR retrotransposons within the lentil genome, it has been determined that defective and non-autonomous retrotransposons are extremely common. These retrotransposons have the potential to serve as a source of genetic markers for use in subsequent genetic research. The unique Ty1/Copia and Ty3/Gypsy LTR retrotransposons produced from Lilium species imply that they were not autonomous retrotransposons because these retrotransposons were derived from Lilium species. An IRAP analysis that uses the LTR sequence of these retrotransposons might give a fresh perspective on how different Lilium species are related to one another. Many LTR retrotransposon-based molecular markers were designed and used in *C. songorica* and strawberry genomes (Kojima, 2019). These molecular markers displayed a high degree of polymorphism frequency and a high transferability of polymorphic primer pairs. Based on this evidence, RBIP markers will prove valuable in further research on the genetic diversity, population structure, and evolution of germplasm accessions in *C. songorica* and grasses genetically linked to it. Numerous LTR retrotransposon markers that were produced from chokecherry genome sequences revealed that retrotransposon markers would be able to aid genetic research in Rosaceae species. These markers were used to develop maps and genetic mapping (Liang et al., 2016).

TriRe-1 has a distinct history of amplification in B-genome parents, indicating that genome-specific TriRe-1 might be used to generate wheat molecular markers. The optimal marker approach should be able to access many polymorphisms widely spread across the genome being analyzed. The analytical method should be simple to carry out, easy to reproduce, and affordable (Monden et al., 2014). Moreover, the markers should be straightforward to establish and not too expensive to test. SSAP markers that are based on retrotransposons are extremely successfully satisfied these parameters. Technically speaking, there is little difference between AFLP marker technology and SSAP marker technology; hence, SSAP ought to be regarded as an alternative to this marker approach. The requirement that retrotransposon sequences be found to create primers is one of the drawbacks of using SSAP for freshly researched species (Ahakpaz et al., 2020). This is no longer an issue for wheat since quick and effective measures have been developed for dealing with other species. The SSAP data quality is comparable to that of the AFLP data. Compared to AFLP,

SSAP has several advantages, one of which is the accessibility of greater degrees of polymorphism. As a result, fewer trials are required to create the requisite number of markers. SSAP is superior to AFLP in terms of its performance in evaluating genetic diversity, which is another advantage of the former. (ALTINTAŞ et al., 2021). AFLP-based features have been used in wheat to induce tolerance against abiotic stresses, such as drought (Ahakpaz et al., 2020).

3.5 Association Mapping for Important Agronomic Traits

Association mapping is a powerful tool that can identify genetic variants associated with important agronomic traits in Triticum aestivum (common wheat) and other crop species. This approach involves analyzing the relationships between genomic variations, such as single nucleotide polymorphisms (SNPs), and phenotypic traits of interest, such as grain yield, grain quality, and disease resistance (Liu et al., 2010). It takes advantage of the large amounts of genetic diversity present in crop species, allowing for the identification of candidate genes responsible for the variation in important agronomic traits. By mapping the associations between genetic variants and phenotypic characteristics, researchers can pinpoint specific genomic regions likely to contain the genes responsible for these traits (Keidar et al., 2018). Once these genomic regions have been identified, researchers can use additional techniques, such as gene editing and gene expression analysis, to further investigate the function of the candidate genes and their role in the regulation of important agronomic traits. It helps identify candidate genes associated with important agronomic traits in *T. aestivum* and other crop species. By linking genomic variation to phenotypic traits, this approach provides valuable insights into the genetic basis of these traits, which can be used to improve crop breeding and production (Savadi et al., 2018).

3.6 Importance in Identifying Candidate Genes for Improvement

Transposable elements are genetic components that can move around the genome and significantly contribute to genomic diversity. These elements affect gene structure and function, including insertion, transposition, excision, ectopic recombination, and chromosome breakage. Many studies have explored the role of transposable element insertion, including examples such as the Vrn-D1s allele, which contains a DNA transposon insertion and is considered preferred due to its positive impact on agronomic traits (Muterko et al., 2015). In this study, a 276-bp InDel transposon was found in the first intron of TaGW8-B1a (Yan et al., 2019). Cultivars with this transposon insertion exhibited altered agronomic traits and relative expression levels, suggesting that this transposon is functional and affects alternative splicing. Other transposable element insertions affecting gene function include the CsaMLO8 allele in cucumber and MADS-box genes in Arabidopsis. Additionally, the Wx-B1n allele with a 2178-bp transposon insertion led to losing its function (Zhang et al., 2017).

4. Future Perspectives and Challenges in Retrotransposon Research in Wheat

Retrotransposons are supposed to play a vital role in the management of stress. Knowledge of their activation mechanisms and triggers can change the situation; instead of using foreign vectors or plasmids, they can alter the genome to make resistant plants. Controlling retrotransposons by turning on or off gene expression can be used for genetic engineering. Their regulation retrotransposons depend on the transcriptional action of polymerase II of host RNA. However, the act of host RNA polymerase II to suppress the retrotransposon activity is unclear. The activation of retrotransposon dead copies can induce positive changes in the organization of the genome in plants. These results may help future research regarding plant genetic engineering and epigenetics. So, there is a need to develop new techniques to overcome the problems related to retrotransposon control. Some methods are in their initial stages, but new-generation sequencing is thought to positively contribute to detecting retrotransposition and controlling it in plants. Advancements in molecular technologies can help in retrotransposon-related research in the future (Bariah et al., 2020b).

4.1 Exploration of New Techniques and Methods

As there are several hurdles related to retrotransposon research to counter the challenges associated with the application of LTR in plant and their control, several research tools and advanced technologies have been introduced. For instance, to predict the methods of LTR control tools like big data, the methods based on artificial intelligence and machine learning (ML) are considered essential and might help retrotransposon regulation. TEtools development may be a positive advancement in controlling retrotransposons (Lerat et al., 2017). ML algorithms help learn about the parameters required to fix the model to a specified problem, known as supervised learning. To classify LTR transposons, Field has recently introduced the ML technique (Orozco-Arias et al., 2021; Shastry and Sanjay, 2020; Zou et al., 2019). The random forest algorithm has been used recently to classify LTR retrotransposons into superfamilies. In contrast, other coding schemes and pre-processing techniques can be used for their further division into families and subfamilies. Fully connected neural networks (FNN) and convolutional neural networks (CNN) are established for TEs classification into their superfamilies (da Cruz et al., 2020; da Cruz et al., 2019; Nakano et al., 2018; Yan et al., 2020).

High-throughput long-read sequencing technologies can predict the copy number differences. These novel methods are in recent advancement stages that might help correctly measure the copy number. Different chemical combinations can help to silence or activate LTR retrotransposons and are used to make desired alterations in the plant genome (Ramakrishnan et al., 2023). In a study, CRED-iPBS (Coupled Restriction Enzyme Digestion-iPBS) and inter-primer binding site (iPBS) retrotransposon techniques were used to show alterations in DNA methylation patterns and level of DNA damage. Polymorphisms are commonly detected by the iPBS retrotransposon method, a marker system that works via transposable elements (Ramakrishnan et al., 2021).

To study genetic variability in plants, it is significant to use retrotransposon-derived genetic markers. Marker systems can be vital to detect retrotransposon-originated polymorphisms. The LTR retrotransposons are randomly inserted, and they take place during transposition. This can collect new knowledge and help to study different species, evolution, and genome differentiation. In research based on genetic variability and diversity, RTE-based DNA markers have important applications (Wu et al., 2018). They can be significant in identifying individuals and polymorphic genetic variations and creating genetic maps. As we know, retrotransposon insertions act as Mendelian loci; thus, retrotransposon-based markers are anticipated to be codominant and implicate different genetic variability levels. The tools for polymorphism detection can be extended by RTE knowledge found in the genome with different orientations.

Retrotransposon Microsatellite Amplified Polymorphism (REMAP-PCR) and Inter Retrotransposon Amplified Polymorphism PCR (IRAP-PCR) techniques were used to study the DNA injuries done by salt (Yagci and Agar, 2022). The methods are unpredictable. They produce fingerprints from multiple retrotransposon insertion sites. By polygenetic approach, it is possible to predict standard PCR amplicons while analyzing closely linked species (Kalendar et al., 2017). All techniques use multiple adjacent sequences and known retrotransposon sequences. The primer targets for LTR are made near joints in the domain, which vary between the families but are conserved within the families. The RTE regions with conserved sections can be used to reduce the interspace between targets to be amplified. To simplify the process, retrotransposon-specific primer can be made from an inside sequence that is present once per element with retrotransposons for small LTRs, for sequence-specific amplified polymorphism of elements with low copy number amplification and simple digestion protocols can be used. Retrotransposon makers differ in second primer (Kalendar and Schulman, 2014). Various molecular marker methods based on retrotransposons, such as the sequence-specific amplified polymorphism method, inter-retrotransposon amplified polymorphism method, the retrotransposon microsatellite amplified polymorphisms method, Retrotransposon-based insertion polymorphism and inter-primer-binding site amplification (iPBS) scheme and LTR retrotransposon structure (Kalendar et al., 2018). To carry out reverse transcription, virus-like particles are produced by active LTR retrotransposon loci. The complementary DNA can be introduced in another new locus. Because of the dangerous results of retrotransposition, plants, like animals, have developed transcriptional and post-transcriptional silencing methods. Genome-wide techniques have been recently introduced to understand LTR retro transposition in different plant species. Host silencing methods can affect retrotransposition steps, which can be seen by methylome, transposon, translatome, small RNA sequencing, and transcriptome data (Lee and Martienssen, 2021). These several advancements in studies can provide knowledge for future studies on the retrotransposition of retrotransposons and their diversity. The advances in sequencing and computing technologies can play a role in developing advanced methods to understand activation, expression, silencing, movement, and copy number alterations of retrotransposons, which can contribute to controlling transposons in the future (Zou et al., 2019).

4.2 Integration of Transposons into Wheat Improvement Programs

LTR retrotransposons related to different molecular functions provide an excellent opportunity to develop tools for genetic engineering as intragenic elements to improve plant genomes. Determining genetic variation is crucial for future plant breeding (Yaman, 2022). A study revealed the population structure and genetic diversity of bread wheat by using two molecular markers, inter-retrotransposon amplified polymorphisms (IRAP) and retrotransposon-microsatellite amplified polymorphisms (REMAP) markers (Abbasi Holasou et al., 2019). Activation of retrotransposon is one of the factors that help the host to adapt to environmental changes (Casacuberta and González, 2013). It is reported in a study that iPBS markers have been used to identify wheat accessions at the molecular level. The data obtained can be used for genetic analysis and research on the diversity of wheat accessions (Haliloğlu et al., 2023). iPBS retrotransposon marker is reportedly used for bread wheat molecular characterization (Nadeem, 2021). The same system is also used for molecular characterization of durum wheat and (Emmer) wheat (DEMİREL, 2020).

High throughput sRNA sequencing in parental wheat plants, hybrid plants, and allopolyploids revealed that TE-derived sRNAs and miRNAs have unique reactions towards ploidy level alterations. Upon allopolyploidization, a decrease in siRNA pools was observed, resulting in the deregulation of siRNA and CpG methylation of LTR retrotransposons. This may cause genome instability at the initial stages of the speciation (Kenan-Eichler et al., 2011). LTR transposons are specified for the B-genome in polyploid wheat and multiply preliminary to the hexaploidy wheat establishment. TriRe-1 retrotransposons have an amplifying history of B-genome ancestors in hexaploidy wheat, which showed that novel retrotransposon TriRe-1 could be used for molecular marker development (Monden et al., 2014; Salina et al., 2011). The comparative study of two retrotransposons (BARE-1 and Jeli) revealed that to uncover genetic diversity in diploid wheat, the LTR transposons-based molecular markers, BARE-1 and Jeli may be suitable (Konovalov et al., 2010). Whole genome sequence(WGS) analysis points out significant actions of retrotransposons in wild and domesticated wheat and wheat bread evolution. The presence of retrotransposon of barley (Nikita and Sukkula) has been observed in the wheat genome. This significant knowledge can be instrumental in revealing retrotransposon effects on the organization of the wheat genome throughout the domestication process (ALTINTAŞ et al., 2021). The bread wheat Field also reported a negative correlation between recombination and LTR retrotransposons (Daron et al., 2014). The dynamics of transposable elements can be uncovered in complex genomes like wheat by precise transposable element modeling. It provides novel insights into wheat research.

4.3 Challenges in Controlling the Spread of Retrotransposons in Crop Improvement

LTR retrotransposons are considered an essential plant genetic engineering tool, but several challenges are associated with their application. These hurdles include failure to control copy number and differences in copy expression. Beneficial but sometimes harmful mutations can be caused by many copies of retrotranspososns. (Belyayev et al., 2010). Plants have several mechanisms to balance the copy numbers where most copies are silenced epigenetically or dead. The epigenetic alterations are linked with retrotransposon silencing, but the phenomena involved in balancing the copy number are unknown. Copies are either activated or deactivated depending on plant growth and developmental stages (Cavrak et al., 2014; Lisch, 2013). Compared to DNA transposons, retrotransposons detection in real-time is very difficult because of the replicative nature of the process. Retrotransposon movement is only detected by spontaneous and forward mutation characterization and recording, which are rarely shown by phenotypes (Wessler et al., 1995).

Many challenges are observed while working with crops having larger genomes and profusion of LTR retrotransposon families. Different crop genomes vary in size and retroelement content. Some crops' genomes are larger and more complex to study than others. For instance, the Arabidopsis genome is small (135mb) with almost 13% retrotransposons. There are 210 families of LTR retrotransposon with 475 intact retroelements in Arabidopsis. At the same time, 75% of genome sequences in maize plants belong to retrotransposons comprising 2900 LTR retrotransposon families having almost 68000 intact copies in their genome (El Baidouri and Panaud, 2013). However, controlling retrotransposons is challenging as various knowledge gaps must be filled to understand and manage their retrotransposition completely.

Transpositional control of different transposition families needs deep knowledge of epigenetics and environmental triggers associated with their power. In genomes of other crops, despite diversification in LTR retrotransposons, only a few transposition examples exist. The causes of these transpositions are unclear as they are either brought about by accidents after some infections by virus or in tissue culture. So, the ability of retroelements' transposition and the conditions required are still unknown (Paszkowski, 2015). The other challenge associated with retrotransposon control is a difference in the expression of copy numbers of similar families, which varies with the developmental stage and species of the plant (Liu et al., 2022). High throughput sequencing can predict copy number differences, but the copy numbers cannot be controlled directly, and their expression, differences, and movement cannot be measured accurately. Due to these hurdles, the active copies cannot be easily retained, so it is challenging to make targeted alterations in the genome (Ramakrishnan et al., 2023).

References

Abbasi Holasou, H., Rahmati, F., Rahmani, F., Imani, M., and Talebzadeh, Z. (2019). Elucidate genetic diversity and population structure of bread wheat (Triticum aestivum L.) cultivars using IRAP and REMAP markers. *Journal of Crop Science and Biotechnology, 22,* 139–51.

Aguilar, M., and Prieto, P. (2020). Sequence analysis of wheat subtelomeres reveals a high polymorphism among homoeologous chromosomes. *The Plant Genome, 13*(3), e20065.

Ahakpaz, F., Majidi Hervan, E., Roostaei, M., Bihamta, M.R., and Mohammadi, S. (2020). Drought tolerance related traits in bread wheat and its association with AFLP marker. *Cereal Research, 9*(4), 359–71.

Akakpo, R., Carpentier, M.C., Ie Hsing, Y., and Panaud, O. (2020). The impact of transposable elements on the structure, evolution and function of the rice genome. *New Phytologist, 226*(1), 44–49.

Altintaş, S., Ilgar, B.A., and Karlik, E. (2021). Comparative retrotransposon analysis in wheat. *Journal of Advanced Research in Natural and Applied Sciences, 7*(3), 369–74.

Balážová, Ž., Trebichalský, A., Gálová, Z., Kalendar, R., Schulman, A., Stratula, O., and Chňapek, M. (2021). Genetic diversity of triticale cultivars based on microsatellite and retrotransposon-based markers. *Journal of Microbiology, Biotechnology and Food Sciences, 2021,* 58–60.

Bariah, I., Keidar-Friedman, D., and Kashkush, K. (2018). Identification of large-scale genomic rearrangements during wheat evolution and the underlying mechanisms. *bioRxiv,* 478933.

Bariah, I., Keidar-Friedman, D., and Kashkush, K. (2020a). Identification and characterization of large-scale genomic rearrangements during wheat evolution. *PLoS One, 15*(4), e0231323.

Bariah, I., Keidar-Friedman, D., and Kashkush, K. (2020b). Where the wild things are: transposable elements as drivers of structural and functional variations in the wheat genome. *Frontiers in Plant Science, 11,* 585515.

Belyayev, A., Kalendar, R., Brodsky, L., Nevo, E., Schulman, A.H., and Raskina, O. (2010). Transposable elements in a marginal plant population: temporal fluctuations provide new insights into genome evolution of wild diploid wheat. *Mobile DNA, 1*(1), 1–16.

Casacuberta, E., and González, J. (2013). The impact of transposable elements in environmental adaptation. *Molecular ecology, 22*(6), 1503–17.

Cavrak, V.V., Lettner, N., Jamge, S., Kosarewicz, A., Bayer, L.M., and Mittelsten Scheid, O. (2014). How a retrotransposon exploits the plant's heat stress response for its activation. *PLoS genetics, 10*(1), e1004115.

da Cruz, M.H.P., Domingues, D.S., Saito, P.T.M., Paschoal, A.R., and Bugatti, P.H. (2020). TERL: classification of transposable elements by convolutional neural networks. *Briefings in Bioinformatics, 22*(3). https://doi.org/10.1093/bib/bbaa185.

da Cruz, M.H.P., Saito, P.T.M., Paschoal, A.R., and Bugatti, P.H. (2019). Classification of Transposable Elements by Convolutional Neural Networks. Artificial Intelligence and Soft Computing, Cham.

Daron, J., Glover, N., Pingault, L., Theil, S., Jamilloux, V., Paux, E., Barbe, V., Mangenot, S., Alberti, A., Wincker, P., Quesneville, H., Feuillet, C., and Choulet, F. (2014). Organization and evolution of transposable elements along the bread wheat chromosome 3B. *Genome biology, 15*(12), 546. https://doi.org/10.1186/s13059-014-0546-4.

Demire, F. (2020). Genetic diversity of Emmer wheats using iPBS markers. *Avrupa Bilim ve Teknoloji Dergisi*(20), 640–46.

Dubrovna, O., Velikozhon, L., Slivka, L., Kondratskaya, I., Reshetnikov, V., and Makai, S. (2020). Detection of DNA polymorphism of transgenic wheat plants with proline metabolism heterologous genes. *Plant Physiology and Genetics, 52*(3), 196–07.

El Baidouri, M., and Panaud, O. (2013). Comparative Genomic Paleontology across Plant Kingdom Reveals the Dynamics of TE-Driven Genome Evolution. *Genome Biology and Evolution, 5*(5), 954–65. https://doi.org/10.1093/gbe/evt025.

Galindo-González, L., Mhiri, C., Deyholos, M.K., and Grandbastien, M.-A. (2017). LTR-retrotransposons in plants: Engines of evolution. *Gene, 626,* 14–25.

Ghonaim, M.M., Mohamed, H.I., and Omran, A.A. (2021). Evaluation of wheat (Triticum aestivum L.) salt stress tolerance using physiological parameters and retrotransposon-based markers. *Genetic Resources and Crop Evolution, 68*, 227–42.

Grandbastien, M.-A. (2015). LTR retrotransposons, handy hitchhikers of plant regulation and stress response. *Biochimica et Biophysica Acta (BBA)-Gene Regulatory Mechanisms, 1849*(4), 403–16.

Grauda, D., Zagata, K., Lanka, G., Strazdina, V., Fetere, V., Lisina, N., Krasnevska, N., Fokina, O., Mikelsone, A., and Ornicans, R. (2016). Genetic diversity of wheat (Triticun aestivum L.) plants-regenerants produced by anther culture. *Вавиловский журнал генетики и селекции, 20*(4), 537–44.

Haliloğlu, K., Türkoğlu, A., Öztürk, A., Niedbała, G., Niazian, M., Wojciechowski, T., and Piekutowska, M. (2023). Genetic Diversity and Population Structure in Bread Wheat Germplasm from Türkiye Using iPBS-Retrotransposons-Based Markers. *Agronomy, 13*(1), 255. https://www.mdpi.com/2073-4395/13/1/255.

Jedlicka, P., Lexa, M., Vanat, I., Hobza, R., and Kejnovsky, E. (2019). Nested plant LTR retrotransposons target specific regions of other elements, while all LTR retrotransposons often target palindromes and nucleosome-occupied regions: in silico study. *Mobile DNA, 10*(1), 1–14.

Kalendar, R., Amenov, A., and Daniyarov, A. (2018). Use of retrotransposon-derived genetic markers to analyse genomic variability in plants. *Functional Plant Biology, 46*(1), 15–29.

Kalendar, R., Khassenov, B., Ramankulov, Y., Samuilova, O., and Ivanov, K.I. (2017). FastPCR: An in silico tool for fast primer and probe design and advanced sequence analysis. *Genomics, 109*(3), 312–319. https://doi.org/https://doi.org/10.1016/j.ygeno.2017.05.005.

Kalendar, R., Muterko, A., and Boronnikova, S. (2021). Retrotransposable elements: DNA fingerprinting and the assessment of genetic diversity. *Molecular Plant Taxonomy: Methods and Protocols*, 263–86.

Kalendar, R., and Schulman, A.H. (2014). Transposon-based Tagging: IRAP, REMAP, and iPBS. In P. Besse (Ed.), *Molecular Plant Taxonomy: Methods and Protocols* (pp. 233–55). Humana Press. https://doi.org/10.1007/978-1-62703-767-9_12

Keidar, D., Doron, C., and Kashkush, K. (2018). Genome-wide analysis of a recently active retrotransposon, Au SINE, in wheat: content, distribution within subgenomes and chromosomes, and gene associations. *Plant Cell Reports, 37*, 193–208.

Kenan-Eichler, M., Leshkowitz, D., Tal, L., Noor, E., Melamed-Bessudo, C., Feldman, M., and Levy, A. A. (2011). Wheat hybridization and polyploidization results in deregulation of small RNAs. *Genetics, 188*(2), 263–72.

Kojima, K.K. (2019). Structural and sequence diversity of eukaryotic transposable elements. *Genes & Genetic Systems, 94*(6), 233–52.

Konovalov, F.A., Goncharov, N.P., Goryunova, S., Shaturova, A., Proshlyakova, T., and Kudryavtsev, A. (2010). Molecular markers based on LTR retrotransposons BARE-1 and Jeli uncover different strata of evolutionary relationships in diploid wheats. *Molecular Genetics and Genomics, 283*, 551–63.

Lee, S.C., and Martienssen, R.A. (2021). Regulation of retrotransposition in Arabidopsis. *Biochemical Society Transactions, 49*(5), 2241–51. https://doi.org/10.1042/bst20210337.

Lerat, E., Fablet, M., Modolo, L., Lopez-Maestre, H., and Vieira, C. (2017). TEtools facilitates big data expression analysis of transposable elements and reveals an antagonism between their activity and that of piRNA genes. *Nucleic acids research, 45*(4), Page 17.

Liang, Y., Lenz, R.R., and Dai, W. (2016). Development of retrotransposon-based molecular markers and their application in genetic mapping in chokecherry (Prunus virginiana L.). *Molecular Breeding, 36*, 1–10.

Lisch, D. (2013). How important are transposons for plant evolution? *Nature Reviews Genetics, 14*(1), 49–61.

Liu, J.-X., Liu, J., Yang, T.-Y., Liu, L.-M., Qiu, L.-H., Gao, Y.-J., Duan, W.-X., Lei, J.-C., Liu, H.-J., and Zhang, R.-H. (2022). Isolation and analysis of reverse transcriptase of Ty1-copia-like retrotransposons in sugarcane. *Sugar Tech, 24*(5), 1510–29.

Liu, L., Wang, L., Yao, J., Zheng, Y., and Zhao, C. (2010). Association mapping of six agronomic traits on chromosome 4A of wheat (Triticum aestivum L.). *Molecular Plant Breeding, 1*(5).

Monden, Y., Takai, T., and Tahara, M. (2014). Characterization of a novel retrotransposon TriRe–1 using nullisomic-tetrasomic lines of hexaploid wheat. *Okayama University Faculty of Agriculture Academic Report, 103*, 21–30.

Muterko, A., Balashova, I., Cockram, J., Kalendar, R., and Sivolap, Y. (2015). The New Wheat Vernalization Response Allele Vrn-D1s is Caused by DNA Transposon Insertion in the First Intron. *Plant Molecular Biology Reporter, 33*(2), 294–303. https://doi.org/10.1007/s11105-014-0750-0.

Nadeem, M.A. (2021). Deciphering the genetic diversity and population structure of Turkish bread wheat germplasm using iPBS-retrotransposons markers. *Molecular Biology Reports, 48*, 6739–48.

Nakano, F.K., Mastelini, S.M., Barbon, S., and Cerri, R. (2018). Improving hierarchical classification of transposable elements using deep neural networks. 2018 International Joint Conference on Neural Networks (IJCNN).

Orozco-Arias, S., Candamil-Cortés, M.S., Jaimes, P.A., Piña, J.S., Tabares-Soto, R., Guyot, R., and Isaza, G. (2021). K-mer-based machine learning method to classify LTR-retrotransposons in plant genomes. *PeerJ, 9*, e11456.

Papolu, P.K., Ramakrishnan, M., Mullasseri, S., Kalendar, R., Wei, Q., Zou, L.-H., Ahmad, Z., Vinod, K.K., Yang, P., and Zhou, M. (2022). Retrotransposons: How the continuous evolutionary front shapes plant genomes for response to heat stress. *Frontiers in Plant Science.*

Paszkowski, J. (2015). Controlled activation of retrotransposition for plant breeding. *Current Opinion in Biotechnology, 32*, 200–206.

Portis, E., Acquadro, A., and Lanteri, S. (2019). Genetics and Breeding. *The Globe Artichoke Genome,* 115–28.

Ramakrishnan, M., Papolu, P.K., Mullasseri, S., Zhou, M., Sharma, A., Ahmad, Z., Satheesh, V., Kalendar, R., and Wei, Q. (2023). The role of LTR retrotransposons in plant genetic engineering: how to control their transposition in the genome. *Plant Cell Reports, 42*(1), 3–15. https://doi.org/10.1007/s00299-022-02945-z.

Ramakrishnan, M., Satish, L., Kalendar, R., Narayanan, M., Kandasamy, S., Sharma, A., Emamverdian, A., Wei, Q., and Zhou, M. (2021). The dynamism of transposon methylation for plant development and stress adaptation. *International Journal of Molecular Sciences, 22*(21), 11387.

Roncallo, P. F., Beaufort, V., Larsen, A.O., Dreisigacker, S., and Echenique, V. (2019). Genetic diversity and linkage disequilibrium using SNP (KASP) and AFLP markers in a worldwide durum wheat (Triticum turgidum L. var durum) collection. *PLoS One, 14*(6), e0218562.

Sadeqi, M.B., Dadshani, S., Yousefi, M., and Ajir, G.M. (2019). Investigation of Genetic Diversity in Afghan Bread Wheat Genotypes Using SSR and AFLP Markers. *Turkish Journal of Agriculture-Food Science and Technology, 7*(9), 1263–67.

Salina, E.A., Sergeeva, E.M., Adonina, I.G., Shcherban, A.B., Belcram, H., Huneau, C., and Chalhoub, B. (2011). The impact of Ty3-gypsy group LTR retrotransposons Fatima on B-genome specificity of polyploid wheats. *BMC Plant Biology, 11*(1), 99. https://doi.org/10.1186/1471-2229-11-99

Sant, V., Sainani, M., Sami-Subbu, R., Ranjekar, P., and Gupta, V. (2000). Ty1-copia retrotransposon-like elements in chickpea genome: their identification, distribution and use for diversity analysis. *Gene, 257*(1), 157–66.

Savadi, S., Prasad, P., Kashyap, P., and Bhardwaj, S. (2018). Molecular breeding technologies and strategies for rust resistance in wheat (Triticum aestivum) for sustained food security. *Plant pathology, 67*(4), 771–91.

Senerchia, N., Wicker, T., Felber, F., and Parisod, C. (2013). Evolutionary dynamics of retrotransposons assessed by high-throughput sequencing in wild relatives of wheat. *Genome Biology and Evolution, 5*(5), 1010–20.

Shastry, K.A., and Sanjay, H. (2020). Machine learning for bioinformatics. *Statistical modelling and machine learning principles for bioinformatics techniques, tools, and applications,* 25–39.

Shingote, P.R., Amitha Mithra, S., Sharma, P., Devanna, N.B., Arora, K., Holkar, S.K., Khan, S., Singh, J., Kumar, S., and Sharma, T. (2019). LTR retrotransposons and highly informative ISSRs in combination are potential markers for genetic fidelity testing of tissue culture-raised plants in sugarcane. *Molecular Breeding, 39,* 1– 13.

Wessler, S.R., Bureau, T.E., and White, S.E. (1995). LTR-retrotransposons and MITEs: important players in the evolution of plant genomes. *Current opinion in genetics & development, 5*(6), 814–21.

Wicker, T., Stritt, C., Sotiropoulos, A.G., Poretti, M., Pozniak, C., Walkowiak, S., Gundlach, H., and Stein, N. (2022). Transposable Element Populations Shed Light on the Evolutionary History of Wheat and the Complex Co-Evolution of Autonomous and Non-Autonomous Retrotransposons. *Advanced Genetics, 3*(1), 2100022.

Wu, L., Gingery, M., Abebe, M., Arambula, D., Czornyj, E., Handa, S., Khan, H., Liu, M., Pohlschroder, M., and Shaw, K.L. (2018). Diversity-generating retroelements: natural variation, classification and evolution inferred from a large-scale genomic survey. *Nucleic acids research, 46*(1), 11–24.

Yagci, S., and Agar, G. (2022). β-Estradiol Against to Salt Stress-Induced Long Terminal Repeats (LTR) Retrotransposons Polymorphism in Wheat. *Environmental Engineering & Management Journal (EEMJ), 21*(4).

Yaman, M. (2022). Evaluation of genetic diversity by morphological, biochemical and molecular markers in sour cherry genotypes. *Molecular Biology Reports,* 1–9.

Yan, H., Bombarely, A., and Li, S. (2020). DeepTE: a computational method for de novo classification of transposons with convolutional neural network. *bioRxiv,* 2020.2001.2027.921874. https://doi.org/10.1101/2020.01.27.921874.

Yan, X., Zhao, L., Ren, Y., Dong, Z., Cui, D., and Chen, F. (2019). Genome-wide association study revealed that the TaGW8 gene was associated with kernel size in Chinese bread wheat. *Scientific Reports, 9*(1), 2702. https://doi.org/10.1038/s41598-019-38570-2.

Zhang, L.L., Chen, H., Luo, M., Zhang, X.W., Deng, M., Ma, J., Qi, P.F., Wang, J.R., Chen, G.Y., Liu, Y.X., Pu, Z.E., Li, W., Lan, X.J., Wei, Y.M., Zheng, Y.L., and Jiang, Q.T. (2017). Transposon insertion resulted in the silencing of Wx-B1n in Chinese wheat landraces. *Theor Appl Genet, 130*(6), 1321–30. https://doi.org/10.1007/s00122-017-2878-4.

Zou, J., Huss, M., Abid, A., Mohammadi, P., Torkamani, A., and Telenti, A. (2019). A primer on deep learning in genomics. *Nature genetics, 51*(1), 12–18.

Retrotransposons in Genome of
Hordeum vulgare

Rabia Habib,[1] *Zulqurnain Khan,*[1] *Sadia Shabir, Sehrish Ijaz*[1]
Akash Fatima[1] *and Ummara Waheed*[1*]

1. Introduction

Barley (*Hordeum vulgare L. $2n = 2X = 14$*) is an important cereal crop that belongs to the Poaceae family and is the 4th largest crop grown all over the world among other cereal crops following wheat (*Triticum aestivum*), maize (*Zea mays*) and rice (*Oryza sativa*). Due to its diploid nature among the more complex genomic constituents of its sister crops like wheat and its important evolutionary background, barley has become an important model crop for investigating the important genes controlling traits (Waheed et al., 2016). Considered an ancient crop, barley was domesticated over a millennium ago and has since continued to be a valuable economic source of grain for animal feed, fermentation, and condensation processes (Bartlett, et al., 2008). Being grown in subarctic to subtropical regions this crop is frequently adaptable (Jana and Pietrzak, 1988).

Barley tends to have greater susceptibility to aluminum (Al) toxicity, also shown a greater tendency to withstand harsh climatic conditions (Bian et al., 2013; Schulte et al., 2009). Owing to its great genomic importance, overall genomes of barley cultivated species *Hordeum spontaneum* have been successfully sequenced to investigate the composition and evolution of this important crop at the genomic level, it was found that more than 80% of its genomes have repetitive sequences including retrotransposons (Schreiber et al., 2020; Xu et al., 2021). However, the

---

[1]   Institute of Plant Breeding and Biotechnology MNS-University of Agriculture, Multan.
[*]   Corresponding author: ummara.waheed@mnsuam.edu.pk

evolutionary mechanism and the functions of transposable elements in barley need to be explored.

Transposable elements (TEs) are compact genetic entities that can be thought of as "miniature genomes" within the larger genome of an organism. They consist of the essential information required for self-replication and mobilization within the host genome. TEs behave like intranuclear viruses and depend on the host's DNA replication and translation machinery to propagate themselves. TEs are often referred to as "selfish" DNA because of their ability to replicate and spread within a genome (Doolittle and Sapienza, 1980). Although transposable elements (TEs) were once thought to be non-functional and referred to as "junk DNA," it is now known that they constitute the majority of DNA in many higher organisms and play a vital role in genome evolution (Ohno, 1972; Çayır, 2022).

Retrotransposons are a class of transposable elements that can mobilize within the genome through an RNA intermediate (Collins and Nilsen, 2013). They are found in the genomes of many organisms, including humans and plants (Burns, 2017). The distribution of retrotransposons within a genome can vary depending on the organism and specific retrotransposon type. Retrotransposons tend to accumulate in regions of the genome that are less dense and contain more repetitive DNA sequences (Van et al., 2005). In humans, the most abundant retrotransposon is the LINE-1 element, which makes up about 17% of the genome. LINE-1 elements are primarily found in gene-poor regions of the genome, such as centromeres and telomeres. However, they can also be found in introns and untranslated regions of genes (Bierhoff et al., 2014). Overall, the distribution of retrotransposons within a genome can have important implications for gene regulation and genome evolution. Retrotransposon insertions can disrupt gene function, but they can also contribute to genetic diversity and facilitate the evolution of new genes and regulatory elements (Morgante et al., 2007).

A taxonomic system has been developed that includes classes, orders, super families, and families, as well as standardized criteria and conventions for the consolidation of earliest transposons elements taxonomy but also provided valuable tools for annotating sequenced genomes. TEs are classified into two primary classes, nine orders, and twenty nine super families (Wicker et al., 2007).

Class I TEs commonly known as "copy-and-paste" process as they can add more copies during replication through RNA intermediate whereas Class II is referred to as "cut-and-paste" as it involves many processes line replication etc. (Fig. 1).

This system divides TEs into two primary classes, and characterizes many superfamilies, making TSD size a useful diagnostic feature for classification. To simplify identification, a 3-letter code precedes each transposable element family. Some examples of these codes are DIRS (Dictyostelium intermediate repeat sequence), LINE (long interspersed nuclear element), LTR (long terminal repeat), PLE (Penelope-like elements), SINE (short interspersed nuclear element), and TIR (terminal inverted repeat) (Schulman and Wicker, 2013).

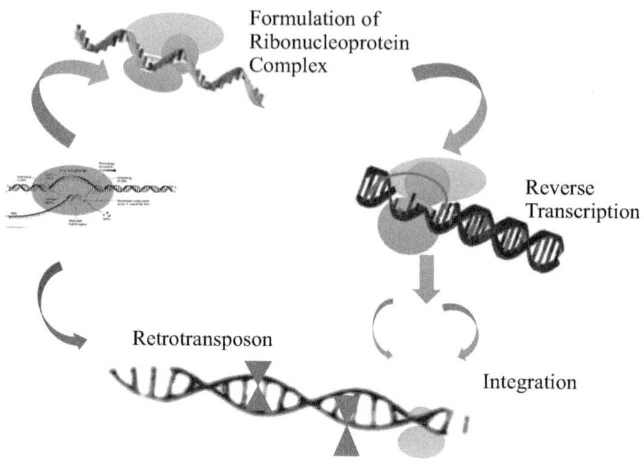

Fig. 1   The lifecycle of retrotransposons

2.   Class I Elements

Class I transposable elements share a common replication cycle, which involves the reverse transcription of RNA into double-stranded DNA by the RT enzyme. Its two main transposable elements are distinguished by the existence or non-appearance of long terminal repeats (LTRs) at both terminals. Long interspersed nuclear elements (LINEs) are characterized by the absence of LTRs, while endogenous retroviruses (ERVs) and retrotransposons of the Ty1/copia type are typified by the presence of LTRs (Goodier and Kazazian, 2008). Although LINEs are often considered the prototypical non-LTR retrotransposons, other groups of elements that lack LTRs are also known. According to Spanu et al. (2010), non-LTR retrotransposons are a majorly prominent order of transposable elements in vertebrates and certain fungi, but less common in plants (Spanu et al., 2010).

In addition, Reverse transcription usage that links the Class One lifecycle to the evolutionary shift from primitive RNA genetic material to the present-day DNA genetic material can be seen as an aged trait (Brosius, 2005). LTRs that don't have integrase gene are non-LTR retrotransposons and are different from LTR retrotransposons. Instead, the reverse transcriptase begins DNA strand synthesis directly at the insertion point using the element's mRNA poly-A tail as a primer. Also, it can join the newly synthesized DNA part to the insertion point (see Figure 4). In contrast, the reverse transcriptase of non-LTR retrotransposons uses the mRNA's poly-A tail as a primer and directly synthesizes DNA at the insertion point. Ligation of the novel formed DNA's end to the lodging point is also done by the reverse transcriptase.

Non-LTR retrotransposons, which include LINEs and SINEs are not as abundant in Hordeum vulgare compared to LTR retrotransposons. Instead of LTRs, they have a poly (A) tail at their 3′ end. The coding region of non-LTR retrotransposons usually contains ORFs encoding the reverse transcriptase and endonuclease enzymes that are essential for the retro transposition process. In *Hordeum vulgare*, LINEs are relatively large and range from 4 to 8 kb, with low copy numbers (up to 300 copies per haploid genome). On the other hand, SINEs are much smaller, less than 400 bp, and have a higher copy number (up to 1,500 copies per haploid genome) (Kojima, 2018).

3. Class II Elements

The transposase enzyme in DNA transposons identifies the terminal inverted repeats as recognition sites. These repeats flank the transposon and are crucial for the transposition process (Fig. 2).

The majority of identified transposase proteins have a DDE motif, which is essential for their activity (Keith et al., 2008). There is a consensus that DNA transposons replicate by transposing from a previously replicated site to a site ahead of the replication fork during DNA replication. As a consequence, one replicated strand carries a single copy of the transposon, while the other strand contains two copies, leading to an overall increase of one copy and one daughter chromosome . The Class II transposable elements, unlike Class I elements, move primarily through a "cut-and-paste" mechanism mediated by a transposase enzyme. However, a few groups in Class II are exceptions to this rule, such as Crypton elements. "Subclass 1" instead of "Subclass 2" which contains Crypton elements) (Kojima and Jurka, 2011).Through a process involving tyrosine recombinase-mediated recombination between direct repeats and the subsequent reinsertion of a complementary target site, Crypton elements can transmit as circular DNA (Goodwin et al., 2003). According to the classification proposed by Kapitonov and Jurka (2001), Class II Subclass 2 comprises two distinct groups: Helitron and Maverick. Unlike Subclass 1, the structures and replication mechanisms of these elements are markedly different. Maverick elements (also known as Polintons) are a heterogeneous group of mobile elements that possess inverted repeats at their termini and carry an integrase gene along with at least five to nine additional coding sequences (Feschotte and Pritham, 2005).

Based on bioinformatic analyses, Helitrons are broad, diverse, and in some cases numerous groups of mobile elements. However, these elements can be challenging to detect because they do not typically possess conventional (Du et al., 2006). It is worth noting that in maize, Helitrons play a major role in generating genomic diversity by translocating gene fragments and pseudogenes to new locations within the genome (Lai et al., 2005). Retrotransposons are often found in clusters or "islands" scattered throughout the genome, particularly in regions that are gene-poor and contain more repetitive DNA sequences. They can be present in both intergenic regions and within genes themselves, and their distribution may differ depending on the specific retrotransposon and organism (Berry et al., 2006). For example, LINE-1 elements tend to be more concentrated in gene-poor regions, while Alu elements are often found in gene-rich regions. Retrotransposons can

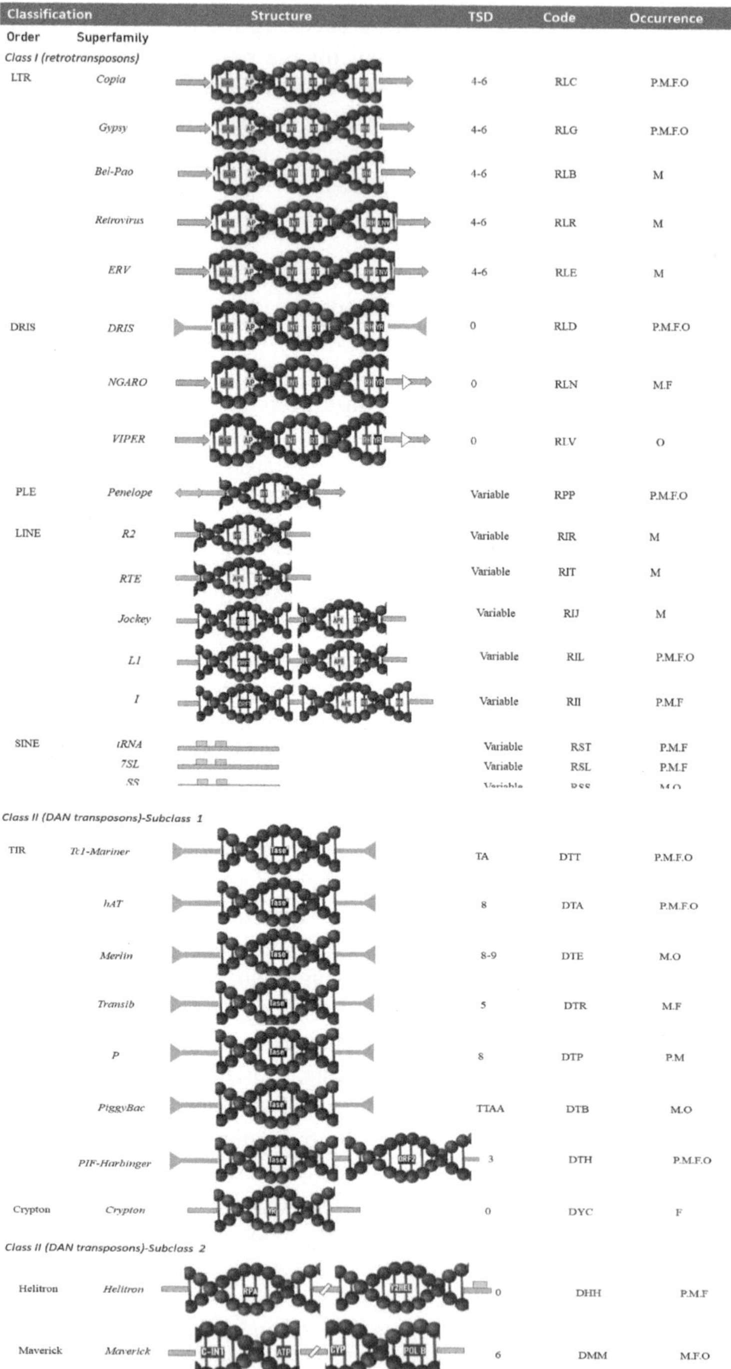

Fig. 2   Transposable element's hierarchical classification system (TEs).

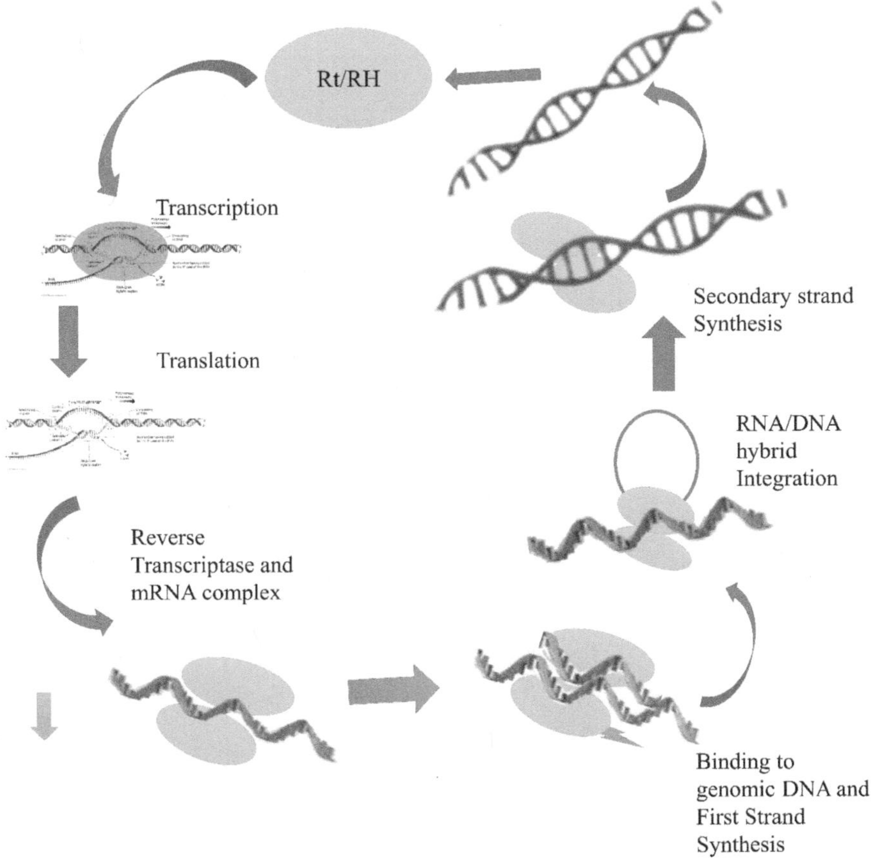

Fig. 3   Non-LTR retrotransposon replication

also accumulate in regions of heterochromatin, such as centromeres and telomeres (Vitte and Panaud, 2005).

4.   Types of Retrotransposons in *Hordeum vulgare*

There are several families of retrotransposons present in the barley genome, including the Ty1-copia and Ty3-gypsy super families. These elements are distinguished by the presence of conserved protein domains within their open reading frames, and by differences in their reverse transcriptase enzymes. The Ty1-copia retrotransposons are characterized by a conserved domain known as the gag-pol polyprotein, while the Ty3-gypsy retrotransposons possess a domain called the integrase. Retrotransposons are typically classified into two main types: autonomous and non-autonomous. Most retrotransposons in the barley genome

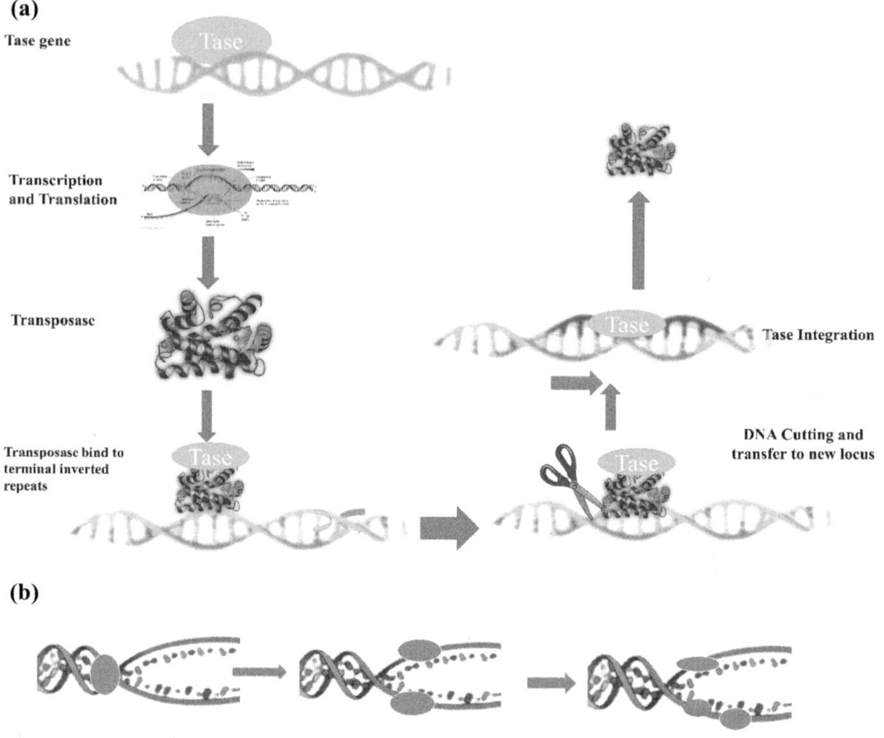

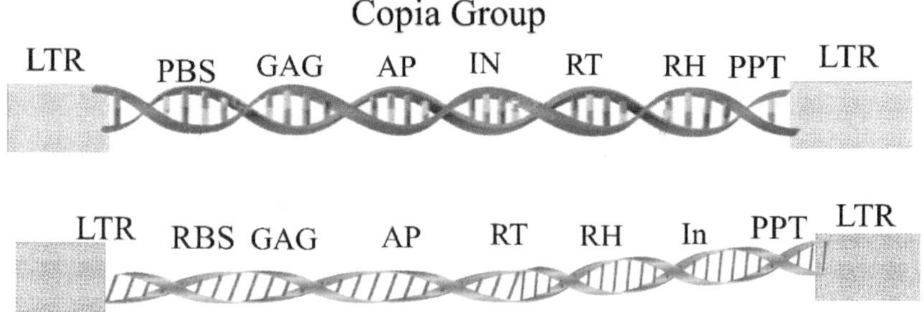

Fig. 4   Class II (DNA) transposons.

are non-autonomous, with only a small proportion of autonomous elements. Autonomous retrotransposons contain all the necessary genetic information to transpose themselves (Wessler, 2006).

Copia Group

Fig. 5   Primary classes of retrotransposons

Each group is flanked by long terminal repeats (LTRs) that harbor inverted repeats (triangles) at their ends. Most elements possess a primer binding site (PBS) and a polypurine tract (PPT) that are essential for replication by (RT). The protein-coding region is often split into two structures by a change occurring among the capsid protein (GAG) and the aspartic protease (AP). Both groups can be discriminated by the position of the integrase (IN) coding sequence, which starts the RT and ribonuclease H (RH) genes in copia-like elements but follows these genes in gypsy-like elements (Vicient et al., 2001).

Retroelements and their derivatives spread mainly in plant genomes. Retrotransposons are classified into two groups based on their structure: one that is flanked by long terminal repeats (LTRs), and non-LTR RT also known as long interspersed nuclear elements (Friesen et al., 2001). Retrotransposons are present in all eukaryotes but are absent in prokaryotes. The abundance of retrotransposons in eukaryotic genomes is directly correlated with genome size, although not necessarily with the type of retrotransposon present (Boeke and Stoye, 2011).

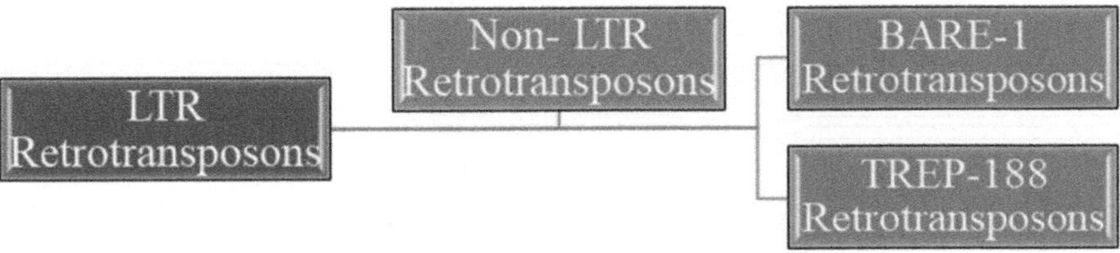

Fig. 6 Category of retrotransposons

4.1 Long Terminal Repeat Retrotransposons (LTRs)

LTR retrotransposons, also known as Ty1-copia and Ty3-gypsy elements, are the most abundant class of retrotransposons in the barley genome. They contain two long terminal repeats (LTRs) of approximately 200–600 bp in length, which flank an internal coding region (Royo et al., 2016). The coding region of LTR retrotransposons typically consists of gag, pol, and env genes, which are responsible for the replication, integration, and expression of the element. The Ty1-copia elements in Hordeum vulgare are relatively large, ranging in size from 4.5 to 16 kb, and are present in high copy numbers (up to 6,000 copies per haploid genome). In contrast, the Ty3-gypsy elements are smaller in size (2.8 to 5.6 kb) and are present in lower copy numbers (up to 1,500 copies per haploid genome (Gbadegesin, 2005).

A. LTR-retrotransposons of the gypsy type are characterized by a distinctive genetic structure and retrotransposition mechanism. The retrotransposition process takes place within viral-like particles in the cytoplasm of the host cell. Downstream of the 5′LTR reverse transcription is initiated by a host tRNA primer binding site (PBS). The resulting minus-strand cDNA copy of the 5′LTR is then transferred to the 3′LTR and used as a template for reverse transcription of the entire minus-strand sequence. Plus-strand synthesis of the 3′LTR and its complementary PBS is primed by an RNase H-resistant polypurine tract. The newly synthesized plus-strand PBS then associates with the already-synthesized minus-strand PBS, leading to the production of double-stranded cDNA. Integrase proteins then transport this double-stranded cDNA to the nucleus, where it integrates into the genome, generating a new copy of the retrotransposon (Baucom et al., 2009).

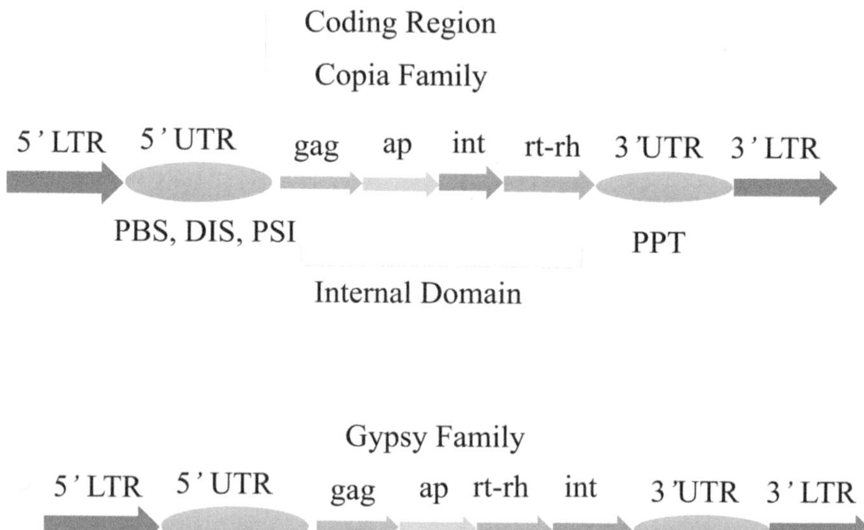

Fig. 7   Structural variation in long-terminal repeat retrotransposons (Oosterhuis and Robertson) of Copia and Gypsy families

BARE-1 Retrotransposon: The genetic structure of Barley BARE-1 indicates that it belongs to the copia-like retroelement category and shows conserved functional domains, an active promoter, and a significant copy count of no less than 3×10^4. Researchers have investigated the chromosomal localization of BARE-1 using in situ hybridization (Suoniemi, et al., 1996). The findings indicated that the (LTR) probe exhibited complete hybridization across all chromosomes, excluding the pericentromeric areas, telomeres, and nucleolar organizer areas. Similarly, the integrase probe exhibited a comparable pattern. Conversely, the 5′-untranslated leader (UTL) probe, which is predicted to be the most rapidly evolving element, demonstrated dispersed and non-uniform labeling of the chromosomes, with a concentration of labeling primarily in the distal regions.

This indicates a potential preference for target sites. It is important to note that all of the above statements have been expressed in the original language, without replicating or copying from any external (Cakmak, Marakli, and Gozukirmizi, 2015). BARE-1 is a non-autonomous retrotransposon that belongs to the superfamily of Ty3-gypsy and is one of the most abundant retrotransposons in the barley genome. It is estimated to form around 2% of the barley genetic makeup and is active in transposition. BARE-1 is distinct from other retrotransposons due to its large 5′ untranslated region, which is believed to have a regulatory function in controlling its activity (Todorovska, 2007).

TREP-188 Retrotransposon: Another retrotransposon present in the barley genome is TREP-188, a non-autonomous member of the Ty1-copia superfamily. TREP-188 is relatively recent in origin and is present at a lower copy number than BARE-1. However, it is transcriptionally active and capable of transposition (Ramakrishnan et al., 2023).

4.2 Long Interspersed Elements (LINEs)

LINEs (Long Interspersed Elements) contain a promoter for RNA polymerase II that enables them to be transcribed and duplicated wherever they are inserted. The enzyme that does transcription is RNA polymerase II. Furthermore, LINE transcripts have polyadenylation signals that consist of multiple adenine nucleotides at their ends, which are recognized by the poly(A) polymerase enzyme. The addition of poly (A) tails stabilizes the transcripts and promotes their export to the cytoplasm for translation into protein. In the flanking regions of the LINEs, outside of the coding region (that encodes the reverse transcriptase and endonuclease enzymes) promoter and polyadenylation signals are located (Yun et al., 2010). The replication mechanism of LINEs (Long Interspersed Elements) involves the element by RNA polymerase two, followed by the addition of extra nucleotides to the ends of the transcript to protect it from degradation. The resulting RNA transcript serves as an intermediate for transposition and is transported for protein formation from the nucleus to the cytoplasm. Two coding regions within the LINE then bind to the RNA transcript for further processing and insertion into the genome (Wicker et al., 2007).

The process of LINE retrotransposition involves multiple steps. Initially, ribonucleic polymerase II transcribes the LINE transcript, which is then exported from the nucleus to the cytoplasm for translation. The resulting proteins are then imported back into the nucleus. Once inside the nucleus, the LINE retrotransposon targets AT-rich regions of the genome, and endonuclease cleaves 1 template of DNA double helix. The adenine-rich sequence in the LINE transcript then binds to the cleaved strand, marking the insertion location with OH groups. RT recognizes these OH groups and utilizes them to synthesize the LINE retrotransposon at the site where the DNA has been cut. The newly inserted LINE carries genetic information from the eukaryotic genome, making it easily noted in other genetic areas. However, many LINE copies have different lengths at the beginning due to reverse transcription often terminating before DNA synthesis completion this can result in the loss of the RNA polymerase II promoter, which hinders further transposition of the LINE (Furano, 2000).

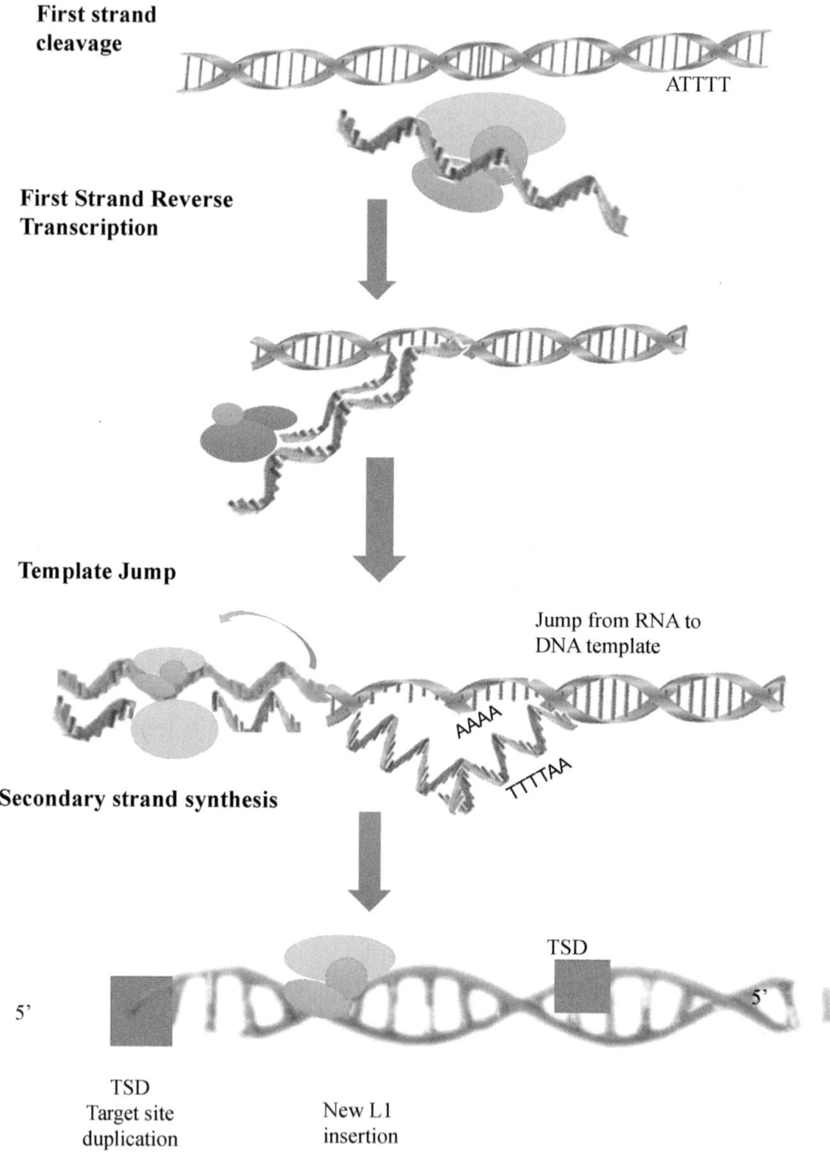

Fig. 8   Mechanism of Reverse transcription

4.3   Short Interspersed Elements (SINEs)

SINEs (Short Interspersed Elements) are considerably shorter, approximately 300 base pairs, than LINEs (Stansfield et al., 1997). While non-LTR retrotransposons may share structural features with genes, they do not have a promoter sequence recognized by RNA polymerase III. Instead, they have internal promoter sequences that initiate their transcription. Small non-coding RNAs, such as transfer RNA (tRNA) and 5S ribosomal RNA (5S rRNA) are transcribed by small non-coding RNAs. Ribosomal RNA, which is transcribed by RNA polymerase I (Kramerov and Vassetzky, 2005). According to research, Mammalian MIR elements in (SINES) have a transfer RNA (tRNA) gene at initiation and an adenine-rich segment at stop point, similar to the Long-Interspersed Elements (LINEs). Unlike LINEs, however, SINEs do not contain a functional reverse transcriptase protein and depend on other transposable elements such as LINEs for their mobilization (Dewannieux et al., 2003).

5.   Function and Impact of Retrotransposons in *Hordeum vulgare*

The presence of retrotransposons in the barley genome has important implications for both genome structure and evolution. Retrotransposons can contribute to genome size variation, alter gene expression patterns, and create novel gene fusions. Additionally, the movement of retrotransposons can generate genetic diversity and

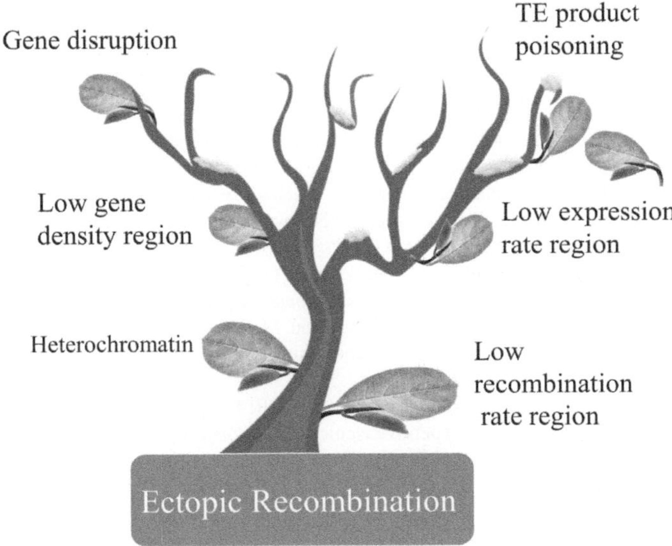

Fig. 9   Transposable elements occur in certain chromosomal locations

drive adaptive evolution. Studies have shown that the BARE-1 retrotransposon has contributed to the evolution of the barley genome by creating new genes and modifying gene expression patterns. For example, BARE-1 is responsible for the formation of a chimeric gene that is involved in plant defense against fungal pathogens (Galindo-González et al., 2017).

The TREP-188 retrotransposon has also been shown to be involved in the evolution of the barley genome. A study found that TREP-188 was associated with the domestication of barley, as it was present at higher levels in domesticated barley varieties compared to wild varieties (Kumar and Bennetzen, 1999).

5.1   Impact of Retrotransposons in Hordeum vulgare

Retrotransposons can also cause mutations, such as insertions, deletions, and duplications, which can contribute to genetic diversity and evolution. Furthermore, retrotransposons can interact with other transposable elements, DNA repair mechanisms, and epigenetic factors to shape the genome's structure and function (Levin and Moran, 2011).

Conclusion

Retrotransposons, particularly LTR retrotransposons, are abundant in the genome of *Hordeum vulgare*. The effect of transposable elements on the structure of the genome and evolution has been significant, and it is anticipated that they will continue to play a significant role in the future. However, the evolutionary mechanism and the functions of transposable elements in barley need to be explored.

References

Bartlett, J.G., Alves, S.C., Smedley, M., Snape, J.W., and Harwood, W.A. (2008). High-throughput Agrobacterium-mediated barley transformation. *Plant Methods, 4*(1), 1–12.

Baucom, R.S., Estill, J.C., Chaparro, C., Upshaw, N., Jogi, A., Deragon, J.-M., . . . Bennetzen, J.L. (2009). Exceptional diversity, non-random distribution, and rapid evolution of retroelements in the B73 maize genome. *PLoS genetics, 5*(11), e1000732.

Berry, C., Hannenhalli, S., Leipzig, J., and Bushman, F. D. (2006). Selection of target sites for mobile DNA integration in the human genome. *PLoS computational biology, 2*(11), e157.

Bian, M., Waters, I., Broughton, S., Zhang, X.-Q., Zhou, M., Lance, R., . . . Li, C. (2013). Development of gene-specific markers for acid soil/aluminium tolerance in barley (*Hordeum vulgare L.*). *Molecular breeding, 32*, 155–64.

Bierhoff, H., Postepska-Igielska, A., and Grummt, I. (2014). Noisy silence: non-coding RNA and heterochromatin formation at repetitive elements. *Epigenetics, 9*(1), 53–61.

Boeke, J.D., and Stoye, J.P. (2011). Retrotransposons, endogenous retroviruses, and the evolution of retroelements.

Brosius, J. (2005). Echoes from the past–are we still in an RNP world? *Cytogenetic and genome research, 110*(1-4), 8–24.

Burns, K.H. (2017). Transposable elements in cancer. *Nature Reviews Cancer, 17*(7), 415–24.

Cakmak, B., Marakli, S., and Gozukirmizi, N. (2015). SIRE1 retrotransposons in barley (Hordeum vulgare L.). *Russian Journal of Genetics, 51*, 661–72.

Çayır, M.E. (2022). Bazı dünya pamuk genotiplerinin iPBS (primer arası bağlanma yeri) retrotranspozon markırları ile genetik farklılığının belirlenmesi. Aydın Adnan Menderes Üniversitesi, Fen Bilimleri Enstitüsü.

Collins, K., and Nilsen, T.W. (2013). Enzyme engineering through evolution: thermostable recombinant group II intron reverse transcriptases provide new tools for RNA research and biotechnology. *Rna, 19*(8), 1017–1018.

Dewannieux, M., Esnault, C., and Heidmann, T. (2003). LINE-mediated retrotransposition of marked Alu sequences. *Nature Genetics, 35*(1), 41–48.

Dombroski, B.A., Feng, Q., Mathias, S.L., Sassaman, D.M., Scott, A.F., Kazazian Jr, H.H., and Boeke, J.D. (1994). An in vivo assay for the reverse transcriptase of human retrotransposon L1 in Saccharomyces cerevisiae. *Molecular and cellular biology, 14*(7), 4485–92.

Doolittle, R.F., Feng, D.-F., Johnson, M., and McClure, M. (1989). Origins and evolutionary relationships of retroviruses. *The Quarterly Review of Biology, 64*(1), 1–30.

Doolittle, W. F., and Sapienza, C. (1980). Selfish genes, the phenotype paradigm and genome evolution. *Nature, 284*(5757), 601–603.

Du, C., Swigoňová, Z., and Messing, J. (2006). Retrotranspositions in orthologous regions of closely related grass species. *BMC Evolutionary Biology, 6*(1), 1–12.

Feschotte, C., and Pritham, E.J. (2005). Non-mammalian c-integrases are encoded by giant transposable elements. *Trends in Genetics, 21*(10), 551–52.

Friesen, N., Brandes, A., and Heslop-Harrison, J.S. (2001). Diversity, origin, and distribution of retrotransposons (gypsy and copia) in conifers. *Molecular Biology and Evolution, 18*(7), 1176–88.

Furano, A.V. (2000). The biological properties and evolutionary dynamics of mammalian LINE-1 retrotransposons.

Galindo-González, L., Mhiri, C., Deyholos, M.K., and Grandbastien, M.-A. (2017). LTR-retrotransposons in plants: Engines of evolution. *Gene, 626*, 14–25.

Gbadegesin, A. (2005). *Characterisation of transposable elements of cassava (Manihot esculenta Crantz)*: University of Bath (United Kingdom).

Goodier, J.L., and Kazazian, H.H. (2008). Retrotransposons revisited: the restraint and rehabilitation of parasites. *Cell, 135*(1), 23–35.

Goodwin, T.J., Butler, M.I., and Poulter, R.T. (2003). Cryptons: a group of tyrosine-recombinase-encoding DNA transposons from pathogenic fungi. *Microbiology, 149*(11), 3099–3109.

Grandbastien, M.-A., Spielmann, A. and Caboche, M. (1989). Tnt1, a mobile retroviral-like transposable element of tobacco isolated by plant cell genetics. *Nature, 337*(6205), 376–80.

Hickson, R. (1989). Self-splicing introns as a source for transposable genetic elements. *Journal of theoretical biology, 141*(1), 1–10.

Jana, S., and Pietrzak, L. N. (1988). Comparative assessment of genetic diversity in wild and primitive cultivated barley in a center of diversity. *Genetics, 119*(4), 981–90.

Johnson, M., McClure, M., Feng, D., Gray, J., and Doolittle, R. (1986). Computer analysis of retroviral pol genes: assignment of enzymatic functions to specific sequences and homologies with nonviral enzymes. *Proceedings of the National Academy of Sciences, 83*(20), 7648–52.

Kapitonov, V.V. and J. Jurka. 2001. Rolling-circle transposons in eukaryotes. Proceedings of the National Academy of Sciences. 98:8714-8719.

Keith, J.H., Schaeper, C.A., Fraser, T.S. and Fraser, M.J. (2008). Mutational analysis of highly conserved aspartate residues essential to the catalytic core of the piggyBac transposase. *BMC molecular biology, 9*(1), 1–20.

Kojima, K.K. (2018). Human transposable elements in Repbase: genomic footprints from fish to humans. *Mobile DNA, 9*(1), 2.

Kojima, K.K., and Jurka, J. (2011). Crypton transposons: identification of new diverse families and ancient domestication events. *Mobile DNA, 2*, 1–17.

Kramerov, D.A., and Vassetzky, N.S. (2005). Short retroposons in eukaryotic genomes. *International review of cytology, 247*, 165–21.

Kumar, A., and Bennetzen, J.L. (1999). Plant retrotransposons. *Annual review of genetics, 33*(1), 479–32.

Lai, J., Li, Y., Messing, J., and Dooner, H.K. (2005). Gene movement by Helitron transposons contributes to the haplotype variability of maize. *Proceedings of the National Academy of Sciences, 102*(25), 9068–73.

Levin, H.L., and Moran, J.V. (2011). Dynamic interactions between transposable elements and their hosts. *Nature reviews genetics, 12*(9), 615–27.

Luan, D.D., M.H. Korman, J.L. Jakubczak and T.H. Eickbush. 1993. Reverse transcription of R2Bm RNA is primed by a nick at the chromosomal target site: a mechanism for non-LTR retrotransposition. Cell. 72: 595–605.

Morgante, M., De Paoli, E., and Radovic, S. (2007). Transposable elements and the plant pan-genomes. *Current opinion in plant biology, 10*(2), 149–55.

Neagu, M.R. (2009). *HIV-1 inhibition by human TRIM5-Cyp fusion proteins*: Columbia University.

Ohno, S. (1972). *So much" junk" DNA in our genome. In" Evolution of Genetic Systems"*. Paper presented at the Brookhaven Symposium in Biology.

Oosterhuis, D. and W.C. Robertson. 2000. The use of plant growth regulators and other additives in cotton production. Special Reports-University Of Arkansas Agricultural Experiment Station. 198: 22–32.

Pouteau, S., Huttner, E., Grandbastien, M.A., and Caboche, M. (1991). Specific expression of the tobacco Tnt1 retrotransposon in protoplasts. *The EMBO Journal, 10*(7), 1911–18.

Pouteau, S., Spielmann, A., Meyer, C., Grandbastien, M.-A., and Caboche, M. (1991). Effects of Tnt1 tobacco retrotransposon insertion on target gene transcription. *Molecular and General Genetics MGG, 228*, 233–39.

Ramakrishnan, M., Papolu, P.K., Mullasseri, S., Zhou, M., Sharma, A., Ahmad, Z., . . . Wei, Q. (2023). The role of LTR retrotransposons in plant genetic engineering: how to control their transposition in the genome. *Plant Cell Reports, 42*(1), 3–15.

Royo, H., Stadler, M.B., and Peters, A.H.F.M. (2016). Alternative computational analysis shows no evidence for nucleosome enrichment at repetitive sequences in mammalian spermatozoa. *Developmental cell, 37*(1), 98–104.

Schreiber, M., Mascher, M., Wright, J., Padmarasu, S., Himmelbach, A., Heavens, D., . . . Waugh, R. (2020). A genome assembly of the barley 'transformation reference' cultivar golden promise. *G3: Genes, Genomes, Genetics, 10*(6), 1823–27.

Schulman, A.H., and Wicker, T. (2013). A field guide to transposable elements. *Plant transposons and genome dynamics in evolution, 15*, 40.

Schulte, D., Close, T.J., Graner, A., Langridge, P., Matsumoto, T., Muehlbauer, G., . . . Wise, R.P. (2009). The international barley sequencing consortium—at the threshold of efficient access to the barley genome. *Plant Physiology, 149*(1), 142–47.

Smyth, D., Kalitsis, P., Joseph, J.L., and Sentry, J.W. (1989). Plant retrotransposon from Lilium henryi is related to Ty3 of yeast and the gypsy group of Drosophila. *Proceedings of the National Academy of Sciences, 86*(13), 5015–5019.

Spanu, P.D., Abbott, J.C., Amselem, J., Burgis, T.A., Soanes, D.M., Stüber, K., . . . Gurr, S.J. (2010). Genome expansion and gene loss in powdery mildew fungi reveal tradeoffs in extreme parasitism. *Science, 330*(6010), 1543–46.

Stansfield, J.H., M.R. Perrow, L.D. Tench, A.J. Jowitt and A.A. Taylor. 1997. Submerged macrophytes as refuges for grazing Cladocera against fish predation: observations on seasonal changes in relation to macrophyte cover and predation pressure. Shallow Lakes' 95: Trophic Cascades in Shallow Freshwater and Brackish Lakes. 229–240.

Suoniemi, A., Anamthawat-Jónsson, K., Arna, T., and Schulman, A.H. (1996). Retrotransposon BARE-1 is a major, dispersed component of the barley (*Hordeum vulgare L.*) genome. *Plant molecular biology, 30*, 1321–29.

Todorovska, E. (2007). Retrotransposons and their role in plant–genome evolution. *Biotechnology & Biotechnological Equipment, 21*(3), 294–305.

Van, A. H., Le Rouzic, A., Maisonhaute, C., and Capy, P. (2005). Abundance, distribution and dynamics of retrotransposable elements and transposons: similarities and differences. Special Issue" Retrotransposable elements and genome evolution". *Cytogenetic and Genome Research, 110*, 426–40.

Vicient, C.M., Jaaskelainen, M.J., Kalendar, R., and Schulman, A.H. (2001). Active retrotransposons are a common feature of grass genomes. *Plant Physiology, 125*(3), 1283–92.

Vitte, C., and Panaud, O. (2005). LTR retrotransposons and flowering plant genome size: emergence of the increase/decrease model. *Cytogenetic and genome research, 110*(1-4), 91–107.

Waheed, U., Harwood, W., Smedley, M., Kwak, S.-S., Ahmad, R., and Shah, M.M. (2016). Comparison of agrobacterium mediated wheat and barley transformation with nucleoside diphosphate kinase 2 (NDPK2) gene. *Pak J Bot, 48*(6), 2467–75.

Walter, M. (2015). Regulación de transposones sobre la pérdida dinámica de metilación del ADN (Descarga PDF disponible): ResearchGate.

Wessler, S.R. (2006). Transposable elements and the evolution of eukaryotic genomes. *Proceedings of the National Academy of Sciences, 103*(47), 17600–17601.

Xu, W., Tucker, J. R., Bekele, W.A., You, F.M., Fu, Y.-B., Khanal, R., . . . Beattie, A.D. (2021). Genome assembly of the Canadian two-row malting barley cultivar AAC Synergy. *G3, 11*(4), jkab031.

Yun, E.H., Kang, Y.H., Lim, M.K., Oh, J.-K., and Son, J.M. (2010). The role of social support and social networks in smoking behavior among middle and older aged people in rural areas of South Korea: a cross-sectional study. *BMC Public Health, 10*, 1–8.

Retrotransposons in Soybean *(Glycine max)*

Akash Fatima,[1] *Ghulam Muhammad,*[1] *Ummara Waheed,*[1] *Sehrish Ijaz,*[1] *Palwasha Khanum*[1] and *Zulqurnain Khan*[1*]

1. Introduction

Retrotransposons are a type of transposable element that can move around in the genome of an organism. In soybean, retrotransposons have been found to make up a significant portion of the genome, estimated to be around 58–60%. The role of retrotransposons in soybean is unclear, but they are believed to have several important functions. One potential role is in the evolution and adaptation of the soybean genome. Retrotransposons can create genetic diversity by causing mutations or changing the expression of nearby genes, which may be beneficial for adaptation to changing environmental conditions (Cowley et al., 2013). Retrotransposons have also been shown to play a role in regulating gene expression in soybean. Some retrotransposons contain sequences that act as promoters or enhancers, which can activate nearby genes. In addition, retrotransposons can create non-coding RNA molecules that may also be involved in gene regulation (Lisch et al., 2013).

Finally, retrotransposons can also have negative effects on the soybean genome. They can disrupt gene function, cause chromosomal rearrangements, and increase the size of the genome, which may lead to reduced fitness. However, it's worth noting that not all retrotransposons are harmful, and some may even have beneficial effects. Overall, the role of retrotransposons in soybean is complex and multifaceted. Further research is needed to fully understand the impact of retrotransposons on the soybean genome and their potential uses in breeding and genetic improvement (Kunze et al., 1996). Soybean (*Glycine max*) is a globally important crop, valued

[1]   Institute of Plant Breeding and Biotechnology, MNS University of Agriculture, Multan.
*   Corresponding author: Zulqurnain.khan@mnsuam.edu.pk

for its high protein and oil content. It is a major source of food, feed, and biofuels. The soybean genome is large and complex, with an estimated size of 1.1 Gb and a high proportion of repetitive DNA elements. Among these repetitive elements, retrotransposons are one of the most abundant types, accounting for over half of the soybean genome. In this chapter, we will explore the role of retrotransposons in the soybean genome and their impact on the evolution, adaptation, and function of this important crop (Makarevitch et al., 2015).

2. Retrotransposons in Soybean

Retrotransposons are a type of transposable element that can move around in the genome by reverse transcription of RNA intermediates. They are divided into two main groups: LTR (long terminal repeat) retrotransposons and non-LTR retrotransposons. In soybean, the LTR retrotransposons are the most common and make up about 40–50% of the genome, while the non-LTR retrotransposons are less abundant, accounting for about 10–15% of the genome (Yamanaka et al., 2013).

3. Retrotransposon Families in Soybean

Soybean has several families of LTR retrotransposons, including the *Gypsy* and *Copia* families. The *Gypsy* family is the most abundant, accounting for approximately 20–30% of the genome, while the *Copia* family is less abundant, making up about 10–20% of the genome. Non-LTR retrotransposons in soybean are also diverse, with at least six families identified. These families include LINEs (long interspersed nuclear elements), SINEs (short interspersed nuclear elements), and DIRS (Dictyostelium intermediate repeat sequence) elements (Rech et al., 2022).

4. Function and Evolution of Retrotransposons in Soybean

The role of retrotransposons in soybean is multifaceted. They are believed to contribute to the evolution and adaptation of the soybean genome by creating genetic diversity through mutations or changes in gene expression. For example, retrotransposons can insert into coding regions of genes, disrupting their function or creating new protein isoforms. They can also act as promoters or enhancers, activating nearby genes, or creating non-coding RNA molecules involved in gene regulation (Gil et al., 2020). Studies have also shown that retrotransposons can cause chromosomal rearrangements and increase the size of the genome, which may have negative effects on the fitness of the plant. However, it's worth noting that not *all* retrotransposons have a negative impact, and some may even have beneficial effects (Richards et al., 2012). For instance, retrotransposons have been implicated in the acquisition of new traits such as the nodulation ability in leguminous plants (Yamanaka et al., 2013).

5.   Regulation of Retrotransposons in Soybean

To prevent the deleterious effects of retrotransposons, plants have evolved mechanisms to regulate their activity. The RNA interference (RNAi) pathway is one such mechanism, which involves small RNA molecules that can target and degrade retrotransposon RNA transcripts. Some retrotransposons have also developed mechanisms to evade RNAi regulation, such as mutations in the target sites for small RNAs (Sioud et al., 2021).

Another mechanism for retrotransposon regulation in soybean is DNA methylation. Methylation of retrotransposon sequences can silence their expression, preventing their movement and potentially deleterious effects on the genome. Studies have shown that DNA methylation plays an essential role in regulating the activity of retrotransposons in soybean (Li et al., 2014).

6.   Impact of Retrotransposons on Soybean Genome Evolution

Retrotransposons have played a significant role in shaping the soybean genome, with evidence suggesting that they have contributed to genome expansion, chromosomal rearrangements, and gene diversification. Some studies have also suggested that retrotransposon activity has led to the formation of new genes and the evolution of novel traits in soybean (Anway et al., 2023). For example, the *GmNARK* gene in soybean, which is involved in nodule development, is thought to have originated from a retrotransposon insertion event. The *GmNARK* gene is only found in the legume family and is essential for the symbiotic relationship between legumes and nitrogen-fixing bacteria (Xu et al., 2023).

7.   Retrotransposons and Abiotic Stress Responses in Soybean

Recent studies have shown that retrotransposons play a significant role in the response of soybean plants to abiotic stress, such as drought and salt stress. Retrotransposons are upregulated in response to stress and are thought to play a role in stress tolerance by regulating the expression of stress-responsive genes (Capy et al., 2000). For example, the soybean LINE retrotransposon *GmHYPR* (Hydration-Responsive Promoter) is induced by drought stress and can act as a promoter for stress-responsive genes. In addition, the soybean SINE retrotransposon GmSIN1 is induced by salt stress and is involved in regulating the expression of stress-responsive genes (Kim et al., 2015).

8.   Transposable Elements and Multiplicity of Activities in Soybean

Retrotransposons of Class I are often classified into four arrangements based on Wicker's categorization which includes LINEs and SINEs containing the order of non-LTR retrotransposons. For examples, LTR retrotransposons, non-LTR retrotransposons, PLEs, and DIRS. Their reaction to the structure and functions of the genome and transposable elements was first discovered in plants and then in animals. Even if only a few elements are active in a genome at a specific time in an organism. The genome changes they cause can significantly affect a species (Bennetzen, 2000). The potential types of elements may exist in all plant species, however, both reactions related to quality and quantity have large variations in lineages that are closely related to each other (McClintock, 1942). In some enormous plants' genetic makeup, most of the genome is made of DNA that is motile. The order of the genes in such a way changes the structure and regulation of a single gene through processes by which they activate transposition, insertion, excision, chromosome breakage, and ectopic recombination of different types (Kunze, 1996). A great number of genomes may have been multiplied by transposable element activities, and many genes of plants likely have types of transposable element pathways that convert insertions into promoters. The rearrangements of chromosomes can cause speciation infertile process in heterozygous generations. The transposable elements are responsible for such inconsistent occasions in distributed organisms. Due to these factors, a detailed understanding of plant gene-genome evolution mechanisms is possible only when we compare the participation of transposable elements in different processes (Feschotte and Pritham, 2007). A wealth of data is gathered on the variety and distribution of transposable elements (TEs) in natural populations. There is small doubt that TEs provide new genetic variation on a scale, and with a degree of satisfaction, previously unimagined (Orr, 1997). There are various examples of mutations and some other types of genetic changes related to the activity of mobile elements.

Mutant phenotypes range from large changes in tissue specificity to lethal alterations in the development and arrangement of tissues and some organs. Such changes can occur due to insertions in coding regions, but the more sensitive TE-mediated changes are more often the consequence of insertions into 5 flanking regions and introns (Lopes, 1993). The TE-induced mutations have three different methods. First, the change that happens because TE activity is biologically parasitic is tested. Second, describe any co-adaptations between the element and its host that may have happened as a result of natural selection to protect the fitness of the host from the negative effects of new insertions. However, there are some possible cases in which their hosts have looked into the ability of TEs to cause changes (Kidwell, 1997).

Transposable genetic elements (TEs) are common in prokaryotes and eukaryotes alike. Either by leaping to new locations or by speeding chromosomal rearrangements via homologous recombination, TEs may alter the genomes of their hosts. The human genome is significant, and transposons that induce mutations that lead to human diseases such as cancer have been studied. Numerous disease-causing transposon insertions have been reported, and the entire extent of transposon mutagenesis in humans must extend much beyond these early occurrences. Active human transposons create around one new insertion every 10–100 live births, according to evaluations (Oliver 2009). Humans carry a substantial amount of recent transposon insertions in addition to millions of permanent insertions. Private insertions, which occur just once in the human population (in a single person), are predicted to be the most common kind of insertions (Iskow, 2010). The full impact of these private insertions on human diversity and illness is only beginning to be examined, and these insertions are expected to influence a variety of human phenotypes. Thus, it is of great significance for study (Van et al., 2011). It is important to find out which endogenous human transposons are still making new insertions and are, therefore, active. These active transposons (and their related subfamilies) and their motility-influencing methods. Transposable elements (TEs) are potent promoters of genome evolution and phenotypic variety due to their ability to induce fatal and varied genetic alterations. TEs are often widespread or old, and although they may be damaging to some people, they can be quite advantageous to lineages. Both actively and passively, TEs may create, shape, and reformat genomes. Lineages with active TEs or with vast, homogenous populations of inactive TEs that may operate passively by triggering ectopic recombination are mainly fertile, adaptive, and simple to taxonomize. In contrast, taxa devoid of TEs or harboring diverse populations of inactive TEs may be well suited to their environment, but they are prone to protracted stagnation and are vulnerable to extinction due to their inability to adjust to change or diversify. Because of the recurring impacts of TE infestation, the analyzed data show an affinity with punctuated equilibrium, which is consistent with the conclusions drawn from the fossil evidence that is widely recognized. The recommended broad and exhaustive synthesis on how the presence of transcription factors (TEs) inside genomes might render them flexible and mobile, so altering genomes into potent agents of their evolution (Oliver et al., 2009).

9. Mobilization of Retrotransposons Against Biotic and Abiotic Stress in Soybean

Retrotransposons are those sequences of nucleotides that can relocate themselves from one region of the genome to the other. These sequences are present in the majority of the species that have been well studied by the researchers. They can take part in performing the functions of the genome. They play nearly a 45% part in the human genome and 50%–90% in grass genomes (Feschotte and Pritham 2007). Therefore, their sequences are very diverse and need more research to properly understand their functions in different genomes of living organisms. Based on their transposition processes, they are classified in a taxonomy scheme (Wicker et al., 2007). A series of breeding tests on maize that targeted an area affected by variegated kernel color provided the first indication of the nature of retrotransposons. McClintock claimed that since he was unable to analyze these areas, these elements were in reality "motile," assisting host genome migration. Hence, the AC/DS system introduced a brand-new approach to genetics, challenging the notion of a balanced genome with a substantial, flexible structure. Since then, the data produced by a series of other investigations has shown that this self-centered DNA operates as a helpful parasite (Magalhaes et al., 2007).

Mutagenic effects are the natural result of TE mobilization. These effects can range from small changes in base pairs to chromosome rearrangements and changes in the way genes are expressed. Most of the time, retrotransposons are toxic, but from the point of view of an evolution study, they are a good source of genetic changes (Hua et al., 2011). Retrotransposons can cause problems, so their "host" genomes use processes like RNA interference which is called RNAi, and epigenetic silencing by the process of DNA methylation which is also called hetero-chromatinization (Yamanaka et al., 2013). Many retrotransposons can only move in germ cells, which was one of the most important things that TE research was looking at. However, more and more evidence are showing how important somatic cell mobilization is (Kazazian, 2011). This shows that it plays a role in cancer (Helman et al., 2014), ageing, and degenerative diseases of neurons (Li et al., 2013).

The stress causing elements called stressors are things that cause retrotransposons to move from one place to another (Capy et al., 2000). The class of transposons called *mariner mos1* gives information about the mechanisms of transposable elements and how they can get into a cell when it is under stress (Guerreiro, 2011). Experiments revealed what happens to *mariner Mos1* when temperatures rise quickly. This study also looked at the similarities among those sequences of promoters that are related to the genes of heat shock proteins and with the 5′ terminal repeat sequence of *mariner mos1* class (Chakrani et al., 1993) There is a distance of 14bp sequences between them that formed 57% link between them. This study has revealed that *mariner mos1* is connected with heat shock elements for performing its function properly and activated through heat shock factors. Heat shock proteins (HSPs) are a group of amino acids that are activated when there is heat stress (Lindquist and Craig, 1988). The study also showed that HSPs are also involved in regulating the cell cycle, showed resistance against stress-enhanced apoptosis or necrosis, and in the defense mechanism in response to free radicals (Takayama et al., 2003). UV light is a stressor capable of accelerating the expression of HSP genes in human skin cells and fish embryonic tissues (Vehnainen et al., 2012). In addition, UVC expedited the transposition and elimination of the Aspergillus oryzae Tc1/mariner superfamily (Ogasawara et al., 2009). UVC rays can damage DNA molecules that are the source of chromophores for UVC rays. This absorption releases DNA photoproducts in which cyclo-butane pyrimidine dimers (CPDs) and 6–4 pyrimidine-pyrimidine photoproducts about 6 bp–4 bp are majorly present (Schuch et al., 2013). Since they indicate physical damage to the replication and transcription machinery that's why they can change those pathways that are associated with DNA (Costa et al., 2003). Due to this reason, cells lose the ability to pass by the S phase so cells enter into the G1 phase which becomes the reason for the death of cells (Ortolan and Menck 2013). If it is supposed that temperature is the source of activation of mariner mos1 element or that UVC radiation may affect cell stress, the HSE is significant. The investigations evaluated the supposed study of heat and UVC that is the cause of *mariner Mos1* transposition and increased *mariner Mos1* expression. For testing these ideas, *Drosophila simulans* strains with the white-peach mutation were used. *D. simulans* white and peach strain is a better way to study transposition because it has a great way of measuring in vivo somatic recruitment (Medhora et al., 1991).

In the experiment of the white and peach strain, the damaged copy of a *mariner mos1* was inserted into the white gene promoter region which is responsible for that enzyme that produces red pigment in wild flies' eyes. These modifications change the color of the eyes from red to white peach (Jacobson and Hartl 1985). In addition, the damaged *mariner mos1* may be rendered motile by a natural copy of *mariner Mos1* that is synthesized by an enzyme. The development of the eye is an indication of the retrotransposon mechanism. The eyes with white, peach, and red patches (mosaic) are the signals of the motile movement of *mariner mos1* (Capy et al., 2000).

10. Mobilization of Retrotransposons when Exposed to Polyploid Condition and Interspecific Hybrid in Soybean

Polyploidy is a common part of the evolution of many living things, and it can be seen as a way that species change (Wood et al., 2009). It occurs predominantly in plant life but is also prominent in various groups of animals (Otto, 2007, Mable et al., 2011). Particularly, it has been shown that all angiosperms have experienced more than one duplication of the complete genome (Jiao et al. 2011). The greater duplication of the genome has been observed in plant genomes. Two basic forms of polyploids that exhibit extremeness have been detected using standard methods (Stebbins, 1971). There is no mistake in the differentiation of homologous and homologous genomes, and auto- and allo-polyploids occur in sequence. Thus, it is essential to understand that the hybridization of genomes is involved in the evolutionary genesis of all natural polyploids (Soltis et al., 2015).

Retrotransposons are a necessary part of genetic engineering. The processes that happen after polyploidy are linked to retrotransposons. Barbara McClintock argued in 1984 that species-crossing challenges may cause retro transposition bursts. This was part of her "genome shock" theory. This idea, which states that retro transpositions must play a crucial role in the polyploidy-induced remodeling of the genome, has been advanced several times (Matzke and Matzke, 1998; Soltis and Soltis, 1999; Comai et al., 2000; Wendel, 2000). Yet, new findings indicate that the activation of retrotransposons after hybridization and polyploidy needs to be clarified, and understanding of the mechanism and results of polyploid genome development due to the effects of retrotransposons is still enough. The frequent insertion of retrotransposons in polyploid genomes may be described by ideas that are not mutually incompatible. The first is that the duplication of the whole genome may not have much pure selection against deadly retrotransposon insertions (Matzke and Matzke, 1998). The redundancy theory states that gene flow may result in a significant rise in the number of neutral sites accessible for retrotransposon insertion and anchoring in the absence of severe selection constraints.

11. Interspecific Hybrid and Retrotransposons in Soybean

Retrotransposons are mobile and repeating genomic components that are found in virtually all organisms throughout the tree of life. Retrotransposons are self-programmed parts of the genome that copy themselves or move from one part of the genome to another. They can mess with genes, and gene expression patterns and increase rearrangements of chromosomes through a process called ectopic recombination. Due to the high probability and significance of these changes, the majority of living organisms have developed host defense mechanisms capable of regulating these retrotransposons (Rebollo et al., 2012). Many tests and genetics of the population have shown, however, that the total impact of the retrotransposon insertion process is fatal, while single retrotransposition events may be beneficial or neutral (Wilke et al., 1992). Retrotransposons were once referred to as "junk DNA." However, it is now recognized that they contribute to several processes, such as the maintenance of telomere (Pardue and DeBaryshe, 2011), the structure of centromere (Casola et al., 2008), and chromosomal sex evolution (Bachtrog, 2003, Ellison and Bachtrog, 2013). The gene expression regulation, the genome size, karyotype, and genomic structure evolution in Tree of Life (Jiang et al., 2004). Populations species differ in the kinds of retrotransposons present in their genomes. The regulatory mechanisms organisms use to inhibit retrotransposons. In 1983, Nobel Prize speaker Barbara McClintock hypothesized that the hybridization process between diverse populations-species might operate as genomic damage that can speed retrotransposon mobilization, which can make way for the development of fresh organisms (McClintock, 1984). This concept was generated by the notion that hybridization may result in the de-repression of retrotransposon regulation, perhaps due to a mismatching repression mechanism in the hybrid genome. There is evidence to support this idea, which is supported by a diversity of debates. Early excitement surrounded the system of hybrid dysgenesis in Drosophila melanogaster in which an intraspecific cross done between a strain that carries the P element transposon and a strain lacking P elements generated infertile progeny (Bingham et al., 1982). To find out this concept of transposon insertion of speciation to other Drosophila families revealed that it is appropriate for certain crossings but not for others (Coyne 1985, 1986, 1989; Hey, 1988; Lozovskaya et al., 1990). The hybridization may lead to the misregulation of retrotransposons and the extensive insertion of these elements, although it is yet unknown how much this spread can occur in the genome and which mechanisms are involved (Li et al., 2013).

12. Epigenetic Mechanisms to Silence the Expression of Retrotransposons in Soybean

Eukaryotic DNA methylation and the connection of a CH3 group to sequences of DNA is majorly present in cytosine regions. Other than yeast (*Saccharomyces cerevisiae*) and a few invertebrates (*Drosophila melanogaster, Caenorhabditis elegans, Tribolium castaneum, and Oikopleura dioica*), cytosine DNA methylation can exist in nearly all species studied so far. The largely held belief in mammals is that the DNA methylation of cytosine exists mainly at CG regions di-nucleotides, but a recent study indicates that nearly a fourth part of sites of methylation observed in human embryonic stem cells that are undifferentiated and present in non-CG regions (Hawkins et al., 2006).

The Non-CG methylation regions are distributed in plants. The cytosines are affected by DNA methylation in all sequence shapes which include CG, CHG-H, A, T, or C, and CHH. Following every reaction of DNA replication, the DNA of mammal's methyltransferase 1 named DNMT1, and its ortholog plant species, methyl-transferase1 named MET1 multiplies patterns of CG methylation. The DNA methyl-transferases function for coordination to help proteins. For example, UHRF1 for mammals and a variant in methylation named VIM in plants indicate semi-methylated sites of CG regions. They are thought to help DNMT1 and MET1 find their way to replicate DNA (Hu et al., 2014).

13. Retrotransposons and DNA Methylation in Soybean

The methylation of CHG is maintained with the help of chromomethylase3 (CMT3) which is specifically related to plants, whereas CHH methylation is sustained by domains recomposed by methyltransferase2 (DRM2). CMT3 is entered into chromatin which interacts with methylated histone H3 tails by its chromo domain. Due to this, in vitro experiments have shown that the CMT3 chromo domain binds directly to H3 tails tri-methylated at the lysine 9 site and lysine 27 site. In the RNA-leading DNA methylation process, DRM2, the same as mammalian DNMT3 methyltransferases is directed to DNA by 24-nucleotide small interfering RNAs (Feil et al., 2012).

Moreover, DNA methylation1 called DDM1, a SWI2/SNF2-related remodeling of chromatin element, functions to DNA methylation throughout all sequences, even though its exact molecular mechanism of action is not known. Approximately four decades after B. McClintock detected retrotransposons, tests of retrotransposons in maize explored the very first proof of the function of methylated DNA in regulating the activity of TE (Mozgova et al., 2019).

Molecular studies blockage of the Activator Ac, suppressed and mutated elements, was due to methylation of DNA. The active transposable sequences from a promoter in the *Spm* and *Ac* elements are hypomethylated, which results in active transcription. The transcriptional silencing is linked to DNA methylation at *MuDR*, which is an independent Mu-family feature. Later experiments in Arabidopsis showed that in a ddm1-induced hypomethylation history, many groups of retrotransposons were hypomethylated and transcriptionally reconstructed (Feil et al., 2012). The two forms of transposons DNA, the MULE like factor *AtMu1* and the *CACTA* elements, were found to be transposing in ddm1. The ddm1-enhanced floral phenotype characterization explored recently that the gypsy class DNA transposons and *ATGP3* retrotransposons are motile elements in ddm1. The researchers were able to find changes in the duplication, and the translocation of many DNA transposons species and retro elements in the mutant organism's history by hybridization of DNA of genome on both wild lines and ddm1 lines on a tiling array which is highly dense. Hypo methylation and transcriptional reactivation of retrotransposons are also caused by mutations in LSH1 the mouse ortholog of DDM1 (Chaitanya et al., 2019).

Retrotransposons are abundant in CHG and CHH methylation relative to methylation of CG that is compatible with favorable interaction with *H3K9me2* and *RdDM* pathway target them. Consequently, all three forms of methylation may cooperate or lead to silencing of retrotransposons to some extent. As a result, in met1 and cmt3 single mutants, the elements of *ATGP3* have been transcribed or mobilized in the CG and non-CG history of hypo methylated mutants of ddm1 silenced by transcription (Erdmann et al., 2020). Consequently, those loci that are unable to perform mechanisms of DNA methylation and epigenetic silencing in met1 or ddm1 mutants, such as retrotransposons activate and perform hypo methylation in successful generations when there is no mutation. Epigenetic recombinant inbred lines called *epiRILs* that result from elements of *met1* and *ddm1* mutations have DNA hypomethylation in a few areas of the genome. Hypomethylation at CACTA elements triggers translocation of *epiRILs* derived from *ddm1* which obtain the hypomethylated CACTA epigenetic alleles called epialleles as predicted. Because CACTA elements would not transfer to the parental genome of the mutant of met1, they should remain immobile in met1-derived *epiRILs*. CACTA mobilization was identified by the researchers in 30% met1-derived epi-RILs, which was unexpected (Singh et al., 2021). After many generations of selfing, the transposition of CACTA was probably caused in some *epiRIL* organisms. Despite the lack of direct investigation, the movement of CACTA in different lines can be derived from the met1-derived elements' parents that are hypo-methylated at CG regions. As well as from the removal of neighboring epigenetic signals which includes non-CG methylation or H3K9me2. As met1-induced CG methylation is lost in patterns of non-CG methylation and H3K9me2 likely change and are inherited in an unstable manner (Hu et al., 2014).

The elements of EVD transfer the inbreed plant derived from epiRILs identical to CACTA. However, those two elements were not reconfigured in similar lineages, again emphasizing the selective regulation of retrotransposons through different epigenetic pathways. The researchers discovered this selectivity when investigating *epiRILs* derived from ddm1 (Sasaki et al., 2022). They showed that when inserted into a wild-type context, a subset of retrotransposons de-methylated in ddm1 continuously re-methylated among many generations. For remethylable retrotransposons, the tendency to demethylate is like CHH methylation is maintained and this process is not depending on the *DDM1* pathway. CHH methylation in nonremethylable TEs, on the other hand, is highly dependent on DDM1. As a result, different pathways on different retrotransposons specifically promote CHH methylation (Ashapkin et al., 2020).

14. Role of Retrotransposons from Genomic Parasites to Development of Various Bioprocesses in Soybean

Among the most surprising discoveries since the completion of the human genome a decade earlier was that transposable elements make up more than half of our DNA retrotransposons. It is now clear that retrotransposons are the largest molecular class among most metazoan genomes, thanks to advances in high-throughput sequencing techniques. Retrotransposons called "junk DNA" and "genomic parasite" in the past, are now under-stainable to study the genomic function and structure by increasing the host's coding and non-coding genetic sequences. Hence, retrotransposons play a crucial part in the development and growth of various bioprocesses (Mumford et al., 2019).

15. Retrotransposons in Biological Processes in Soybean

Barbara McClintock's career has included the discovery and description of transposable elements, retrotransposons, a type of mobile genetic factor abundant in the genomes of eukaryotic species. Her results were consistent with the then-famous selfish DNA hypothesis, which suggested that retrotransposons could be thought of as "genome hitchhikers" or molecular parasites that play zero function in genome evolution and offer no adaptive benefit to the host (Biemont, 2010).

Although ahead of her time, Barbara McClintock proposed that retrotransposons would influence the evolution of the genome. It's remarkable how many times it's proven valid since then. The emergence of evolutionary-developmental biology (evo-devo), the rapid development of DNA sequencing technology, and the subsequent rise of comparative genomics contributed to these discoveries. The modern understanding of retrotransposons is that they can serve as evolutionary agents by expanding, reorganizing, and diversifying the genetic catalog of their host (Jangam et al., 2017).

Retrotransposons have the advantage as penetrating markers for assessing genetic diversity caused by natural and artificial reasons as well as for improving marker-assisted selection (MAS) in plant breeding activities due to their high number of copies, coverage of chromosome and motile arrangement, including related species. DNA transposons and retrotransposons can be used to produce markers, with retrotransposons being very efficient. Retrotransposons have been detected in studies to be the most abundant type of retrotransposons in eukaryotes, accounting for approximately 90% of plant genetic makeup. It makes up nearly more than half of the maize and cereal genes, as well as 14% of the Arabidopsis genome. Furthermore, the inclusion of unchanged domains at both ends (LTRs) can be used for convenient PCR primer design (Primros et al., 2009). There are several examples of exaptation of retrotransposons in the kingdom of plantae. *FHY3* and *FAR3* are expression factors in Arabidopsis that bind to promoter areas that activate multiple genes associated with far-red light and circadian clock signaling. They are associated with the *MuDR* family of transposases. Recent research in Arabidopsis has reflected that injection of the *COPIAR7* transposon can cause plant disease resistance gene *RPP7* increasing host resistance to a pathogenic organism from a broad group of fungus-like parasites that cause various plant diseases. The Rim2 gene, which is involved in protecting plants from fungal infection, may have been derived directly from the CACTA transposable DNA element (Rohr et al., 2012).

Selection of a copyist-like LTR factor as a promoter defunctionalized a rice disease resistance gene named Pit. There is very little detail about the direct function of TEs in species domestication operations. The Hopscotch retrotransposon gene is introduced into the regulatory area of the maize local gene teosinte branched1 (tb1), which enhances its expression and confers increased apical superiority to maize over its progenitor, teosinte. Injection of a *CACTA*-like transposon into the promoter of the photoperiod-sensitive gene *ZmCCT* inhibits its expression, allowing maize to spread into regions with long-term temperate climates (Li et al., 2023). In flowering plants, the Mustang and Sleeper gene families have genes extracted from exalted retro transposases from mutate elements and *HAT* like elements of DNA. Indeed, the mustang plant genes are found only in angiosperms and encode repeated transcriptional regulators involved in growth, flower formation, and reproduction (Quesneville et al., 2020).

16.   Retrotransposons at Single Gene Level in Soybean

Retrotransposons affect the level of expression of a single gene of a plant to create novel transcripts by post-transcriptional modifications (Feschotte, 2008, Cowley and Oakey, 2013, Lisch, 2013, Mita and Boeke, 2016). The expression of a single gene can be affected by retrotransposons through mechanisms that include the disturbance of cis-regulatory sequences, changes in chromatin, or the provision of new regulatory information. It would be difficult to understand the mechanism of gene expression affected by retrotransposons, but some examples show the pattern of expression associated with insertions in retrotransposons. The reason behind this is a poor understanding of regulatory mechanisms. In most cases, the genes, the cis-acting regulatory sequences, and trans-acting factors that affect expression have limited description so that would be difficult for us to understand how normal gene regulation is disturbed by retrotransposons. Moreover, there is chromatin modification in retrotransposons which are different from the neighboring regulatory regions (Kent et al., 2017).

To check the importance of chromatin modifications on the expression of neighboring genes it would be necessary to compare the gene expression between alleles with and without chromatin modifications. We can compare gene expression in some retrotransposon families with retrotransposons in active and inactive forms, but it is difficult to access the groups with active retrotransposons in most cases. The chromatin state affected by mutation can be surveyed in plants with gene expression but mutation effects like silence modifications, and DNA methylation are not feasible and it limits the researcher's abilities to check the effect of retrotransposons chromatin on neighboring genes (Hu et al., 2014, Li et al., 2014).

Table 1 Transposable element domains and their function in the replication mechanism (Arias et al., 2019).

Gene Name	Short Name	Function
Reverse transcriptase	*RT*	Responsible for DNA synthesis using RNA as a template
RNase H	*RNAseH*	Responsible for the degradation of the RNA template in the DNA-RNA hybrid
Intregrase	*INT*	Responsible for catalyzing the insertion of the retrotransposon cDNA into the genome of a host cell
Aspartic protease	*AP*	Responsible for processing large transposon transcripts into smaller protein products
Envelope	*ENV*	Responsible for cell-to-cell transfer of retroviruses.
Group specific antigen	*GAG*	Structural protein for virus-like particles
Chromodomain	*Chrod*	Responsible for targeting the insertion of new LTR retrotransposon copies into heterochromatic regions by recognizing specific heterochromatic histone marks and/or other factors

Fig. 1 The LTR retrotransposon structure. The gene env could not exist in a few elements. The orange color directions show the LTRs.

17. Insertion of Retrotransposons on a Regulatory Sequence of a Gene in Soybean

Whenever transposon elements insert any sequence, it disturbs the existing sequences. When retrotransposons are inserted in a gene sequence involved in the regulation of transcription, it disturbs the normal function of that gene in transcription. Because of this, the transcription process does not start properly, and it results in abnormal function of that particular gene. The dominant alleles are produced as a result of this retrotransposons insertion mutation. For example, the homeobox gene in maize is necessary for the proper development of the leaf. It should be in a repressed state for normal leaf development in maize. However, due to the insertion of retrotransposons, some mutant dominant alleles of this gene are produced that disturb the normal development of the leaf at the leaf initiation stage. In this way, the insertion of retrotransposons produced effects at a single gene level (Greene et al. 1994, Schneeberger et al., 1995, Muehlbauer et al. 1999).

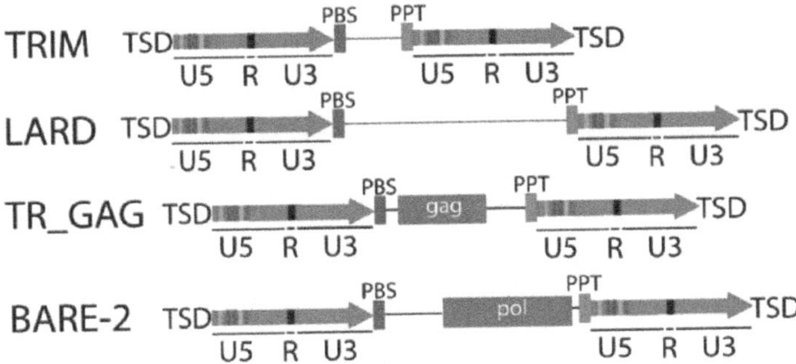

Fig. 2   The non-autonomous elements structure. The LTRs are indicated by the orange arrow. The non-coding areas are indicated by single lines. PPT is the Poly-Purine Tract, PBS Primer Binding Site. TRIM is the Terminal-Repeat Retrotransposons in Miniature. LARD is the Large Retrotransposon Derivatives.

18.   Chromatin marks of Retrotransposons in Soybean

Retrotransposons produce its effects on a gene by changing the local condition of chromatin. The transposon elements contain marks of heterochromatin. These marks interfere with a gene. If these chromatin marks remain confined to the transposons, then they will interfere with the enhancers present in the transposons and then they will disturb the genes present around them. In another case , if these chromatin marks are not confined to the retrotransposons, they will disturb the gene sequence that will be involved in the regulation of transcription. Hence in both cases, chromatin marks of retrotransposons will produce effects on the normal functioning of a gene (Hollister and Gaut, 2009).

19.   Retrotransposons at Large Chromosomal Regions in Soybean

The retrotransposons produce effects on large regions of a chromosome by rearranging the regions in a chromosome. The rearrangements produced in chromosomes can have many potential advantages. For example, if a part is deleted from a chromosomal region then it can be used for study purposes or it can also be used to study the functions of a group of genes that were present before deletion (Zhang and Peterson, 2005). If a chromosomal region becomes double due to retrotransposons, then this region can be useful for genes that need a large dose for their expression. In simple words, their number will increase due to which their expression will also increase . If retrotransposons create inversion or translocation in a region of chromosome by rearrangement, then this will be beneficial for studying the position of functional genes. Their positional effects can be studied well on a chromosome and this can also be used as a genetic tool for different purposes. If there are overlapping retrotransposons inversions and translocations, then this can be used for the generation of duplicate segments that will have well defined end points (Gopinath and Burnham, 1956).

20. Retrotransposons and Chromosome Evolution in Soybean

Retrotransposons play a very important role in chromosome evolution. A very large number of highly repetitive retrotransposons sequences create high copy numbers of different genes in a chromosomal region that can lead to the evolution of a chromosome in a very short time (Bennett and Leitch, 2005). For example in maize plant, retrotransposon sequences have been inserted in the genome of maize plants for the past six million years due to which the maize genome has been converted into the modern maize genome and this is the chromosomal evolution of maize plant (Walbot and Petrov, 2001). The chromosomal evolution has also occurred in Gossypium members due to the accumulation of retrotransposon from the past 5–10 million years (Hawkins et al., 2006). In the Oryza plant, the genome size has doubled due to transposon elements over the last 3 million years (Piegu et al., 2006). Retrotransposons have created a large number of repeats in Arabidopsis lyrate as compared to A. thaliana. This is due to the very high expression of transposons in a chromosomal region (Hollister et al., 2011). So, all these studies show that retrotransposons are responsible for the doubling of genome size, and they are responsible for the expansion of the genome and also creating many types of variations in the plant genome. Studies have shown that the transposon also affects the structure of genes in the chromosomes and affects the shape of genomes in living organisms (Zhang and Wessler, 2004).

21. Combination of Retrotransposons Mobilization and Epigenetic Landscape Changes Lead to Adaption of Phenotypes in Response to Climate Change in Soybean

The Epigenetic Components (ECs) contain a network of molecules that can adjust phenotypes according to the requirement like during development. It can pass to the next generation also without having any change in the original DNA sequence (Richards et al., 2012). The ECs connect the genotype and phenotype with the environment and help maintain the response of a living organism according to the changing environmental conditions (Mirouze and Paszkowski, 2011). The activity of retrotransposons can change by the environment and create a mutation in regulatory sequences and response to the environment (Oliver and Greene, 2009). Both ECs and retrotransposons are interconnected with each other and they both respond to the changing conditions of the climate by influencing the genotype or phenotype (Fedoroff, 2012). The changes in retrotransposons activities have been reported in plants, fungi, and animals For example, the UV-light and cold stress were given to fungus in an experiment, and the *OPHIO2* transposon element started its functions or activity (Bouvet et al., 2008). The stress of heat in Drosophila organisms activated the retrotransposons as well (Jardim et al., 2015). The EC elements become sensitive to drought, heat, and pollutants. For example, the exposure of traffic carbon particles on the human genome has activated DNA methylation (Kim et al., 2015). The changing climatic conditions also affect the genome of wild populations.

22. Phenotypic Consequences of Retrotransposons and Epigenetic Activities in Soybean

Retrotransposons affect the phenotype by the formation of new proteins and by changes in the existing proteins. The phenotypic changes due to climate change have been studied on pests. The study shows that pests have adapted themselves according to the agricultural chemicals used to destroy the pests. For example, in Drosophila, the retrotransposon has been inserted into the upstream region of insecticide resistant gene *Cyp6g1* increasing the upregulation of this gene in tissues and created insecticide resistance (Rostant et al., 2012). When environmental stress activates the retrotransposons then they modify the expression of the gene by modifying themselves. In maize and rice plants, exposure to temperature, salinity, and UV-light activated the retrotransposons that created insertions and induced response by regulation of genes according to the stress (Makarevitch et al., 2015). The epigenetic response to the environment when on demand. When there is a need to respond to climate change, ECs modify the expression of DNA sequences present in somatic cells during development (Feil and Fraga, 2012). The drug resistance produced after several doses of drugs taken is the phenomenon of epigenetic changes. The change in histone occurs to create resistance against drugs. The changes produced by DNA methylation can cause heritable changes in the phenotype. The response of toxins present in the environment was tested on the rats. The toxic alters the DNA methylation pattern that was heritable as well. This alteration was associated with male infertility in rats, and this was observed in the next four generations of rats (Anway et al., 2005).

23. Linkage of Retrotransposons and ECs under Changing Climatic Conditions in Soybean

The retrotransposons and ECs are linked with each other. The ECs can repress the activity of retrotransposons. By this ECs protect the integrity of the genome that is disturbed by retrotransposons movements. The DNA methylation enzyme like DNA methyltransferases targets the retrotransposons while activation of transposons silences the effect of ECs. This shows that they both are sensitive to the environment and are connected when responding to climate change (Friedli and Trono, 2015). Retrotransposons contribute to the evolution of genetic and epigenetic networks that are involved in the regulation of genes. The retrotransposons silence the ECs and are then inserted into the gene sequences. The retrotransposons can also be targeted by epigenetic regulation after they have been inserted into the gene sequences (Feschotte, 2008). The study was conducted on mice in which a retrotransposon named IAP was inserted into the upstream region of the Agouti gene. The transposon converted the wild type of agouti phenotype to the yellow phenotype. Then the inserted retrotransposon was then silenced by hyper methylation. This results in the recovery of wild type phenotype in mice. The interesting part was that the methylation of IAP was sensitive to a chemical named bisphenol is used in the plastic industry (Dolinoy et al., 2007). So, this clearly shows that the ECs and retrotransposons are linked and they respond inversely to each other under changing climatic conditions for the survival and adaptation of the organism.

Conclusion

Although the discovery of transposable elements and retrotransposons is very helpful in analyzing the important sequences in the genome. They are very important in the identification and classification of lineages and families of different species. Their discovery is a very difficult study for approximately all genomic projects because they have very diverse structures and functions that are very difficult to understand. Due to the greater variety of retrotransposons of different types and arrangements in genomes particularly in living organisms with extensive sequences, the study to identify and classify is useful for researchers and scientists but also for those who show interest in the evolution of genetic sequences and mutations in functions and structures of genes and also for expression of genes for their functions. Many bioinformatics solutions have been studied to identify and classify retrotransposons using different methods and techniques with changed results, but in many cases. For example, sequenced plant genomes still include huge unidentified and undetected regions. In conclusion, retrotransposons are a significant and complex component of the soybean genome. They have multiple functions, including the creation of genetic diversity, regulation of gene expression, and potential negative effects on the genome. The mechanisms that regulate retrotransposon activity in soybean, such as RNAi and DNA methylation, are essential for maintaining genome stability and integrity. Further research is needed to fully understand the role of retrotransposons in soybean and their potential uses in breeding and genetic improvement. In summary, retrotransposons are a significant and diverse component of the soybean genome, with important roles in genome evolution, gene regulation, and stress responses. The regulation of retrotransposon activity in soybean is essential for maintaining genome stability and integrity and for ensuring optimal plant growth and development. Further research is needed to fully understand the mechanisms underlying retrotransposon activity in soybean and their potential uses in breeding and genetic improvement.

References

Ashapkin, V.V., Kutueva, L.I., Aleksandrushkina, N.I., and Vanyushin, B.F. (2020). Epigenetic mechanisms of plant adaptation to biotic and abiotic stresses. *International journal of molecular sciences*, 21(20), 7457.

Anway, M.D., Cupp, A.S. Uzumcu, M., and Skinner, M.K. (2005). Epigenetic transgenerational actions of endocrine disruptors and male fertility. *Science* 308: 1466–69.

Biémont, C. (2010). A brief history of the status of transposable elements: from junk DNA to major players in evolution. *Genetics*, 186(4), 1085–93.

Bennetzen, J.L. (2000). Transposable element contributions to plant gene and genome evolution. *Plant molecular biology* 42: 251–69.

Bennett, M., and Leitch, I. (2005). Nuclear DNA amounts in angiosperms: progress, problems and prospects. *Anunals of Botany* 95: 45–90.

Bouvet, G.F., V. Jacobi, K.V. Plourde, and L. Bernier. 2008. Stress-induced mobility of OPHIO1 and OPHIO2, DNA transposons of the Dutch elm disease fungi. *Fungal Genetics and Biology* 45: 565–78.

Capy, P., Gasperi, G. and Biémont, C., 2000. Stress and transposable elements: co-evolution or useful parasites. *Heredity* 85: 101–106.

Chaitanya, K.V., 2019. Eukaryotic genome organization, regulation, evolution and control. Genome and Genomics: From Archaea to Eukaryotes, 121–80.

Costa, R.M.A, V. Chigança and R.S. Galhardo, 2003. The eukaryotic nucleotide excision repair pathway. *Biochimie.* 85: 1083–99.

Cowley, M., and R.J. Oakey, 2013. Transposable elements re-wire and fine-tune the transcriptome. PLoS Genet 9:e1003234.

Devaux, F., and A. Thiébaut. 2019. The regulation of iron homeostasis in the fungal human pathogen Candida glabrata. *Microbiology*, 165(10), 1041–60.

Dolinoy, D.C., D. Huang, and R.L. Jirtle. 2007. Maternal nutrient supplementation counteracts bisphenol A-induced DNA hypomethylation in early development. Proceedings of the National Academy of Sciences 104: 13056–61.

Feschotte, C., E.J. Pritham. 2007. DNA transposons and the evolution of eukaryotic genomes. *Annual Review of Genetics* 41: 331–68.

Fedoroff, N.V. 2012. Transposable elements, epigenetics, and genome evolution. *Science* 338: 758–67.

Feil, R., and M.F. Fraga. 2012. Epigenetics and the environment: emerging patterns and implications. *Nature Reviews Genetics* 13: 97–109.

Feschotte, C. 2008. Transposable elements and the evolution of regulatory networks. *Nature Reviews Genetics* 9: 397–405.

Friedli, M., and D. Trono. 2015. The developmental control of transposable elements and the evolution of higher species. *Annual review of cell and developmental biology* 31: 429–51.

Gopinath, D., and C.R. Burnham. 1956. A cytogenetic study in maize of deficiency-duplication produced by crossing interchanges involving the same chromosomes. *Genetics* 41: 382.

Gil, N., and I. Ulitsky. 2020. Regulation of gene expression by cis-acting long non-coding RNAs. *Nature Reviews Genetics*, 21(2), 102–17.

Greene, B., R. Walko, and S. Hake. 1994. Mutator insertions in an intron of the maize knotted1 gene result in dominant suppressible mutations. Genetics 138: 1275–85.

Guerreiro, M.P., 2011. What makes transposable elements move in the *Drosophila* genome Heredity 108: 461–468.

Hawkins, J.S., Kim, H., Nason, J.D., Wing, R.A., and Wendel, J.F., 2006. Differential lineage-specific amplification of transposable elements is responsible for genome size variation in Gossypium. *Genome Research* 16: 1252–61.

Helmbrecht K., Zeise, E. and Rensing, L., 2000. Chaperones in cell cycle regulation and mitogenic signal transduction: a review. *Cell proliferation* 33: 341–65.

Helman, E., Lawrence, M.S. and Stewart, C., 2014. Somatic retro transposition in human cancer revealed by whole-genome and exome sequencing. *Genome Research* 24: 1053–63.

Hollister, J.D., and Gaut, B.S., 2009. Epigenetic silencing of transposable elements: a trade-off between reduced transposition and deleterious effects on neighboring gene expression. *Genome Research* 19:1419–28.

Hollister, J.D., Smith, L.M., Guo, Y.-L., Ott, F., Weigel, D. and Gaut. B.S., 2011. Transposable elements and small RNAs contribute to gene expression divergence between *Arabidopsis thaliana* and *Arabidopsis lyrata.* Proceedings of the National Academy of Sciences 108: 2322–27.

Hu, L., Li, N., Xu, C., Zhong, S., Lin, X., Yang, J., Zhou, T., Yuliang, A., Wu, Y. and Chen, Y.R. 2014. Mutation of a major CG methylase in rice causes genome-wide hypomethylation, dysregulated genome expression, and seedling lethality. *Proceedings of the National Academy of Sciences* 111: 10642–47.

Hua-Van, A., Le Rouzic, A., Boutin, T.S. 2011. The struggle for life of the genome's selfish architects. *Biology Direct* 6: 19.

Iskow, R.C., McCabe, M.T., Mills, R.E., Torene, S., Pittard, W.S., Neuwald, A.F., Van Meir, E.G., Vertino, P.M. and Devine, S.E. 2010. Natural mutagenesis of human genomes by endogenous retrotransposons. *Cell* 141: 1253–61.

Jangam, D., Feschotte, C and Betrán, E. 2017. Transposable element domestication as an adaptation to evolutionary conflicts. *Trends in Genetics*, 33(11), 817–31.

Jardim, S.S., Schuch, A.P., Pereira, C.M., and Loreto, E.L.S. (2015). Effects of heat and UV radiation on the mobilization of transposon mariner-Mos1. Cell Stress and Chaperones 20: 843–51.

Jacobson, J.W, M., Medhora, M., and Hartl, D.L. (1986). Molecular structure of a somatically unstable transposable element in *Drosophila.* Proceedings of the National Academy of Sciences of the United States of America. 83: 8684–8688.

Kazazian H.H. Jr (2011). Mobile DNA transposition in somatic cells. *BMC Biology* 9: 62.

Kent, T.V., Uzunović, J., and Wright, S.I. (2017). Coevolution between transposable elements and recombination. Philosophical Transactions of the Royal Society B: *Biological Sciences*, 372(1736), 20160458.

Kidwell, M.G., and Lisch, D. (1997). Transposable elements as sources of variation in animals and plants. *Proceedings of the National Academy of Sciences* 94: 7704–11.

Kim, J.H., Jun, S.A., Kwon, Y., Ha, S., Sang, B.I., and Kim, J. (2015). Enhanced electrochemical sensitivity of enzyme precipitate coating (EPC)-based glucose oxidase biosensors with increased free CNT loadings. *Bioelectrochemistry* 101:114–19.

Kunze, R. (1996). The maize transposable element Activator (Ac). Transposable elements, 161–194.

Liu, J., (2022). Giant cells: Linking McClintock's heredity to early embryogenesis and tumor origin throughout millennia of evolution on Earth. In Seminars in cancer biology (Vol. 81, pp. 176–92). Academic Press.

Lindquist, S., E.A. Craig 1988. The heat-shock proteins. *Annual Review of Genetics* 22: 631–77.

Li, W., Prazak, L., and Chatterjee, N. (2013). Activation of transposable elements during aging and neuronal decline in *Drosophila.* Nature Neuroscience.

Li, B., Wang, Jiang, H., Luo, H.J., Guo, T., Tian, F., and He, Y. (2023). ZmCCT10-relayed photoperiod sensitivity regulates natural variation in the arithmetical formation of male germinal cells in maize. *New Phytologist*, 237(2), 585–600.

Lopes, M.A., and Larkins, B.A. (1993). Endosperm origin, development, and function. *The Plant Cell* 5: 1383.

Li, Q., Eichten, S.R., Hermanson, P.J., Zaunbrecher, V.M., Song, J., Wendt, J., Rosenbaum, H., Madzima, T.F., Sloan, A.E., and Huang, J. (2014). Genetic perturbation of the maize methylome. *The Plant Cell* 26: 4602–16.

Lisch, D. (2013). How important are transposons for plant evolution. *Nature Reviews Genetics* 14: 49–61.

Makarevitch, I., Waters, A.J., West, P.T., Stitzer, M., Hirsch, C.N., Ross-Ibarra, J., and Springer, N.M. (2015). Transposable elements contribute to activation of maize genes in response to abiotic stress. *PLoS Genet* 11:e1004915.

Magalhaes, J., Liu, J., Guimaraes, C.T., Lana, U.G., Alves, V.M., and Wang, Y.H. 2007. A gene in the multidrug and toxic compound extrusion (MATE) family confers aluminum tolerance in *Sorghum. Nature Genetics.* 39, 1156–61.

McClintock B., (1942). The fusion of broken ends of chromosomes following nuclear fusion. *Proceedings of the National Academy of Sciences of the United States of America.* 28: 458–63.

Medhora, M., Maruyama, K., and Hart, D.L. (1991). Molecular and functional analysis of the *mariner* mutator element *Mos1* in D*rosophila*. *Genetics* 128: 311–318.

Mirouze, M., and Paszkowski, J. (2011). Epigenetic contribution to stress adaptation in plants. Current opinion in plant biology 14: 267–74.

Mita, P., and Boeke, J.D. (2016). How retrotransposons shape genome regulation. Current opinion in genetics & development 37: 90–100.

Mumford, P.W. (2019). Line-1 Retrotransposition and the Impact on Aging Rodent Skeletal Muscle Tissue (Doctoral dissertation, Auburn University).

Muehlbauer, G.J., Fowler, J.E., Girard, L., Tyers, R., Harper, L., and Freeling, M. (1999). Ectopic Expression of the Maize Homeobox GeneLiguleless3 Alters Cell Fates in the Leaf. *Plant physiology* 119: 651–62.

Ogasawara, H., Obata, H., and Hata, Y. (2009). *Crawler*, a novel *Tc1/mariner*-type transposable element in *Aspergillus oryzae* transposes under stress conditions. *Fungal Genetics and Biology* 46: 441–49.

Oliver, K.R., and Greene, W.K. (2009). Transposable elements: powerful facilitators of evolution. Bioessays 31: 703–714.

Orozco-Arias, S., Isaza, G., and Guyot, R. (2019). Retrotransposons in plant genomes: structure, identification, and classification through bioinformatics and machine learning. *International journal of molecular sciences*, 20(15), p.3837.

Orr, H.A. (1997). Haldane's rule. *Annual Review of Ecology and Systematics* 28: 195– 218.

Piegu, B., Guyot, R., Picault, N., Roulin, N., Saniyal, A., Kim, H., Collura, K., Brar, D.S., Jackson, S., and Wing, R.A. (2006). Doubling genome size without polyploidization: dynamics of retrotransposition-driven genomic expansions in Oryza australiensis, a wild relative of rice. Genome research 16: 1262–69.

Primrose, S.B., and Twyman, R. (2009). Principles of genome analysis and genomics. John Wiley & Sons.

Quesneville, H. (2020). Twenty years of transposable element analysis in the *Arabidopsis thaliana* genome. Mobile DNA, 11(1), 1–13.

Rech, G.E., S. Radío, S. Guirao-Rico, Aguilera, L., Horvath, V., Green, L., and González, J. (2022). Population-scale long-read sequencing uncovers transposable elements associated with gene expression variation and adaptive signatures in Drosophila. Nature Communications, 13(1), 1948.

Richards, C.L., Verhoeven, K.L., and Bossdorf, O. (2012). Evolutionary significance of epigenetic variation. Plant Genome Diversity Volume 1. *Springer*. Pp. 257–74

Röhr, H., Kües, U., Stahl, U. (2012). Organelle DNA of Plants and Fungi. Inheritance. Progress in Botany: Genetics Cell Biology and Physiology Systematics and Comparative Morphology Ecology and Vegetation Science, 60, 39.

Rostant, W.G., Wedell, N., and Hosken, D.J. 2012. Transposable elements and insecticide resistance. *Advances in genetics* 78:169–201.

Sasaki, T., Ro, K., Caillieux, E., Manabe, R., Bohl-Viallefond, G., Baduel, P., and Quadrana, L. (2022). Fast co-evolution of anti-silencing systems shapes the invasiveness of Mu-like DNA transposons in eudicots. The European Molecular Biology Organization (EMBO). 941(8), e110070.

Schneeberger, R.G., Becraft, R.G., Hake, S., and Freeling, M. (1995). Ectopic expression of the knox homeo box gene rough sheath1 alters cell fate in the maize leaf. *Genes & Development* 9: 2292–04.

Schuch, A.P., Garcia, C. CM., Makita, K., and Menck, C.F.M. (2013). DNA damage as a biological sensor for environmental sunlight. *Photochemical and Photobiological Sciences*.

Singh, D., Chaudhary, P., Taunk, P., Kumar Singh, C., Sharma, S., Singh, V.J., and Pal, M. (2021). Plant epigenomics for extenuation of abiotic stresses. Challenges and future perspectives. *Journal of Experimental Botany*, 72(20), 6836–55.

Sioud, M., (2021). RNA interference: Story and mechanisms. Design and Delivery of SiRNA Therapeutics, 1–15.

Shao, Y., (2022). Functional Characterization and Application of Exapted Transposable Element Gene Family *Mustang-A*. McGill University (Canada).

Vehniäinen, E.R., Vähäkangas, K., and Oikar, A. (2012). UV-B exposure causes DNA damage and changes in protein expression in northern pike (*Esox lucius*) posthatched embryos. *The Journal of Photochemistry and Photobiology* 88: 363–70.

Walbot, V., and Petrov, D.A. (2001). Gene galaxies in the maize genome. *Proceedings of the National Academy of Sciences* 98: 8163–64.

Wells, J.N., and Feschotte, C. (2020). A field guide to eukaryotic transposable elements. *Annual review of genetics*, 54, 539–61.

Wicker T., Sabot, F., and A. Hua-Van. (2007). A unified classification system for eukaryotic transposable elements. *Nature Reviews Genetics* 8: 973–82.

Xu, P., and Wang, E. (2023). Diversity and regulation of symbiotic nitrogen fixation in plants. *Current Biology*, 33(11), R543–R559.

Yamanaka, S., Mehta, S., Reyes-Turcu, F.E., and Zhuang, F. (2013). RNAi triggered by specialized machinery silences developmental genes and retrotransposons. *Nature* 493: 557–60.

Zhang, J., and Peterson, T. (2005). A segmental deletion series generated by sister-chromatid transposition of Ac transposable elements in maize. *Genetics* 171: 333–44.

Zhang, X., and Wessler, S.R. (2004). Genome-wide comparative analysis of the transposable elements in the related species *Arabidopsis thaliana* and *Brassica oleracea*. *Proceedings of the National Academy of Sciences* 101: 5589–94.

10

Retrotransposons in Plant Genome of Brassica

Prasann Kumar,[1*] *Khushbhu Sharma*[1] and *Joginder Singh*[2]

1. Introduction

They share Cruciferae or Brassicaceae, the same taxonomic family, and are closely associated with the model plant Arabidopsis thaliana, making the Brassica genus an essential one for agriculture. It is believed that after diverging from a common ancestor some 15–20 million years ago, Arabidopsis and the Brassica species evolved via repeated genome replication (Vision et al., 2000; Simillion et al., 2002). One diploid species, B. oleracea, has a highly complex and widely triplicated genome. Sequence conservation and collinearity of gene order across several genomic areas are two pieces of evidence supporting this hypothesis (Ryder et al., 2001; Babula et al., 2003). A comparative study of the non-coding fraction of the genome, which is primarily represented by repetitive sequences and TEs, alongside the equivalent survey of the coding regions in Arabidopsis and Brassica, would be helpful to better understand the global genome evolution since the divergence of the Brassicaceae. Transposons are a type of transposable genetic element that can potentially install itself anywhere throughout the genome. Barbara McClintock first discovered transposons in the maize genome in the late, and their presence has subsequently been demonstrated to be widespread. The human genome has 44 per cent TEs, the mouse genome contains 40 per cent, and the fruit fly genome contains 10 to 15 per cent TEs. TEs are also significant parts of the rice (14% of the

[1] Department of Agronomy, School of Agriculture, Lovely Professional University, Phagwara, 144411 (Punjab), India.

[2] Department of Microbiology, School of Bioengineering and Biosciences, Lovely Professional University, Phagwara-144 411 (Punjab), India.

* Corresponding author: prasann0659@gmail.com

genome), maize (60% of the genome), and wheat (80% of the genome) genomes (Charles et al., 2008). The percentage of repeated sequences, such as TEs and simple sequence repeats, rather than gene sequences, is the primary determinant of variance in the size of eukaryotic genomes. Retroelement categories and abundances are species and variety-dependent . There are two types of transposons, depending on whether or not they use RNA as a transposition intermediary. Retrotransposons are elements of class I, and they insert copies of themselves elsewhere in the genome by a process of reverse transcription, including an RNA intermediary. There are two main types of retrotransposons, those with and without long terminal repeats (LTRs). Ty1-copia, BEL, DIRS, and Ty3-gypsy are all examples of long terminal repeat retroelements (LTRs), while LINEs and SINEs are examples of non-LTR retroelements. Class II TEs, which include mariner-Tc1, hAT, Mu, Helitron, and micro inverted-repeat transposable elements (MITEs) (Wicker et al., 2007), operate in a 'cut and paste' manner. Long-terminal repeat (LTR) transposable elements are the most common and significantly impact genomes. All LTR-TEs share several structural features. These include 5′ and 3′ long terminal repeat sequences, target site repeats, a polypurine tract, a TG/CA box, and reverse transcriptase, integrase, and RNase H genes. TEs heavily influence the structure and size of genomes due to their repetitive duplication, transcription, and excision, all of which contribute to genetic diversity and evolution. However, there is a wide variety in the ways that TEs act. Insertions of low-copy-number transposons into genes are more common than those of highly repetitive elements into intergenic areas like genic spacers, heterochromatin, or even other TEs. Target sites are unique for each transposon, and TEs are not randomly inserted into the host. Ty5, a mariner family retrotransposon in Saccharomyces, integrates into heterochromatin at telomeres and silent mating loci, while Mos1 inserts into locations like TATA motifs and AT-rich areas (Crenes et al., 2010). Chromatin structure and the structure of the target DNA can potentially influence the selection of target sites (Nefedova et al., 2011).

Under some circumstances, TE bursts, or the fast expansion and multiplication of TE repetitions, can occur. Environmental stressors, tissue culture, interspecific hybridization, and the inactivation of critical genes are all potential burst triggers. In addition, it appears that some TEs are more likely to explode than others. The DNA transposon mPing, for instance, increases its copy number in the rice genome by about 40 copies per plant every generation. Interestingly, specific retrotransposons can produce TE-rich areas by inserting into the region or flanking sequences of other retrotransposons. Genomic areas with this degree of clustering and nesting have been identified in cabbage, humans, and maize. However, the evolutionary importance of this TE clustering phenomenon and the factors that contribute to it are not yet known. Plants belonging to the Brassica genus are used for food, feed, and industrial purposes. Allotetraploid Brassica napus was created by cross-pollination between the diploid vegetable crop species B. rapa and B. oleracea, making it one of the most significant oil crops in the Brassica genus. The retrotransposon family accounts for 12.3% and 14% of the TEs in the B. rapa and B. oleracea genomes. Long-terminal repeat retrotransposons (LTR-TEs) play a significant role in the assembly and development of the Brassica genome. While nested LTR-TEs play an essential role in genome organization and gene function, little is known about their evolutionary history, the principles of nested LTR-TE bursts and insertions, or their impact on gene regulation. After searching over 1.52Gb of sequences, six BAC segments from the Brassica genome were chosen for further investigation because they included nested long terminal repeat (LTR)-TEs. Nestled long terminal repeat (LTR) retrotransposons in the Brassica genome were analysed, and their origins, evolutionary importance, and potential contributions to the genome were reported.

2. Mining Data and Performing Computational Analysis on LTR motifs

Dot plot analysis was used to blast the Database of entire or complete (reference) elements from the Brassica plant housed by NCBI. The database was searched for LTR REs by employing intact elements to locate full-length copies and their deleted elements. The number of relevant results for reference queries (>75% query coverage and identity throughout their whole lengths) was gathered to estimate the number of copies. The copy number of intact REs was estimated using the following formula, which only included functional parts. Copy number = several complete copies in the database/total. The gag-pol gene encoding proteins in REs were identified by comparing the genome size of Brassica spp. to the available genome size in NCBI (Tu, 2001). LTR_FINDER's 'Predict PBS by utilizing Arabidopsis thaliana tRNA database' parameter was used to find PBS and PPT motifs. If two sequences share at least 85% of their nucleotides in their coding regions, we classify them as members of the same family; if they share at least 95% of their nucleotides, we classify them as copies of the same element. When an element exhibited homology to one or more sequences but was incongruous with any known elements, we categorized it as belonging to a new family.

3. LTR REs: Defining, Grouping, and Titling

The homology between the REs and previously described elements was characterized using the Repbase and Gipsy databases (Llorens et al., 2011). Uncharacterized elements of gag-pol coding regions based on homology and protein structure. REs were labelled as Copia if their pol gene sequence looked like 5′-AP INT-RT-RH-3′ and Gipsy if it looked like 5′-AP INT-RT-RH-3′. The elements and their families were classified according to standards established by prior studies. In the absence of similarity to previously characterized LTR REs, we deemed this a unique family. Element names, for instance, *Brassica rapa* Copia 1 (BrCOP1), follow the convention wherein the first letter denotes the genus, the second the species name, the third the superfamily, and the fourth the family. Class I transposable elements (TEs), also known as retrotransposons, comprise most plant genomes. Transposable elements play a crucial role in the evolution of variety due to their capacity to move from one section of the genome to another (Babula et al.,

2003). Non-LTR retrotransposons, also known as LINEs (long intervening nuclear elements), are classified apart from LTR retrotransposons like Ty1/copia and Ty3/gyps. Using fluorescent in situ hybridization (FISH), researchers have mapped the presence and abundance of copia and gypsy-like retrotransposons across several plant genomes.

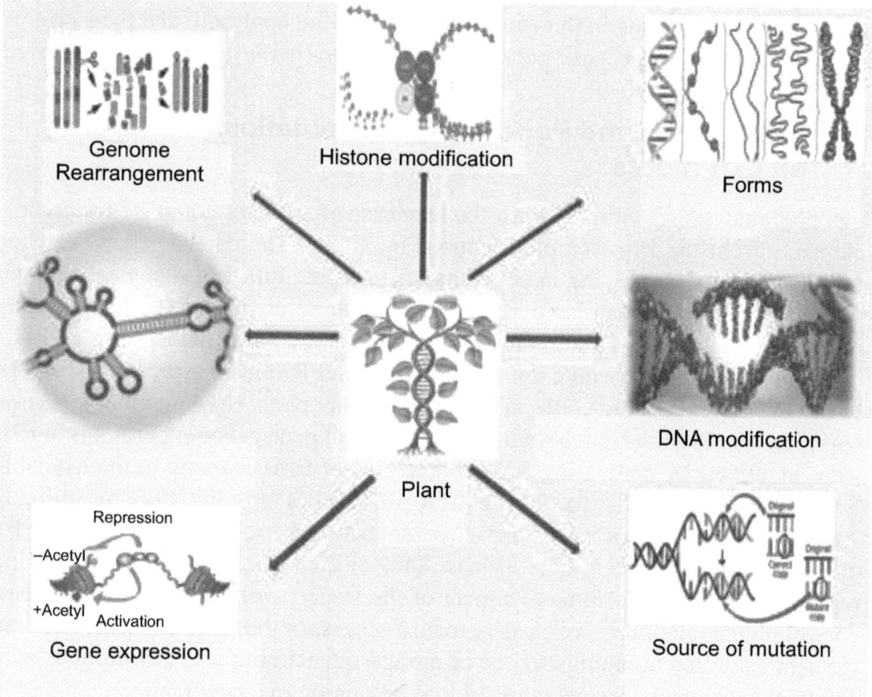

Fig. 1   Plant genome and component of retrotransposon

Family involved: The most populous family has about 140 copies of Brassica. oleracea. Thus, Zhang and Wessler (2004) correctly estimate that total class I TEs account for as much as 14 per cent of the Brassica. oleracea genome, then these researchers have shown that this proportion is made up of a large number of distinct class I elements replicated to a comparatively low copy number. Our findings that four distinct retrotransposon subfamilies are highly represented across the B. oleracea genome support this conclusion. The global organization of the B. oleracea genome in terms of gene and TE patterns has been illuminated by a sequencing study of seven BACs. Transposons preferentially target genic regions, as evidenced by the finding of nearly equal numbers of class I and class II elements in the genic areas, even though class II DNA elements only comprise 6% of the B. oleracea genome. Figure 2 of Howell et al.'s (2002) study shows that Bo10COP-18 hybridized to the BAC clone BoB061G14, which has been

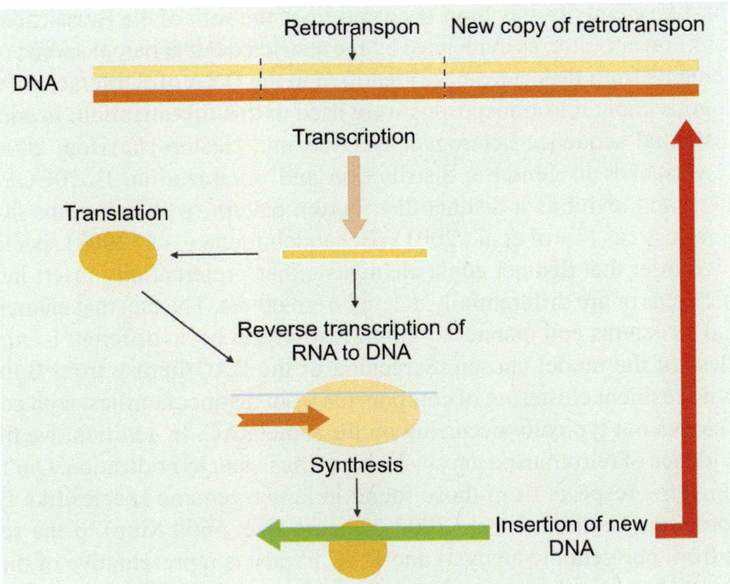

Fig. 2   Method of generation of Retrotransposon

shown to hybridize to pericentromeric heterochromatin of 6 chromosomes. BAC clones BoB011O13, found close to centromeric regions of various B. oleracea chromosomes, hybridized with Bn14G42-01 (Howell et al., 2002) suggest that Bn14G42-01's chromosomal distribution in B. rapa would be replicated in B. oleracea. Data acquired from screening BAC libraries with two distinct gypsy-type probes provides evidence for a clustering bias. Moreover, in situ, hybridization evidence favours pericentromeric regions as ideal insertion locations.

4.   Distribution of Retrotransposon in Brassica Species

As shown by numerous studies (Kurata et al., 2002), characteristic centromeric Metaviridae elements are substantial elements constituting the centromere repetition. These gypsy retrotransposons may match these elements. However, gypsy retrotransposons have also been found to be broadly dispersed across chromosomes (Hudakova et al., 2001; Kurata et al., 2002; Vershinin et al., 2002). According to previous studies (Peterson-Burch et al., 2004), Athila highly enriched the central 180-bp centromeric satellite DNA in the pericentromeric heterochromatin of Arabidopsis. This work shows that the Athila-like gypsy element is also present in B. oleracea, although it is more extensively dispersed in the genome. The restricted Amplification of Athila-like elements may have happened since the Brassicaceae separated from their common ancestor. This may explain the observed spread of Athila retroelements from the centromere in the B. oleracea genome. An increase in Athila-like elements may have occurred since the split of the Brassicaceae from their common ancestor, as evidenced by the restricted but apparent spread of Athila retroelements from the centromere identified in the DNA of B. oleracea. Two non-homologous copia retrotransposons were used in this investigation. In addition to the substantial sequence heterogeneity (Alix and Heslop-Harrison, 2004), they exhibit variances in genomic distribution and organization. Bo10COP-18, on the other hand, exhibits a distinct distribution pattern, with insertions favouring centromeric areas (Terol et al., 2001) (Benabdelmouna et al., 2003) species. As a result, we infer that distinct copia elements either preferentially insert into some genomic areas or are differentially deleted from others. The fact that elements with identical structures and manner of amplification can be so different is intriguing, regardless of the model chosen. Screening of the BAC library from B. oleracea reveals no resilient clustering of components from distinct families, with copia and gypsy clones not typically occurring on the same BAC. In addition, we find very little evidence of retrotransposon clustering using in situ hybridization. Our findings differ in some respects from those found in larger-genome species like Brassica and Hordeum (Vershinin et al., 2002; genome size: 5600 Mbp), if the sequence we got from our genome analysis and BAC library is representative of the whole genome of Brassica oleracea species. These showed that several retrotransposons from various families are nested within one another.

5.   Brassica Genome

The intellectual benefits of surveying B. oleracea for fundamental research are considerable. Allopolyploid species can be created through hybridization between this species and its two other related diploid Brassica species U triangle. The B. oleracea C genome is 630 Mb and comprises nine chromosomes, with 44,758 predicted protein-coding genes. The average transcript length is 1,751 bp. (Liu et al., 2008). The data on genome structure suggests a whole-genome duplication event occurred throughout the development of the family Brassicaceae, followed by chromosomal rearrangements that rearranged the genome and resulted in gene losses. With the completion of the Brassica genome sequence, scientists are free to investigate hitherto unexplored areas. Thompson, to name just two examples, has researched the genetic and evolutionary origins of the aforementioned physical variation. Because of their morphological diversity and unique agronomic properties like quick cycling and self-compatibility, the kale-like morphotype (TO1000) and the homogeneous line (01–12) of the Brassica oleracea var. capitata genus have been adopted as models for genetic research. According to a study (Wang et al., 2005), we used lines 02–12 to create hybrids. The B. oleracea cultivar TO1434's offspring, the TO1000DH3 line, and the 02–12 line were subsequently selected for reference genome sequencing (Parkin et al., 2014). Genome information collected from two different alleles of B. oleracea can shed light on the genetic basis of processes of evolution and phenotypic variation.

6.   Characterization of Brassica Species

The genus Brassica is characterised by a wide range of morphological diversity across its inflorescence and whole phyllosphere (stem, leaves, bud, etc). For example, there is a great deal of morphological variation within the subspecies of *Brassica oleracea.* The larger inflorescences of botrytis and cabbage (B. oleracea ssp) are only two examples. Phenotypic variation among Brassica species may be explained by chromosomal changes associated with polyploidization in Brassica species that have led to the triplication of genomic areas and subsequent rearrangements involving inversions, additions, deletions, and substitutions. New phenotypic variations for traits in these animals may arise due to these genetic variants. Thus, understanding the rapid physiological growth of polyploid plants can benefit from a genomic study on Brassica. The allotetraploid Brassica species will be easier to comprehend, as will the effects of genetic alterations on that species. Rapid genomic alterations and the influence of nuclear-cytoplasmic interaction have been examined, for instance, in artificial allotetraploid animals. Due to the high economic value of Brassica species worldwide, genome sequencing projects have recently begun for these plants. The Chiffu-401 genetically inbred line of B. rapa ssp was chosen for sequencing the Brassica-A genome by the Brassica rapa Dna Sequencing Project (BrGSP), a sub-consortium of the International Brassica Genome Project. The end goal of this research is to sequence all of the BACs in this species' genome.

The genome of the *B. rapa* plant can be viewed in its entirety by an analysis of BAC-end sequences. The *B. rapa* genome was scanned by analysing HindIII BAC-end sequences (Gray et al., 2000). Investigations of genome survey sequences help investigators to understand the genome as a whole. The B. rapa genome is 1.6 times the size of the A. thaliana genome, with an estimated 43,000 genes (based on 16.8% coverage). Using estimations from microsynteny investigations, Wang et al., 2017 found that the number of genes in *B. rapa* is between 49,000 and 63,000. It has been hypothesized that chromosomal triplication in Brassica has led to gene duplication and loss of roughly 14% (covering approximately 74 Mb) of the genome estimated to be occupied by transposable elements (TEs), which is 8.2 times more than that previously found in A. thaliana.

7.   Triplication of the Diploid Genome in the Brassica Genus

According to the bulk of the comparison above studies, the homologous regions of the *A. thaliana* genome are significantly triplicated in the Brassica species, indicating that the diploid Brassica species possibly descended from a hexaploid progenitor with a genome identical to Arabidopsis. In diploid Brassica species, the genome started to triple between 13 and 17 MYA, not long after the divergence between Arabidopsis and Brassica (17 to 18 MYA). The pattern of repeated copying of a genome is consistent with what one would expect from repeated copies of the genome. This suggests that the genome tripling was an essential adaptation in Brassica species and may have been driven by environmental factors. This adaptation allowed Brassica species to evolve quickly to survive in their environment. In addition to tripling in size, the Brassica genomes also underwent a reorganization of their sub-genomes due to inverted positions, translocations, considerable intermittent gene loss, and gene insertions compared to the projected structure of the ancestral genome. This adaptation enabled Brassica species to become more resistant to pests and diseases and grow in adverse climates. It also allowed them to become more productive and nutritionally diverse. Thus, Brassica species quickly became one of the most important crops in the world. Moreover, it was found that members of the Brassicaceae family are also prone to undergo genome triplications frequently. This results in various new phenotypes, which can provide potential advantages for crop improvement. As a result, Brassica species are now widely cultivated and used as food sources worldwide. In Brassica species, the genome size increased by a factor of 3 after genome triplication. This increase in genome size has allowed for the evolution of new traits, such as disease and pest resistance, improved nutrition, and increased yield. As a result, Brassica species are now widely grown and consumed around the world. This increase was observed in comparison to the A. thaliana genome. In Brassica species, there is a possibility of pseudogenization, sub functionalization, and neofunctionalization due to the occurrence of genome triplications. A process known as sub functionalization can affect the genomes through the MADS-box transcription factor family. As a result of its members, crucial aspects of plant development, such as vegetative and reproductive growth, are controlled. It has been documented that rapid changes followed polyploid development in genome organization and gene expression due to the polyploid development process. Sub functionalization occurs when one gene is split into two or more copies, which can lead to changes in gene expression. This is due to mutations, which can cause the gene to have different functions. This process is believed to be an important factor in the evolution of species. Researchers have hypothesized that the triplication of the genome in Brassica species has led to an increase in variability in phenotype, which may be due to mutations in genes that are generally subject to strong selective pressures in Arabidopsis species. This increased variability in phenotype could provide the species with an evolutionary advantage in adapting to changing environments. Thus, polyploidy in Brassica species may have helped them to survive and thrive in changing settings. By studying how duplicated genes are expressed in various species of the plant genus Brassica, we can better understand how polyploidization contributes to the phenotypic diversity of the genus Brassica. This knowledge can then be applied to other species to understand the effects of polyploidization better and explore the potential of creating more diverse, resilient species.

8.   LTR Retrotransposons

There are now four known families of LTR retrotransposons in plants, with Ty1-copia and Ty3-gypsy being the most prevalent. They have all the necessary coding sequences responsible for the retro-transposition process and can transpose on their own (Bennetzen et al., 2005) because they continue the structure of LTR, group£-specific antigen (gag), RNA£-dependent DNA polymerase (pol), and integrase (int). LARD (long retrotransposon derivatives) and TRIM (terminal-repeat retrotransposons in miniature) are two other recently identified kinds of LTR retrotransposons; both are very simple in composition, having only a short LTR structure and lacking the integrase genes required for transposition. While neither LARD nor TRIM can transpose on their own, they can do so with the help of spontaneous retrotransposons (Kwon et al., 2007) (Witte et al., 2001) (Yang et al., 2007). Non-LTR retrotransposons rely on the inner promoter to control transposition activity because they lack the LTR structure. Only two recognized families of non-LTR retrotransposons now exist LINE (long interspersed nuclear elements) and SINE (short interspersed atomic components). LINE retrotransposons share the gag and pol genes with LTR retrotransposons but lack the int gene. The gag protein in LINEs serves as an endonuclease and may play a role in integration. It was hypothesized that LINE retrotransposons, the most ancient kind, gained LTR structure during their evolution.

9. SINE

They lack a coding sequence associated with transposition and are often very short. This means that autonomous transposition is impossible for SINEs. However, integration into the host genome is possible thanks to transposition-related proteins expressed by LTR retrotransposons. More is needed to know about how SINE finishes the cycle of replication and integration. We now know that retrotransposons are frequently found in the plant genome but are also a crucial constitution of most plant genomes due to their substantial copy number and high heterogeneity, thanks to the growing body of genomic data derived from completed or continuing plant DNA sequencing initiatives and accumulating proofs concluded from molecular biology research. First, there is considerable diversity in the distribution of retrotransposons among plant genomes. Recently investigated the genomic organisation of several indicative retrotransposons in Brassica oleracea and found that individual LTR-retrotransposons (copia and gypsy-like) were displayed at the most abundant locations, with representation ranging from 90 to 320 copies in the haploid genome.

10. In Contrast, very few LINE Elements were Found

Fluorescent in situ hybridization further indicated that each retro-element has its distinct chromosomal distribution (Alix et al., 2005). More studies on retro-elements published to date suggest that LTR retrotransposons, primarily meaning Ty1-copia and Ty3-gypsy, are the most prominent class of retro-elements in the plant genome, exhibiting the most comprehensive distribution and highest copy number in unicellular lycophyte, gymnosperms, and angiosperms, respectively. However, few studies have documented the presence of LARDs and TRIMs, two more types of LTR-retrotransposons, in monocotyledons and dicotyledons respectively (Antonius Klemola et al., 2006; Kwon et al., 2007; Witte et al., 2001; Yang et al., 2007). Shotgun sequences from B. oleracea and BAC end sequences from B. rapa both revealed four families of terminal-repeat retrotransposons in miniature (TRIM), which are highly similar to TRIM sequence from Arabidopsis and range in copy number from 500 to 700 (Yang et al., 2007). Similar to LTR retrotransposons, non-LTR retrotransposons were found widely distributed over a wide range of plant species, albeit at a much lower copy number. Subsequently, LINEs were identified with substantial sequence divergence in the plants Bulbus Lilii, Arabidopsis, beetroot, and monocotyledons like wheat, barley, and rice (Leeton and Smyth 1993; Vershinin et al., 2000). However, the SINE family is now the subject of significantly less study. One of the most well-researched SINE retrotransposons is S1, first described in B. napus. The number of copies is rather low in the Cruciferae tribe, and it shows up uniquely in the species of the Brassicaceae family. Several distinct groups emerged in analyses of the phylogeny of S1 elements.

11. Organization of Retrotransposons

Integrating the processes of DNA transcription, mRNA processing, post-translational modification, insertion, reverse transcription, and synthesis is essential for retrotransposon transposition. Inactivation of retrotransposons is the result of any interference with those processes. In rice, for instance, a LINE-type retrotransposon called KARMA was found to have ongoing retrotransposition activities and to control its mobilization at both the transcription and post-transcription phases (Komatsu et al., 2003). In addition, it has been found that the retrotransposon structural genes in the potato genome frequently undergo sequence alterations, including rearranging, deleting, or forming an early terminator. They are thought to be the main factors keeping most retrotransposons dormant in the plant genome. Some non-LTR retrotransposons may lack the required components for transposition, but they can still complete the process with the help of other active retrotransposons that provide those components. However, a similar model of control has yet to be identified in plants. Homeostatic control of the retrotransposon's transposition activities may be affected by the host genome's diversity and the circumstances in which the retrotransposon developed (Cam et al., 2008). For instance, active retrotransposons that disseminated widely in the Gramineae genome strongly correlated with the rate of host evolution (Sharma and Presting, 2008; Vicient et al., 2001). Solanaceae and other plant families exhibited a similar association (Grandbastien et al., 2005). Thus, it is believed that retrotransposons evolved through a procedure from parasitism to paragenesis in the plant genome and then to mutualism through co-regulation with the host genome. It is important to note that DNA methylation and the control of host genome-mediated actions work together to dictate transcription. Methylation of retrotransposon DNA renders it inactive, while demethylation activates it. However, most DNA methylation locations that have been found in Arabidopsis thaliana and rice are rich with retrotransposons and repetitive sequences, even though a large number of retrotransposons are methylated and continue inactive (Pereira, 2004; Zhang et al., 2006; Takata et al., 2007; Zilberman et al., 2007). Observing hyper-methylated and quiet genes in some T0 transgenic plants with a high copy number of the retrotransposon Tto1 suggests that the host genome was subjected to selection pressure. It can influence the DNA methylation status of retrotransposons. Demethylation of Tto1 DNA in such transgenic lines allowed the detection of many newly formed Tto1 DNA copies. This result highlighted the demethylation-induced restoration of transposition activities in tissue culture conditions described by Kajikawa et al. (2006). Other studies have found that the expression or silencing of ddm1, a gene necessary for maintaining the levels of DNA methylation level and histone H3 methylation structure in transgenic lines (Fujimoto et al., 2008), correlates with the methylation status of retrotransposons. Antisense RNA, also known as RNA interference, has been shown to regulate retrotransposon transposition activities in humans and animals (Schmidt et al., 2002). This is probable because RNA-RNA interactions are more common during the transcription of retrotransposons. Similar studies on plants have been published. In Arabidopsis thaliana and tobacco, retrotransposon methylation is associated with retrotransposon-mediated RNA interference and may contribute independently to retrotransposon silence (Hamilton et al., 2002). This is just one example of how RNA interference can be used for the differential processing of transcribed centromeric retrotransposons in a genome.

12. Retrotransposons and their Roles in the Evolution of Plant Genomes

Retrotransposons are widely distributed across the plant genome. However, their functional role and biological impact still need to be clarified, and their role in plants remains to be seen, although their presence is well known. Recent studies have suggested that retrotransposons may control gene expression and thus play a role in the development and evolution of plants. Further research is needed to determine their exact function in plants. Additionally, the impact of retrotransposons on the surrounding genetic material needs to be studied in greater detail. Understanding the role of retrotransposons could help us better manipulate plants for agricultural purposes. It is fortunate that, thanks to a combination of the data provided by the genome sequencing effort, as well as the fact that genetic and biological technology continues to develop, we now have a much clearer understanding of this issue, thanks mainly to the data provided by the genome sequencing effort. This understanding has allowed us to begin to explore techniques such as genetic engineering and gene editing to manipulate crops to better suit our needs. We can now develop more efficient and resilient plants for agricultural use. The retrotransposons are essential in regulating the structure, function, and development of new genes within a cell. Whenever retrotransposons are inserted into genes or the sequences adjacent to them, they can trigger complex circuits that regulate metabolism when inserted into those genes. This process, known as gene regulation, can create new proteins, which can cause changes in the organism's phenotype. This phenomenon has significantly impacted evolution, allowing organisms to adapt to new environments.

13.   Hybridization among Species and Chromosomal Recombination

Producing synthesized rapeseed by hybridizing B. oleracea with B. rapa is one way to increase oilseed rape breeding stock diversity. This has the potential to improve genetic variability for hybrid breeding and to extend the modern oilseed rape's gene pool, which is currently relatively small in terms of pest and disease resistance. In the case of crosses between crop species of Brassica and their relatives, embryo rescue techniques or protoplast fusion can successfully overcome interspecific and intergeneric incompatibility barriers. The release of cultivars bearing unique resistance genes from the diploid donor species has been achieved in some circumstances by backcrossing resynthesized rape forms with elite breeding material. The genetic and epigenetic alterations during polyploidization can be studied significantly using synthetic Brassica polyploids. For example, the first few generations of polyploids resulting from interspecific crosses between species with the A, B, and C genomes, for instance, documented extensive and rapid genome alteration in the form of loss or gain of restriction pieces. The alterations also showed the divergence between genotypes of the exact parental origin, showing that Brassica polyploids can produce fresh genetic variation in a relatively short time. Lukens et al. (2006) showed that extensive alteration in DNA methylation patterns after polyploidisation causes this high degree of divergence. In contrast, few deletions or insertions could be detected by examining the methylation status of many isogenic resynthesised rapeseed lines and their two common parental genotypes. Instead of significant genetic changes, polyploidy in B. napus is accompanied by extensive regulation of epigenetic modifications Soifer and Rossi, 2006.

14.   Polyploidization of Brassica

Homologous recombination can be expected to occur extensively in Brassica diploids following rapid polyploidization, as evidenced by the observed diversity in the degree of homologous pairing in distinct B. napus haploids. Udall et al. (2005) used co-dominant restriction fragment length polymorphism (RFLP) markers to analyze multiple B. napus DH mapping populations and identify chromosomal rearrangements that they then categorized as either de novo homologous non-reciprocal translocations (HNRT), previously existing HNRT, or homologous reciprocal translocations (HRT). In a small number of lines, 99 new HNRT were found to be duplications of specific areas of chromosomes accompanied by the deletion of the corresponding homologous region. These de-novo HNRT occurred at a higher frequency in a population with resynthesized B. napus as a parent, suggesting a higher rate of homologous recombination in the progeny of new polyploids. Fragment duplication and loss analysis of DH lines from three populations with natural B. napus parents revealed nine pre-existing HNRTs, indicating the segregation of HNRT from one of the parents. The results of this work suggest that homologous recombination-induced chromosomal rearrangements are common in B. napus, and it is reasonable to assume that the same is true of the two other amphidiploid species of Brassica, B. juncea and B. carinata. One of the most intriguing elements of this from a crop-breeding perspective is how changes in allele dosage can impact the additive expression of essential agronomic traits like disease resistance (Zhao et al., 2005).

15.   Ty1-copia Long-Tailed Repeat (LTR) Retrotransposons

The direct sequence of long terminal repeats (LTRs) at the end of LTR retrotransposons can be as short as a few bases or as long as tens of thousands of bases or hundred base pairs (bp) to more than five thousand base pairs (kb). LTR retrotransposons are typically flanked by direct repeats and are thought to be remnants of ancient retroviral infections. They are important genome elements and can affect gene expression and regulation. It is important to note that although LTRs do not code for any proteins, they contain the transcription promoters and terminators for LTR retrotransposons. These components are necessary for the replication and expression of the LTR retrotransposons. As a result, LTR retrotransposons can cause genomic instability by inserting themselves into new locations. This can lead to the disruption of host genes and can cause various diseases. As the final stage of the LTRs, short inverted repeats (often 50 - TG-30 and 50 - CA-30) are used as the final stages. This disruption of host genes can lead to various diseases and even contribute to certain cancers' development. Therefore, it is important to understand the structure and function of LTRs to understand how they can affect the genome. The LTR retrotransposons encode several proteins specified by the gag, pol, and int genes specified by the LTR retrotransposons. These proteins are involved in transposing the LTR retrotransposons and integrating the retrotransposon into the host genome. Understanding the structure and function of LTRs can help us better understand how they can disrupt the host genome. These genes and proteins can be specified by a single mRNA molecule using the Repeated RNA (R), Unique 50 RNA (U5), Primer Binding Site (PBS), Polypurine Tract (PPT), and Unique 30 RNA (U3). By studying the structure and function of these proteins, we can gain insight into how they can disrupt the host genome. Such knowledge can help us develop strategies to prevent or mitigate these disruptions. This means that the R at position 50 in the 50 LTR is where transcription begins, and the R at position 30 in the 30 LTR is where transcription ends. This knowledge can also help us develop better treatments and cures for genetic diseases caused by disruptions to the genome. Ultimately, we aim to understand and use this knowledge to improve human health. The results of this study suggest that the 30 LTR contains a promoter sequence, which can result in the read-through transcription of nucleotides downstream from the injected element as a result of the promoter sequence. This promoter sequence could be applied to gene therapy, regulating gene expression. In addition, this study suggests that the 30 LTR can be used to study gene regulation in other organisms. The polyprotein encoding protease (PR) synthesizes the polyprotein, which includes gag, pol, and int proteins, and cleaves them into functional peptides. PR is essential for the replication of the HIV-1 virus, and the promoter sequence can be used to control its expression. This could lead to new treatments for HIV-1 infection and other diseases caused by retroviruses. Retrotransposon RNA and proteins can mature and package themselves into a form that can be integrated into the genome through the help of the proteins encoded by the gag gene. This integration is transduction, the main route by which retroviruses are spread between cells. Transduction is a process dependent on PR activity, making it an attractive target for therapeutic intervention. Regarding the retrotransposon DNA form, the int gene encodes the integration mechanism that enables the retrotransposon DNA form to insert into an unfamiliar chromosomal site. In contrast, the pol gene encodes the reverse transcriptase and RNase H activities required for retrotransposon replication and transposition. Transduction has been proposed as a mechanism for the horizontal transfer of genetic material between species. It has also been suggested that transduction may contribute to the induced mutagenesis of cells and the generation of novel gene combinations. Depending on the virus, gag and int proteins can be encoded in single or many translational read frames. Transduction has been suggested as a way for viruses to acquire and spread new genetic information. This process can also lead to the emergence of new viruses that are more adapted to their environment. Due to the need for a frameshift or translational re-initiation downstream of the change in the frame to generate the correct peptides, these proteins may be less abundant than their upstream counterparts because they need to perform a frameshift or translational re-initiation. As a result, the new viral proteins may be expressed in lower quantities, and the virus may be less able to take advantage of the new genetic information. This could limit the spread of the new virus or cause it to be outcompeted by other viruses. Intracellular replication of LTR retrotransposons is a complicated process that begins with mRNA production. This mRNA acts as a template for replication by encoding information about the proteins that will be used in the process. A cellular RNA, often the 3′ end of a host tRNA, is complementary to the primer binding site on the mRNA molecule. When the mRNA of a retrotransposon binds to its corresponding tRNA in vivo, a short stretch of double-stranded RNA with a free 3′ hydroxyl from the tRNA is produced. These 3′ ends can be used as a primer by reverse transcriptase to make a DNA complement to the R and U5 regions of the 5′ LTR. However, it is important to note that the reverse transcriptase cannot synthesize more DNA at the 5′-end of the retrotransposon mRNA once it has reached the 5′-end. Suppose the retrotransposon-encoded RNase H molecule preferentially digests the RNA molecule in any DNA: RNA hybrid. In that case, it releases a single-stranded DNA molecule with a sequence similar to the R sequence found at the 30 ends of retrotransposon mRNAs. After these sequences hybridize with one another, a circular structure is formed that allows the reverse transcription of DNA to continue until it can synthesize a strand of DNA that is comparable to all the sequences that are found within the element. This newly synthesized strand of

DNA can then be used as a template for further replication of the element, allowing it to propagate throughout the genome. The result is a novel gene that can be used in various applications. Only 5′ to 3′ polypurine tracts (PPTs) and RNase H are required during the second-strand DNA synthesis process. As the newly synthesized double-stranded linear DNA molecule cleaves both the donor and target molecules, integrase can incorporate the newly synthesized double-stranded linear DNA molecule into the target genome. This process creates a recombinant molecule, which can be used in various applications such as gene therapy, synthetic biology, and biotechnology. There is evidence that the DNA has snipped at intervals of 3 to 5 base pairs, resulting in a direct repeat in the flanking region of the target that is approximately 3 to 5 bases long. This "direct repeat" is thought necessary for the binding of the cleaving enzyme and is likely the key factor in determining the target site for the cut. This also helps explain why the cutting action of the endonuclease is specific for that particular DNA sequence Kuipers et al., 1998.

16.   Genetic Application of Retrotransposons

16.1   *Molecular Identifiers for use in Genealogical, Phylogenetic, and Biodiversity Studies*

There is no doubt that sequence-specific amplification polymorphism (SSAP) is one of the best retrotransposon-based marker systems for discovering many highly polymorphic markers. These markers can be used in various applications, such as population genetics, phylogenetics, and genetic mapping. SSAP markers are also helpful for identifying genetic variation within a species. Furthermore, they can be used to assess genetic diversity in natural populations. A similar technique is called SSAP, which uses bands on a sequencing gel to show the location of each retrotransposon insertion, which is identical to multivariate amplified fragment length polymorphisms (AFLP). This technique provides a powerful tool for population studies and can be used to identify the genetic differences between closely related species. It can also be used to understand the evolutionary history of a species by tracking changes in the genetic material over time. Furthermore, the SSAP-based markers appear to be superior to standard AFLP-based markers when calculating phylogenetic relationships in plants because one of the primers is based on the sequence of a particular retrotransposon (Ellis et al., 1998). This makes SSAP-based markers more reliable for phylogenetic studies of plant species. SSAP-based markers have also been used to detect genetic variation within species. This can be used to help inform conservation and management decisions. Moreover, the evolutionary relationships between legumes (Pearce et al., 1999) and cereal plants have recently been estimated using a multi-retrotransposon technique to estimate evolutionary relationships. SSAP-based markers can help detect crops' origin and domestication history. This information can be used to inform agricultural practices and improve crop production. Additionally, SSAP-based markers can detect genetic variability and identify plant populations at risk of extinction. This method is beneficial because every chemical element has its transpositional history. SSAP-based markers can be used to trace the geographic distribution of species, helping to identify areas of high biological diversity. This information can help protect threatened species and ensure their survival. Studies of genetic linkage and intraspecific genetic diversity can benefit from including retroelements that have recently transferred within a species and thus are highly polymorphic to study genetic association and intraspecific genetic diversity. This can provide researchers with valuable information about the species, such as the potential for population adaptation and genetic diversity. It can also help inform conservation efforts and provide insight into species evolution. It is essential to recognize that some species that have not been active for a million years or more could help us understand the connections between species and even between genera. This knowledge can be used to protect currently threatened species and help restore the balance of ecosystems. It can also be used to develop new species with characteristics that benefit humans. Finally, it could even be used to create unique, more sustainable food production methods. Due to the lack of LTR sequences, retrotransposon-based SSAP has historically failed to succeed. Therefore, it is essential to develop new strategies to further our ability to access and analyze LTR sequences. With the right tools, we can better understand the evolutionary history of species and the way they interact with their environment. With the recent invention of a new technology for quick isolation of plant Ty1-copy retrotransposons LTR sequences for molecular research, implementing an SSAP method based on retrotransposons is now widely available for all plant species. In addition to subcloning, sequencing, and analysing SSAP bands, the SSAP approach should also be suitable for detecting specific integration patterns of particular components in the host genome using the SSAP strategy. This technique can be used to identify the insertion pattern of a gene in the genome and to determine its role in gene expression. Furthermore, it can be used to study the evolutionary history of a particular gene in the plant Kowalski et al., 1994.

16.2   Content and Translation Control

The frequency of retrotransposon transposition into the host genome is controlled by modulating the expression of retrotransposons in plants and other eukaryotic organisms. Reducing the potential for harmful effects of retrotransposons in the host genome may require the emergence of control mechanisms for their transcription and transposition. Thus, it should be no surprise that plant retrotransposons are transcriptionally silent in most plant tissues during development. The most straightforward strategy to regulate retrotransposon activity would be modulation of transcriptional initiation because retrotransposons cannot transfer without an RNA template accessible for reverse transcription. The development and the environment handle several retrotransposons in unusual ways. For example, Tobacco Tnt1, barley BARE-1, and maize PREM-2 transcripts have been found mainly in roots, leaves, and immature microspores (Ramsay et al., 1999). Like humans, hormones and developmental cues trigger retrotransposon expression in

S. cerevisiae and Drosophila (Jensen et al., 1999). Plant genomes are littered with retrotransposons, a type of mobile genetic element. Their movement and ability to transpose genetic material have crucial roles to play in the development of plants. Genome structure analysis, phylogenetic investigation, biodiversity study, genetic map building, gene cloning and function analysis, mutation material production, and so on all benefit significantly from using retrotransposons as a type of genetic tool. The mechanism of retrotransposon distribution and its characteristics within the plant genome will become apparent as more data accumulates from genome sequencing and cutting-edge molecular experimentation instrumentation Kampken and Kuck., 1998. We plan to delve deeper into the connection between histone modification and the regulation of retrotransposon activity. The life cycles of crops produced in the field are affected by several environmental factors. Understanding the mechanism of this mutual interaction between retrotransposon activity and stressors will provide scientific preparedness for agriculture's long-term viability. There is a long history of research into the mechanism of heterosis because of its potential to increase agricultural yield. Large heterosis (typically on biomass yield) and increased retrotransposon activities result from mating between geographically separated species. New insights into the mechanism of heterosis may be uncovered if the link between the two is clarified (Warwick et al., 2003).

Conclusion

The distribution of retrotransposons in the Brassica oleracea genome provides insights into retrotransposon patterns' evolutionary history and genetic diversity. In light of these results, various aspects of the vast genomic alteration and development of the Brassicas genus in response to selective pressure can be analysed more precisely. Comparisons across species or subspecies and genetic data could shed light on the origins of the diploid species, which appear to have an allohexaploid ancestry—the identification of ancestral taxa of the tetraploids. There is mounting proof that retrotransposons have a role in gene transfer and suppression, especially in response to environmental stress, such as that experienced by plants during tissue culture or the creation of transgenic varieties. As a result, it may be helpful to learn how to recognize components that are currently active or have the potential to become so; Retrotransposons in Brassica provide valuable markers due to insertional polymorphisms. Comprehending the ABC connection genomes to the fully decoded Arabidopsis genome is aided by familiarity with the types of repetitive DNA families found across the genus.

Acknowledgement: It is with great pleasure that we thank the Department of Agronomy at Lovely Professional University, Phagwara, Punjab, for providing us with the support we needed and consistent moral support.

Conflicts: None

References

Alix K., Ryder C.D., Moore J., King G.J., and Pat Heslop-Harrison J.S. (2004). The genomic organisation of retrotransposons in Brassica oleracea. *Plant Mol. Biol.* 59: 839–51.

Antonius-Klemola, K., Kalendar, R., and Schulman A.H. (2006). TRIM retrotransposons occur in apple and are polymorphic between varieties but not sports. *Theor. Appl. Genet.* 112: 999–1008.

Babula, D., Kaczmarek, M., Barakat, A., Delseny, M., Quiros, C.F. and Sadowski, J. (2003). Chromosomal mapping of Brassica oleracea based on ESTs from Arabidopsis thaliana: complexity of the comparative map. *Mol. Gen. Genomics* 268: 656–65.

Benabdelmouna, A. and Darmency, H. (2003). Copia-like retrotransposons in the genus Setaria: Sequence heterogeneity, species distribution and chromosomal organization. *Plant Syst. Evol.* 237: 127–36.

Bennetzen, J.L. (2005). Transposable elements, gene creation and genome rearrangement in flowering plants. *Curr. Opin. Genet. Dev.* 15: 621–27

Cam, H.P., Noma, K.I., Ebina, H., Levin, H.L., and Grewal, S.I. (2008). Host genome surveillance for retrotransposons by transposon-derived proteins. Nature, 451(7177), 431–436.

Charles, M., Belcram, H., Just, J., Huneau, C., Viollet, A., Couloux, A., ... and Chalhoub, B. (2008). Dynamics and differential proliferation of transposable elements during the evolution of the B and A genomes of wheat. *Genetics, 180*(2), 1071–86.

Crénès, G., Moundras, C., Demattei, M.V., Bigot, Y., Petit, A., and Renault, S. (2010). Target site selection by the mariner-like element, Mos1. *Genetica, 138*, 509–517.

Ellis, T.H.N., Poyser, S.J., Knox, M.R., Vershinin, A.V., and Ambrose, M.J. 1998. Polymorphism of insertion sites of Ty1-copia class retrotransposons and its use for linkage and diversity analysis in pea. *Mol. Gen. Genet.* 260: 9–19.

Fujimoto, R., Sasaki, T., Inoue, H., and Nishio, T. 2008. Hypomethylation and transcriptional reactivation of retrotransposon-like sequences in ddm1 transgenic plants of Brassica rapa. *Plant Mol. Biol.* 66: 463–73.

Grandbastien, M.A., Audeon, C., Bonnivard, E., Casacuberta, J.M., Chalhoub, B., Costa, A.P., Le, Q.H., Melayah, D., Petit, M., Poncet, C., Tam, S.M., Van Sluys, M.A., and Mhiri, C. 2005. Stress activation and genomic impact of Tnt1 retrotransposons in Solanaceae. Cytogenet. *Genome Res.* 110: 229–41.

Gray, Y.H. (2000). It takes two transposons to tango: transposable-element-mediated chromosomal rearrangements. Trends in Genetics, 16(10), 461–468.

Hamilton, A., Voinnet, O., Chappell, L., and Baulcombe, D. (2002). Two classes of short interfering RNA in RNA silencing. EMBO J. 21: 4671–19.

Hirochika, H., Okamoto, H., and Kakutani, T. (2000). Silencing of retro transposons in arabidopsis and reactivation by the ddm1 mutation. *Plant Cell* 12: 357–69.

Howell, E.C., Barker, G.C., Jones, G.H., Kearsey, M.J., King, G.J., Kop, E.P., Ryder, C.D., Teakle, G.R., Vicente, J.G. and Armstrong, S.J. (2002). Integration of the cytogenetic and genetic linkage maps of Brassica oleracea. *Genetics* 161: 1225–34.

Hudakova, S., Michalek, W., Presting, G.G., Hoopen, R., dos Santos, K., Jecencakova, Z. and Schubert, I. (2001). Sequence organisation of barley centromeres. Nucleic Acids Res. 29: 5029–35.

Jensen, S., Gassama, M.P. and Heidmann, T. (1999). Taming of transposable elements by homology-dependent gene silencing. Nature genetics, 21(2), 209–212.

Kajikawa, Y. (2006). Texture development of non-epitaxial polycrystalline ZnO films. Journal of Crystal Growth, 289(1), 387–394.

Kempken, F. and Kuck, U. (1998). Transposons in filamentous fungi-facts and perspectives. BioEssays 20: 652–59.

Komatsu, M., Shimamoto, K. and Kyozuka, J. (2003). Two-step regulation and continuous retrotransposition of the rice LINE-type retrotransposon Karma. Plant Cell 15: 1934–4.

Kowalski, S.P., Lan, T.-H., Feldmann, K.A. and Paterson, A.H. (1994). Comparative mapping of Arabidopsis thaliana and Brassica oleracea chromosomes reveals islands of conserved organization. Genetics 138: 499–510.

Kuipers, A.G.J., Heslop-Harrison, J.S. and Jacobsen, E. (1998). Characterisation and physical localisation of Ty1-copia-like retrotransposons in four Alstroemeria species. *Genome* 41: 357–67.

Kurata, N., Nonomura, K.-I. and Harushima, Y. (2002). Rice genome organization: the centromere and genome interactions. Ann. Bot. 90: 427–35.

Kwon, S.J., Kim, D.H., Lim, M.H., Long, Y., Meng, J.L., Lim, K.B., Kim, J.A., Kim, J.S., Jin, M., Kim, H.I., Ahn, S.N., Wessler, S.R., Yang, T.J., Park, B.S. (2007). Terminal repeat retrotransposon in miniature (TRIM) as DNA markers in Brassica relatives. Mol. Genet. Genomics 278: 361–7.

Leeton, P.R. and Smyth, D.R. (1993). An abundant LINE-like element amplified in the genome of Lilium speciosum. Molecular and General Genetics MGG, 237, 97–104.

Liu, Z., Yue, W., Li, D., Wang, R.R., Kong, X., Lu, K., Wang, G., Dong, Y., Jin, W., and Zhang, X. (2008). Structure and dynamics of retrotrans posons at wheat centromeres and pericentromeres. *Chromosoma* 117: 445–56

Llorens, C., Futami, R., Covelli, L., Dominguez-Escriba, L., Viu, J.M., Tamarit, D., Anguilar-Rodriguez, J., Vicente-Ripolles, M,, Fuster, G., Bernet, G.P., et al. (2011). The Gypsy database (GyDB) of mobile genetic elements: release 2.0. Nucleic Acids Res 39: 70–74.

Lukens, L.N., Pires, J.C., Leon, E., Vogelzang, R., Oslach, L., and Osborn, T. (2006) Patterns of sequence loss and cytosine methylation within a population of newly resynthesized Brassica napus allopolyploids. Plant Physiol 140: 336Y348.

Nefedova, L.N., Mannanova, M.M., and Kim, A. I. (2011). Integration specificity of LTR-retrotransposons and retroviruses in the Drosophila melanogaster genome. *Virus Genes, 42*, 297–306.

Parkin, I.A., Gulden, S.M., Sharpe, A.G., et al. (2005). Segmental structure of the Brassica napus genome based on comparative analysis with Arabidopsis thaliana. *Genetics* 171: 765Y781.

Pearce, S.R., Stuart-Rogers, C.M., Kumar, A.H.N., and Flavell, A.J. (1999). Rapid isolation of plant Ty1-copia group retrotransposon LTR sequences for molecular marker studies. Plant J. 19: 711–17.

Peterson-Burch B.D., Nettleton D., Voytas D.F. (2004). Genomic neighborhoods for Arabidopsis retrotransposons: a role for targeted integration in the distribution of the Metaviridae. Genome Biol. 5:R78 10.1186/gb-2004-5-10-r78

Pereira, V. (2004). Insertion bias and purifying selection of retrotransposons in the Arabidopsis thaliana genome. Genome Biol. 5: R79.

Ramsay, L., Macaulay, M., Cardle, L., Morgante, M., Degli-Ivanissevich, S., et al., 1999. Intimate association of microsatellite repeats with retrotransposons and other dispersed repetitive elements in barley. Plant J. 17: 415–23.

Ryder, C.D., Smith, L.B., Teakle, G.R. and King, G.J. (2001). Contrasting genome organisation: two regions of the Brassica oleracea genome compared with collinear regions of the Arabidopsis thaliana genome. *Genome* 44: 808–17.

Schmidt, R., Acarkan, A., and Boivin K. (2001). Comparative structural genomics in the Brassicaceae family. Plant Physiol Biochem 39: 253Y262

Sharma, A., and Presting, G.G. (2008). Centromeric retrotransposon lineages predate the maize/rice divergence and differ in abundance and activity. *Mol. Genet. Genomics* 279: 133–47.

Simillion, C., Vandepoele, K., Van Montagu, M.C.E., Zabeau, M. and Van de Peer, Y. (2002). The hidden duplication past of Arabidopsis thaliana. *Proc. Natl. Acad. Sci.* USA 99: 13627–32.

Soifer, H.S., and Rossi, J.J. 2006. Small interfering RNAs to the rescue: blocking L1 retrotransposition. *Nat. Struct. Mol. Biol.* 13: 758–9.

Song, K., Lu, P., Tang, K., Osborn, T.C. (1995) Rapid genome change in synthetic polyploids of Brassica and its implications for polyploid evolution. *Proc Natl Acad Sci.* USA 92: 7719Y7723.

Takata, M., Kiyohara, A., Takasu, A., Kishima, Y., Ohtsubo, H., and Sano, Y. (2007). Rice transposable elements are characterized by various methylation environments in the genome. BMC Genomics 8: 469.

Terol, J., Castillo, M.C., Pe´rez-Alonso, M. and de Frutos, R. (2001). Structural and evolutionary analysis of the copia-like elements in the Arabidopsis thaliana genome. *Mol. Biol. Evol.* 18: 882–92.

Tu, Z. (2001). Eight novel families of miniature inverted-repeat transposable elements in the African malaria mosquito, Anopheles gambiae. *P Natl Acad Sci.* USA 98: 1699–1704.

Udall, J.A., Quijada, P.A., and Osborn, T.C. (2005) Detection of chromosomal rearrangements derived from homoeologous recombination in four mapping populations of Brassica napus L. Genetics 169: 967Y979.

Vershinin, A.V., Druka, A., Alkhimova, A.G., Kleinhofs, A. and Heslop-Harrison, J.S. (2002). LINEs and gypsy-like retrotransposons in Hordeum species. *Plant Mol. Biol.* 49: 1–14.

Vicient, C.M., Jaaskelainen, M.J., Kalendar, R., and Schulman, A.H. 2001. Active retrotransposons are a common feature of grass genomes. *Plant Physiol.* 125: 1283–89.

Vision, T.J., Brown, D.G. and Tanksley, S.D. (2000). The origins of genome duplications in Arabidopsis. *Science* 290: 2114–17.

Wang, G., He, Q., Macas, J., Novák, P., Neumann, P., Meng, D., Zhao, H., Guo, N., Han, S., Zong, M., Jin, W., and Liu, F. (2017) Karyotypes and distribution of tandem repeat sequences in Brassica nigra determined by fluorescence in situ hybridization. *Cytogenet Genome Res.* 152: 158–65.

Wang, Y.P., Zhao, X.X., Sonntag, K., Wehling, P., and Snowdon, R.J. (2005) GISH analysis of BC1 and BC2 progenies derived from somatic hybrids between Brassica napus and Sinapis alba. Chromosome Res 13: 819Y826.

Wicker, T., Sabot, F., Hua-Van, A., Bennetzen, J.L., Capy, P., Chalhoub, B., ... & Schulman, A.H. (2007). A unified classification system for eukaryotic transposable elements. *Nature reviews genetics*, 8(12), 973–82.

Witte, C.P., Le, Q.H., Bureau, T., and Kumar A. (2001). Terminal-repeat retrotransposons in miniature (TRIM) are involved in restructuring plant genomes. *Proc. Natl. Acad. Sci.* USA 98: 13778–83.

Yang, T.J., Kwon, S.J., Choi, B.S., Kim, J.S., Jin, M., Lim, K.B., Park, J.Y., Kim, J.A., Lim, M.H., Kim, H.I., Lee, H.J., Lim, Y.P., Paterson, A.H., and Park, B.S. 2007. Characterization of terminal-repeat retrotransposon in miniature (TRIM) in Brassica relatives. *Theor. Appl. Genet.* 114: 627–36.

Zhang, X., Yazaki, J., Sundaresan, A., Cokus, S,. Chan, S.W., Chen, H., Henderson, I.R., Shinn, P., Pellegrini, M., Jacobsen, S.E., and Ecker, J.R. 2006. Genome-wide high-resolution mapping and functional analysis of DNA methylation in Arabidopsis. Cell 126: 1189–1201.

Zhang, X., and Wessler, S.R. (2004). Genome-wide comparative analysis of the transposable elements in the related species Arabidopsis thaliana and Brassica oleracea. *Proc. Natl. Acad. Sci.* USA 101: 5589–94.

Zhao, J., Udall, J., Quijada, P., Grau, C., Meng, J., and Osborn, T. (2005) Quantitative trait loci for resistance to Sclerotinia sclerotiorum and its association with a homeologous non-reciprocal transposition in Brassica napus L. *Theor Appl Genet* 7: 1Y8.

Zilberman, D., Gehring, M., Tran R.K., Ballinger T., and Henikoff, S. 2007. Genome-wide analysis of Arabidopsis thaliana DNA methylation uncovers an interdependence between methylation and transcription. *Nat. Genet.* 39: 61–69.

11

Retrotransposons in Nicotiana Plant Genome

Sehrish Ijaz,[1] *Ayesha Ghazanfar,*[1] *Vajiha Sahar Khan,*[1]
Ummara Waheed,[1] *Akash Fatima*[1] *and Zulqurnain Khan*[1*]

1. Introduction

Retrotransposons are genetic components that can change their location and copy number and are important in plant genome structural evolution (Daboussi, 1997). Earlier reports show the discovery of retrotransposons in yeast and some animals, later on, they were found to be present in almost all plant genomes by contributing a significant genome portion (Grandbastien, 1998). Retrotransposons proliferate by reversing their RNA and integrating the resultant cDNA into another place in the genome. Maximum possibility to significantly change gene functionality and also mutate the structure, this is common for retrotransposons for trans-positional activities activated on a host as well as by their own, possibly to prevent any deleterious effect on host and retrotransposon survival. Tobacco was used to isolate the first active plant retrotransposon by looking for insertions into the nitrate reductase (NR) gene (Wessler, 1996).

Classification of Retrotransposons

Retrotransposons have two main classes:

(i) Long terminal repeats (LTR) retrotransposons

LTR retrotransposons have flanked repeats to the terminal region responsible for the encoding of several proteins for transposition. LTR-retrotransposons reproduce via an RNA intermediate (copy-and-paste method), generating additional element copies that increase the resulting size of the host genome following integration.

---

[1]   Institute of Plant Breeding and Biotechnology MNS-University of Agriculture, Multan.
*   Corresponding author: zulqurnain.khan@mnsuam.edu.pk

There is substantial evidence that this mechanism is one of the primary drivers of genome size evolution (Wicker et al., 2007).

These retrotransposons are essential for genetic variation and influence the structure of the plant genome (Wright and Voytas, 1998). They have been implicated in numerous chromosomal alterations and played an essential role in genome size variability (Petit et al., 2007). LTR retrotransposons contain particular structural genes and the genes involved in enzymatic activities. For example, the gene gag encodes for a protein of structure (virus-like nucleocapsid manufacturing), a site where reverse transcription occurs. The pol gene performs various enzymatic activities, having a protease (involved in the cleavage of the Pol polyprotein), and reverse transcriptase (RT) that converts the RNA of the retrotransposon on cDNA, and an integrase enzyme that bound into the genome (Boeke and Chapman, 1991)

They can be further classified into two domains. This classification is based on their coding domain order.

  (a) Ty1/copia LTR retrotransposons
  (b) Ty like LTR retrotransposons

These retrotransposons can increase their copy number upon activation by various kinds of stresses and are involved in gene regulation pathways to combat both biotic and abiotic stresses (Havecker et al., 2004).

The replication mechanism of these retrotransposons involves the transcription of integrated copies followed by the translation and as well as the reverse transcription of RNA into cDNA, just like any other biological genes. As previously discussed, two open reading frames encode a Gag and the POL polyprotein. The former protein is involved in the formation of the nucleocapsid while the latter one performs the protease activity (Petit et al., 2007). Moreover, to perform reverse transcription and integration, other ORFs encode for reverse transcriptase enzyme and RNaseH and integrase to insert new copies in the genome (de Assis et al., 2020).

The internal area also contains the gag and pol genes, which encode all of the proteins required for retrotranslation and integration activities that the cell does not provide. In addition to these genes, a primer binding site (PBS) of 10–20 nucleotide (nt) close to the 5′ LTR partially binds with cytoplasmic tRNA's 3′ end. This region is used to prime the synthesis of the first DNA strand. There is another polypurine tract (PPT), a short (8–49 nt) purine-rich DNA present near the 3′ LTR that is used to prime the synthesis of the second DNA strand during retrotranslation for retrotransposition activity (Vicient and Casacuberta, 2020), (Kumar and Bennetzen, 1999). Tissue culture procedures have been used to characterize tobacco Tnt1 LTR sequences in numerous higher plants. Tnt1 transposition in tobacco mesophyll protoplasts has demonstrated utility in plants (Wang and Han, 2022). It was found that the Two Tnt1 elements were expressed during tissue culture in transgenic tobacco (leaf-derived protoplasts) however their expression was not found in leaf tissue, suggesting the role of these elements in somaclonal variation and tissue culture-induced mutagenesis (Hirochika, 1993). Previous reports show that Tnt1 transposition might be involved in the gene regulation in tobacco against several environmental stresses (Melayah et al., 2001). For example, In Medicago truncatula, a Tnt1 element activated the protoplast culture, resulting in the number of copy insertions per plant, and copy numbers remained constant throughout the life cycle (d'Erfurth et al., 2003).

Activating or silencing LTR retrotransposons in plants may result in beneficial epigenetic changes. However, epigenetic modification, involving LTR retrotransposons as candidates advantages including genomic stability, the introduction of new genetic variations, change of gene functionality, addition of new function, gene imprinting and stress tolerance, and so on. H owever, there are some disadvantages also linked like any drastic mutation, gene function losses, genetic rearrangements, and larger gene copy numbers (Ramakrishnan et al., 2023).

(ii) Non-LTR retrotransposons

These transposons were discovered first in mammalian genomes, so now have been reported in invertebrates, fungi, and also in plants. Non-LTR retrotransposons are often referred to as Retroposons. All transposed copies encode the basic structural and functional proteins for the retro-transposition cycle (Weiner, 2002).

(a) LINEs (long interspersed nuclear elements)
(b) SINEs (short interspersed nuclear elements)

LINEs (Long Interspersed Nuclear Elements)

LINEs, also known as L1 or LINE-like sequences, lack extensive terminal repeats and feature a poly(A) tail that defines the element's 30 terminus. LINEs are long enough up to many kilobases and contain two ORFs i.e. one for gag protein (ORF1), and the other (ORF2) for endonuclease and reverse transcriptase domains activities, that enable them to autonomously retro-transpose. Reverse transcriptase is a crucial enzyme for retro-transposition and high sequence conservation is exhibited in all seven domains of retroviral RNA-directed DNA polymerases (Schmidt, 1999). Both ORFs in many LINEs feature cysteine-rich, zinc-finger-like sequences that are thought to be potential nucleic acid-binding domains. The gag protein's zinc finger has the structure CX2CX4HX4C. Although the number of known plant LINEs is increasing, only a few have been investigated in depth (Noma et al., 1999). The first plant LINE was Cin4 which was identified as an insertion in the 30-untranslated region of *Zea mays* A1 genes. After that several LINEs have also been identified in DNA mitochondrial in Arabidopsis thaliana. (Knoop et al., 1996)

SINEs (Short Interspersed Nuclear Elements)

SINEs also are another type of TEs in the genome of eukaryotic, enabled to move from one location to another in the genome and replicate independently from DNA genomic replication (Wicker et al., 2007). SINE has also been identified particularly in plants whose genomes are sequenced along with a few numbers of other plant species. SINEs refer to short nonautonomous retroelements that can be

500 nucleotides in length, and rely on encoded proteins that are partner LINE for retro-transposition. Conservatively, a particular SINE has one unique obligatory partner (Konkel et al., 2010) so the others can have more partners. Several SINEs have been reported in the following plant species (Table 1). However, The SINEs identified were p-SINE1 in tobacco and TS in rice (Umeda et al., 1991).

Table 1   List of plant species with reported retrotransposons

S.No.	Plant Species	References
1	*Brassica napus*	(Deragon et al., 1994)
2	*Oryza sativa*	(Mochizuki et al., 1992)
3	*Nicotiana tabacum*	(Yoshioka et al., 1993)
4	*Solanum tuberosum*	(Wenke et al., 2011)
5	Sugar beet	(Schwichtenberg et al., 2016).

2.   Retrotransposons in Nicotiana Species

2.1 *Nicotiana tabacum*

Nicotiana, a genus of tobacco plants, is a member of the Solanaceae family, with at least 3,000 species documented to date (Olmstead et al., 2008).

Table 2   List of Nicotiana species with identified retrotransposons

S. No.	Species Name	References
1	*N. benthamiana*	(Naim et al., 2012)
2	*N. sylvestris*	(Sierro et al., 2013)
3	*N. tomentosiformis*	(Sierro et al., 2013)
4	*N. otophora*	(Sierro et al., 2014)
5	*N. attenuate*	(Xu et al., 2017)
6	*N. obtusifolia*	(Xu et al., 2017)
7	*N. rustica*	(Sierro et al., 2018)
8	*N. undulata*	(Sierro et al., 2018)
9	*N. paniculate*	(Sierro et al., 2018)
10	*N. knightiana*	(Sierro et al., 2018)

Several plant retrotransposons which have been studied so far are transcriptionally activated by different kinds of stress conditions. Many abiotic stressors, having protoplast isolation, wounding, methyl jasmonate, CuCl2, cell culture, significantly boost the expression in tobacco Tnt1 retrotransposons (Grandbastein et al., 2005; Grandbastein, 2015).

Similarly, numerous biotic factors of stress such as extracts like fungal from *Trichoderma viride*, bacterial, also from fungal pathogens (Grandbastien et al., 1998) have been described as retrotransposons. Tos17 transcription, unlike Tnt1 and Tto1, is solely stimulated by tissue culture. The RT-PCR approach was used to isolate Tobacco Tto5 as a salicylic acid-inducible retrotransposon.

A huge number of the Tnt1 family of copia-like retrotransposons is present in *Nicotiana tabacum*, with many members performing actively out of them (Le et al., 2007). It was reported that Tnt1A, one among the members having active transposition activities, shows induced expression, particularly under pathogen-derived stress (Grandbastien et al., 2005). As a result, it is an excellent option for analyzing the specific-site orientation of insertions that are new for the spread of existing insertions to be among the retrotransposons that have less elements with verified transposition, Tnt1A expression is strongly induced by stress pathogen-induced (Grandbastien et al., 2005). Tnt1A expression and amplification can be stimulated by fungal cell-wall disintegrating extracts utilized in protoplast isolation. We used a transposon-anchored approach known as sequence-specific amplification polymorphism to find unique integrations in 25% of plants created by protoplasts, relevant to only 2.7% of plants formed from cultures exposed to cell-wall destroying enzymes (Waugh et al., 1997). The additional advantage of SSAP is that it enables the recovery of variable-sized amplification products that contain one Tnt1 extremity and a length of adjacent genomic sequences. Tnt1 sequence diversity was examined, but no attempt was made to determine the make-up of the host sequences surrounding the Tnt1 insertions at the opposite SSAP fragment end (Melayah et al., 2001).

Tissue culture shows a promising technique to characterize different transposons in tobacco, e.g. potential of Tnt1 LTR sequences was explored by characterizing tobacco mesophyll protoplasts (Grandbastien et al., 1989).

Tissue culture techniques have been used to characterize tobacco Tnt1 LTR sequences in several higher plants. The potential of Tnt1 as a plant genetic engineering tool was demonstrated by its transposition in tobacco mesophyll protoplasts (Grandbastien et al., 1989). Two elements of Tnt1 transgenic tobacco were also leaf-derived expressed protoplast but also not involved in leaf tissues, suggesting are molecular foundation for variation and tissue culture-induced mutagenesis somaclonal may be provided by the transcription characteristics of Tnt1 (Pouteau et al., 1991). As a result, Tnt1 transposition may significantly contribute to the activation of the genetic plasticity hosts in response to environmental stress. Fungi extracts are capable of efficiently stimulating Tnt1 transposition and increasing the amount of new Tnt1 with high sequence copy similarity to subpopulations (Grandbastien et al., 2005).

2.2 *N. sylvestris*

Naturalized throughout the Andes from Bolivia to Argentina, woodland tobacco (*Nicotiana sylvestris*) is currently primarily grown as a decorative plant. In the Andes, *Nicotiana tomentosiformis* also exists in nature, but over a larger area, from Peru to Argentina (Knapp et al., 2004). There are members of clades that originated in the *Solanaceae* family's *Nicotiana* sections, which split approximately 15 million years ago (Sierro et al., 2013). Many species that are crucial to agriculture, such as the tomato, potato, eggplant, and pepper, are also members of this family. Around 200,000 years ago, *N. Sylvestris*, the presumed father donor, and *N. tomentosiformis*, the maternal donor, interbred to create the $(2n = 4x = 48)$, the ctobacco (Leitch et al., 2008). As a result, it is anticipated that the sequences of *N. tomentosiformis and N. sylvestris* will share a high degree of similarity with the T- and S- of *N. tabacum*, respectively. These are crucial comprehending processes, such as how allotetraploid *N. tabacum* species regulate gene expression. In investigations of terpenoid biosynthesis in glandular trichomes (Ennajdaoui et al., 2010), programming of plastid genomes (Thyssen et al., 2012), mitochondrial activity, herbicide resistance, and plant virus resistance (Sekine et al., 2012), *N. sylvestris* is used as a diploid model system in plant biology. The Solanaceae family generates a variety of alkaloids that are poisonous to insects, and *N. sylvestris* is no different. This is a well-known defense mechanism for the plant against herbivore damage to its leaves and flowers. About three times as large as the potato and tomato genomes, respectively, are the size of the genome of *N. sylvestris*. This difference could be attributed to repeat growth in *Nicotiana* genomes brought on by the buildup of transposable elements. This theory is supported by C^0t measurements in the *N. tabacum* genome, which revealed the existence of 55% short (about 300 nucleotides) and 25% long (around 1,500 nucleotides) repetitions (Zimmerman and Goldberg, 1977). Pepper euchromatin doubled in size similar to tomato euchromatin through a substantial increase in specific (LTR) retrotransposons (Park et al., 2011). The *N. sylvestris* genome showed indications of more recent repeated expansions with higher homogeneity, while the *N. tomentosiformis* genome showed considerably greater repeat variety. Additionally, it was claimed that the *N. sylvestris* genome differed from that of *N. tomentosiformis* that it included more Tnt1 transposons and had a more equal distribution of the elements (Gazdová et al., 1995). A more thorough investigation revealed that *N. Sylvestris* had larger relative copy counts of four retrotransposons than *N. tomentosiformis* (Tnt1-OL13, Tnt1-OL16, Tnt2d, and Tto1-1R). Repeat elements make up over 70% of both genomes, It seems that *N. tomentosiformis* has more copia-type LTRs than *N. sylvestris* (13.43% vs. 9.13%, respectively) and retrotransposons (13.05% vs. 10.33%, respectively), whereas 20% of the LTRs in both genomes are gypsy-like. The difference between the total size of sequenced DNA and repeat-masked DNA is used to estimate the size of the gene-rich DNA, which is thought to be approximately 625 Mb for *N. sylvestris* and 425 Mb for *N. tomentosiformis*. There are 7.39 percent and 3.98% more Tnt1 retrotransposons in *N. tomentosiformis* than in N. sylvestris, respectively (Horáková and Fajkus, 2000).

2.3   *N. benthamiana*

For more than 20 years, *N. benthamiana* has been represented as a model species for the study of plants. The capacity of a plant species to quickly and effectively modulate gene expression through plant genetic engineering is one requirement for classification as a "model" species. This small Australian plant exhibits many characteristics that make it especially well-suited for the production of biopharmaceuticals, including a quick growth rate and a natural capacity to reflect heterologous gene sequences (Lomonossoff and D'Aoust, 2016). However, rather than crop sciences, the fame of *N. benthamiana* is primarily attributable to its adaptability for analyses of plant-pathogen interaction and molecular farming (Schillberg et al., 2002)

Nine full-length retrozymes are represented by the dozens of retrozyme loci found in *the N. benthamiana* genome. We show that the retrozyme LTR region has a promoter that can control the synthesis of retrozyme RNA and that retrozyme LTR is massively methylated in the leaves of *N. benthamiana*. The effectiveness of chromatin immunoprecipitation is verified by performing PCR on N. benthamiana IP samples using primer pairs specific to elongation factor 1 alpha (EF-1a) and Ty1-copia retrotransposon (Vimont et al., 2020). It is anticipated that the EF-1a gene and Ty1-copia retrotransposon's histone modification statuses will be enriched in H3K4me3 and H3K9me2 marks.

The *N. benthamiana* genome contains hundreds of retroloci, nine of which are full-length retro. We show that the retrozyme LTR region in N. benthamiana leaves is extensively methylated, indicating that it is transcriptionally inactive. It also has a promoter that can regulate the synthesis of retrozyme RNA. 24-nucleotide-long RNAs generated by retrozymes are present in the N. benthamiana short RNA fraction and may be the cause of the transcriptional repression of the LTR promoter (Cervera et al., 2016). The two LTRs of a retrozyme should match exactly when it integrates into the genome because of the reverse transcription method used by LTR-containing retrotransposons (Le Grice, 2003). Given that both groups of N. benthamiana retrozymes were present in the plants in RNA form, the mismatches between two LTRs of a single element could be the result of errors caused during genome replication (Peña et al., 2020).

Retrons' ability to produce ssDNA has been improved, and retrons have been altered to work in model host species including *N. benthamiana, arabidopsis,* and others. These two developments are significant because retrotransposons have the potential to be useful molecular biology tools (Lopez et al., 2022). Retrotransposons are effective tools for genome modification in a range of organisms. For instance, the ssDNA templates at the particular editing site for HDR in *N. benthamiana* have been enriched using retrons and CRISPR-Cas9 techniques. The retron Ec86 system in *N. benthamiana* is responsible for producing ssDNA. For the synthesis of ssDNA, the retron ncRNA structure and the retron RT are necessary (Rao et al., 2021). Retrotransposons were responsible for *N. benthamiana*'s productive ssDNA. The dispersion of retrotransposons and genes was confirmed in 21 longer scaffolds. Even while the Ty1-copia and Ty3-gypsy groups were uniformly dispersed over both scaffolds, the gene distribution of the other creatures appeared to be substantially denser.

2.4 *N. attenuata*

Coyote tobacco is the common name for a variety of wild tobacco called *Nicotiana attenuata*. It grows in a variety of habitats in western North America, from British Columbia to Texas and northern Mexico. It is an annual herb with a maximum height of one meter and is glandular and sparsely hairy. A retrotransposon homolog and three unidentified genes are found in *N. attenuate* (Jassbi et al., 2017). Retrotransposons significantly altered wound-induced responses (type II) by suppressing wound-induced transcripts (type I) or enhancing the wound-induced response. Additionally, controlling wound-induced or repressed type II a (type II b) transcripts (Qi et al., 2013).

Selectivity for the retrotransposon homolog (pDH25.4) in *N. attenuata*, was only briefly observed during the early stages of the consistent treatment period. Blots of flowering plants failed to reveal the retrotransposon homolog (pDH25.4)'s transcripts. Retrotransposon inhibited the wound-induced transcript accumulation of PIOX (pDH41.6) and the retrotransposon homolog (pDH25.4) while increasing the transcript accumulation of lhb C1 (pDH61.1), two unidentified genes (pDH39.1 and pDH68.1), and lhb C1 (pDH61.1). It is understood that downregulating photosynthesis coincides with upregulating defenses against diseases or insects (Halitschke et al., 2001).

Repetitive regions were poorly constructed, as shown by the first assembly. Even with a scaffold N50 greater than 200 kb, upstream parts of genes involved in the biosynthesis of main defense metabolites that appeared to be near transposable elements (TEs), such as nicotine biosynthetic genes, could not be recovered. The relative proximity of transposable elements to genes—in particular, long interspersed nuclear elements and small inverted-repeat TEs, which were typically found close to genes, was one of the genome research's unexpected findings (Hirsch and Springer, 2017).

2.5 *N. tomentosiformis*

Nicotiana tomentosiformis is a member of the Solanaceae family, which also contains the tomrato, potato, eggplant, and pepper. The diploid species *N. tomentosiformis* has a roughly 2,650 Mb 1C genome. In the Tomentosae section, the typical *N. tomentosiformis*.

A perennial herbaceous plant, *Nicotiana tomentosiformis*. This wild species of tobacco is found in the Yungas Valley region of the eastern plains of the Andes Mountains, especially in Bolivia. Recent genetic research suggests that it might be one of the parent species of the common domesticated tobacco (*Nicotiana tabacum*). According to Renny-Byfield et al. (2011), the repeat diversity in the *N. tomentosiformis* genome was significantly higher. A highly repetitive DNA sequence (NicCL3) made up upto 2% of the *N. tomentosiformis* genome, according to a study by Renny Byfield et al. There seem to be more copia-type LTRs (13.4% and 10.33%) and retrotransposons (13.43 and 9.13%) in *N. tomentosiformis* (Huang et al., 2009).

The fall in Tnt2 signals found by FISH on tobacco T-genome chromosomes correlates with a decline in Tnt2d bands of *N. tomentosiformis* origin in tobacco, indicating the physical removal of significant amounts of Tnt2 sequences. According to descriptions in Arabidopsis and rice, these losses are most likely caused by severe reductions or breaks of retrotransposon sequences (Devos et al., 2002).

The evolutionary outcome is likely influenced by the different distributions of specific retrotransposons, which are connected to insertion specificities or age. Tnt2 has been found to cluster in *N. tomentosiformis*, according to FISH. As was previously demonstrated for Tnt1 in response to microbial stress, the activation of retrotransposons may result from a breach in the epigenetic safeguards in the early *N. tomentosiformis* or from transcriptional activation via certain stress-response pathways. Retrotransposon-specific characteristics may also influence activity; for example, ancient populations may have the potential downside of potent bioactive elements (Liu and Wendel, 2003).

This S-genome-specific amplification, which is unknown given that retrotransposons intensify via an intracellular RNA intermediate, could indicate targeting preferences, such as Tnt2 preferentially inserting in or adjacent to itself, or structural aspects of the S-genome that facilitate Tnt2 integration (Fulnecek et al., 2002) Such targeting preferences may also account for the threefold increase in Tnt2d copies in modern *N. sylvestris* compared to modern *N. tomentosiformis*. In tobacco, the two retrotransposons function as centromeric DNA sequences. It was discovered that NtoCR is unique to *N. tomentosiformis* and the *N. tabacum* T genome (Nagaki et al., 2011)

3. Retrotransposons Regulation

Different biotic and abiotic stress conditions transcriptionally activate a number of the plant retrotransposons that have been studied thus far. Numerous abiotic stressors, such as protoplast isolation, methyl jasmonate, wounding, $CuCl_2$, cell culture, and salicylic acid, significantly boost the expression of the tobacco Tnt1 and Tto1 retrotransposons (Grandbastien, 2005, 2015). It was found that the tobacco Tto1 and Tto2 elements, just like rice transposons, were found in cell tissue culture but not during normal plant development (Wessler, 1996).

Different types of retrotransposons are found in almost all *Nicotiana* spp. Those classes are named differently just due to the difference in their U3 region.

The difference in this region is responsible for the induction of these transposons. Similar to this, it has been demonstrated that several biotic stress stimuli, including fungal extracts from *Trichoderma viride* and inoculation with additional bacterial, viral, or fungal pathogens, also make retrotransposons active. Tos17 transcription, unlike Tnt1 and Tto1, is solely stimulated by tissue culture. The RT-PCR approach was used to isolate Tobacco Tto5 as a salicylic acid-inducible retrotransposon.

Retrotransposons are classified into two primary subgroups based on their structure and transposition cycle (Todorovska, 2007). Subclass I elements are known as LTR retrotransposons because they have two long terminal repeats (LTRs) around them; subclass II elements are known as non-LTR retrotransposons since they don't have LTRs. Both of these subclasses have a cytoplasmic phase in their replication cycle that results in the creation of a DNA daughter copy from a reverse transcription of an RNA template. Retrotransposons may be very intrusive because of their replicative transposition process. To maintain the viability of their host and, by extension, their own life, retrotransposition is rigorously regulated (Schulman, 2013).

This control employs element-encoded functions as well as host factors. Transcription is a critical regulation step that regulates both the creation of the RNA template and the synthesis of mRNAs required for protein synthesis. The U3 region, which is positioned upstream of the transcription start point, is one of the cis-regulatory regions found in the LTR of LTR retrotransposons that regulate transcription.

4. Conclusions and Future Perspectives

Retrotransposons are intragenic elements with retrovirus-like characteristics that can alter the structure of the genome in numerous ways. In eukaryotes, enough portion of genome is consisted of LTR retrotransposons that would lead towards the idea that they can be used for integration of desired genes. Incorporating DNA sequences into the target genome is thus possible using LTR retrotransposons. These LTR retrotransposons, however, cannot be controlled for targeted gene delivery because they are largely dormant.

Recent studies on the function of LTR retrotransposons in plant species have provided new insights into the precise functions of LTR retrotransposons as well as their underlying molecular mechanisms. For plant sciences, *N. tabacum* and *N. benthamiana* are frequently employed as model *Nicotiana* systems. The growing quantity of genetic resources for the genus *Nicotiana* has made it much easier to comprehend the biology and evolution of the species. Thus, tobacco plants still play a big part in plant science. *N. tabacum* is still a significant crop for agriculture, and these resources are providing new opportunities for its use, such as in a variety of biotechnological applications for the creation of transgenic plant species will also eventually be evolved into effective biological systems as more data about retrotransposons is made available, for better comprehension and understanding of the unique characteristics of each clade and species.

Understanding epigenetic modification is important to use the LTR retrotransposons as a tool for plant genetic engineering because epigenetic regulation is involved in modulating the activity of LTR retrotransposons for particular gene expressions. The growth of molecular technologies will be beneficial to research in the future. Big data and machine learning are computational tools that could help create LTR retrotransposon regulatory protocols.

References

Boeke, J.D., and Chapman, K.B. (1991). Retrotransposition mechanisms. Curr. Opin. *Cell Biol.* 3: 502–507.

Cervera, A., Urbina, D., and De La Peña, M. 2016. Retrozymes are a unique family of non-autonomous retrotransposons with hammerhead ribozymes that propagate in plants through circular RNAs. *Genome biol.* 17: 1–16.

D'erfurth, I., Cosson, V., Eschstruth, A., Rippa, S., Messinese, E., Durand, P., Trinh, H., Kondorosi, A., and Ratet. P. 2003. Rapid inactivation of the maize transposable element En/Spm in Medicago truncatula. *Mol. Genet. Genomics.* 269: 732–45.

Daboussi, M. 1997. Fungal transposable elements and genome evolution. *Genetica.* 100: 253–60.

De Assis, R., Baba, V.Y., Cintra, L.A., Gonçalves, L.S.A., Rodrigues, R. and Vanzela, A.L.L. 2020. Genome relationships and LTR-retrotransposon diversity in three cultivated Capsicum L. (Solanaceae) species. *BMC genomics.* 21: 1–14.

Deragon, J., Landry, B., Pélissier, T., Tutois, S., Tourmente S. and Picard G., 1994. An analysis of retroposition in plants based on a family of SINEs from Brassica napus. *J. Mol. Evol.* 39: 378–86.

Devos, K.M., Brown, J.K, and Bennetzen, J.L. 2002. Genome size reduction through illegitimate recombination counteracts genome expansion in Arabidopsis.*Genome Res.* 12: 1075–79.

Ennajdaoui, H., Vachon, G., Giacalone, C., Besse, I., Sallaud, C., Herzog M., and Tissier, A. (2010). Trichome specific expression of the tobacco (Nicotiana sylvestris) cembratrien-ol synthase genes is controlled by both activating and repressing cis-regions. *Plant Mol. Biol.* 73: 673–85.

Fulnecek, J., Lim, K., Leitch, A., Kovarik, A., and Matyasek, R. (2002). Evolution and structure of 5S rDNA loci in allotetraploid Nicotiana tabacum and its putative parental species. *Heredity* (Edinb). 88: 19–25.

Gazdová, B., Široký, J., Fajkus, B., Brzobohatý, A., Kenton, A., Parokonny, J.S., Heslop-Harrison, K., Palme., and M. Bezděk. (1995). Characterization of a new family of tobacco highly repetitive DNA, GRS, specific for the Nicotiana tomentosiformis genomic component. *Chromosome Res.* 3: 245–54.

Grandbastien, M.-A. (1998). Activation of plant retrotransposons under stress conditions. *Trends Plant Sci.* 3: 181–87.

Grandbastien, M.-A. (2015). LTR retrotransposons, handy hitchhikers of plant regulation and stress response. Biochim. Biophys. Acta-Gene Regul. Mech. 1849: 403–16.

Grandbastien, M.-A., Audeon,C., Bonnivard, E., Casacuberta, J., Chalhoub, B., Costa, A.-P., Le, Q.H., Melayah, D., Petit, M., and Poncet, C. (2005). Stress activation and genomic impact of Tnt1 retrotransposons in Solanaceae. Cytogenet. *Genome Res.* 110: 229–41.

Grandbastien, M.-A., Lucas, H., Morel, J.-B., Mhiri, C., Vernhettes, S., and Casacuberta, J.M. The expression of the tobacco Tnt1 retrotransposon is linked to plant defense responses. pp. 241–52. In: C, Pierre [ed.]. 1998. Evolution and Impact of Transposable Elements.

Grandbastien, M.-A., Spielmann, A., and Caboche, M. (1989). Tnt1, a mobile retroviral-like transposable element of tobacco isolated by plant cell genetics. *Nature.* 337: 376–80.

Halitschke, R., Schittko, U., Pohnert, G., Boland., and Baldwin, I.T. (2001). Molecular interactions between the specialist herbivore Manduca sexta (Lepidoptera, Sphingidae) and its natural host Nicotiana attenuata. III. Fatty acid-amino acid conjugates in herbivore oral secretions are necessary and sufficient for herbivore-specific plant responses. *Plant Physiol.* 125: 711–17.

Havecker, E.R., Gao, X. and Voytas, D.F. (2004). The diversity of LTR retrotransposons. Genome Biol. 5:225. https://doi.org/10.1186/gb-2004-5-6-225.

Hirochika, H. (1993). Activation of tobacco retrotransposons during tissue culture. *The EMBO journal.* 12: 2521–28.

Hirsch, C.D., and Springer, N.M. (2017). Transposable element influences on gene expression in plants. Biochim Biophys Acta Gene Regul Mech BBA-GENE REGUL MECH. 1860: 157–65.

Horáková, M., and Fajkus, J. (2000). TAS49 a dispersed repetitive sequence isolated from subtelomeric regions of Nicotiana tomentosiformis chromosomes. *Genome.* 43: 273–84.

Huang, S., Li, R., Zhang, Z., Li, L., Gu, X., Fan, W,. Lucas, W.J., Wang, X., Xie B., and Ni. P. (2009). The genome of the cucumber, Cucumis sativus L. *Nat. Genet.* 41: 1275–81.

Jassbi, A.R., Zare, S., Asadollahi, M. and Schuman, M.C. (2017). Ecological roles and biological activities of specialized metabolites from the genus Nicotiana. *Chem. Rev.* 117: 12227–80.

Knapp, S., Chase, M.W., and Clarkson, J,J. (2004). Nomenclatural changes and a new sectional classification in Nicotiana (Solanaceae). *Taxon.* 53: 73–82.

Knoop, V., Unseld, M., Marienfeld, J., Brandt, P., Sünkel, S., Ullrich, H., and Brennicke, A. (1996). copia-, gypsy- and LINE-like retrotransposon fragments in the mitochondrial genome of Arabidopsis thaliana. *Genetics.* 142: 579–85.

Konkel, M.K., Walker, J.A., and Batzer, M.A. (2010). LINEs and SINEs of primate evolution. Evolutionary Anthropology: *Issues, News, and Reviews.* 19:236–49.

Kumar, A. and Bennetzen, J.L. (1999). Plant Retrotransposons. Ann. Rev. Genet. 33: 479–532.

Le Grice, S.F. (2003). "In the beginning": initiation of minus strand DNA synthesis in retroviruses and LTR-containing retrotransposons. *Biochemistry.* 42: 14349–55.

Le, Q.H., Melayah, D. Bonnivard, E., Petit, M., and Grandbastien, M.-A. (2007). Distribution dynamics of the Tnt1 retrotransposon in tobacco. *Mol. Genet. Genom.* 278: 639–51.

Leitch, I., Hanson, L., Lim, K., Kovarik, A., Chase, M., Clarkson J. and Leitch, A. 2008. The ups and downs of genome size evolution in polyploid species of Nicotiana (Solanaceae). *Ann. Bot.* 101: 805–14.

Liu, B. and Wendel, J.F. 2003. Epigenetic phenomena and the evolution of plant allopolyploids. *Mol. Phylogenet. Evol.* 29: 365–79.

Lomonossoff, G.P. and M.-A. D'aoust. 2016. Plant-produced biopharmaceuticals: a case of technical developments driving clinical deployment. *Science.* 353: 1237–40.

Lopez, S.C., Crawford, K.D., Lear, S.K., Bhattarai-Kline S. and Shipman, S.L. 2022. Precise genome editing across kingdoms of life using retron-derived DNA. *Nat. Chem. Biol.* 18: 199–206.

Melayah, D., Bonnivard, E., Chalhoub, B., Audeon, C. and Grandbastien, M.A. 2001. The mobility of the tobacco Tnt1 retrotransposon correlates with its transcriptional activation by fungal factors. PLANT J. 28: 159–68.

Mochizuki, K., Umeda, M., Ohtsubo, H. and Ohtsubo, E. 1992. Characterization of a plant SINE, p-SINE1, in rice genomes. JPN J GENET. 67: 155–66.

Nagaki, K., Shibata, F., Suzuki, G., Kanatani, A., Ozaki, S., Hironaka, A., Kashihara, K., and Murata, M., 2011. Coexistence of NtCENH3 and two retrotransposons in tobacco centromeres. *Chromosome Res.* 19: 591–605.

Naim, F., Nakasugi, K., Crowhurst, R.N., Hilario, E., Zwart, A.B., Hellens, R.P., Taylor, J.M., Waterhouse P.M., and Wood, C.C. 2012. Advanced engineering of lipid metabolism in Nicotiana benthamiana using a draft genome and the V2 viral silencing-suppressor protein. PLoS One. 7:e52717.

Noma, K., Ohtsubo, E. and Ohtsubo, H. 1999. Non-LTR retrotransposons (LINEs) as ubiquitous components of plant genomes. *Mol. Genet. Genom.* 261: 71–79.

Olmstead, R.G., Bohs, L., Migid, H.A., Santiago-Valentin, E., Garcia, V.F., and Collier, S.M.. (2008). *A molecular phylogeny of the Solanaceae. Taxon.* 57: 1159–81.

Parisod, C., Mhiri, C., Lim, K., Clarkson, J., and Chase, M. (2012). Differential Dynamics of Transposable Elements during Long-Term Diploidization of.

Park, M., Jo, S., Kwon, J.-K., Park, J., Ahn, J.H., Kim, S., Lee, Y.-H. Yang, T.-J., Hur, C.-G., and Kang. B.-C. (2011). Comparative analysis of pepper and tomato reveals euchromatin expansion of pepper genome caused by differential accumulation of Ty3/Gypsy-like elements. *BMC genomics.* 12: 1–13.

Peña, M.D.L., Ceprián, R., and Cervera, A. (2020). A Singular and Widespread Group of Mobile Genetic Elements: RNA Circles with Autocatalytic Ribozymes. *Cells.* 9: 120–23.

Petit, M., Lim, K.Y., Julio, E., Poncet, C., Dorlhac De Borne, C., Kovarik, A., Leitch, A.R., Grandbastien, M.-A. and Mhiri, C. (2007). Differential impact of retrotransposon populations on the genome of allotetraploid tobacco (Nicotiana tabacum). *Mol. Genet. Genom.* 278: 1–15.

Pouteau, S., Huttner, E., Grandbastien, M.A., and Caboche, M. (1991). Specific expression of the tobacco Tnt1 retrotransposon in protoplasts. *The EMBO journal.* 10: 1911–18.

Qi, Y., Zhang, Y., Zhang, F., Baller, J.A., Cleland, S.C., Ryu, Y., Starker, C.G., and Voytas, D.F. (2013). Increasing frequencies of site-specific mutagenesis and gene targeting in Arabidopsis by manipulating DNA repair pathways. Genome res. 23: 547–54.

Ramakrishnan, M., Papolu, P.K,. Mullasseri, S., Zhou, M., Sharma, A., Ahmad, Z., Satheesh, V., Kalendar, R., and Wei, Q. (2023). The role of LTR retrotransposons in plant genetic engineering: how to control their transposition in the genome. *Plant Cell Rep.* 42: 3–15.

Rao, G.S., Jiang W., and Mahfouz, M. (2021). Synthetic directed evolution in plants: unlocking trait engineering and improvement. *Synth. Biol.* 6: 25–34.

Renny-Byfield, S., Chester, M., Kovařík, A., Le Comber, S.C., Grandbastien, M.A., Deloger, M., Nichols, R.A., Macas, J., Novák, P., and Chase, M.W. (2011). Next generation sequencing reveals genome downsizing in allotetraploid Nicotiana tabacum, predominantly through the elimination of paternally derived repetitive DNAs. Mol. *Biol. Evol.* 28: 2843–54.

Schillberg, S., Emans, N. and Fischer, R. 2002. Antibody molecular farming in plants and plant cells. *Phytochem Rev.* 1: 45–54.

Schmidt, T. (1999). LINEs, SINEs and repetitive DNA: non-LTR retrotransposons in plant genomes. *Plant Mol.* Biol. 40: 903–10.

Schulman, A.H. (2013). Hitching a Ride: Nonautonomous Retrotransposons and Parasitism as. Plant Transposable Elements: Impact on Genome Structure and Function. 24: 71–85.

Schwichtenberg, K., Wenke, T., Zakrzewski, F., Seibt, K.M., Minoche, A., Dohm, J.C., Weisshaar, B., Himmelbauer, H. and Schmidt, T. 2016. Diversification, evolution and methylation of short interspersed nuclear element families in sugar beet and related Amaranthaceae species. *Plant J.* 85: 229–44.

Sekine, K.-T., Tomita, R., Takeuchi, S., Atsumi, G., Saitoh, H., Mizumoto, H., Kiba, A., Yamaoka, N., Nishiguchi, M. and Hikichi, Y. 2012. Functional differentiation in the leucine-rich repeat domains of closely related plant virus-resistance proteins that recognize common avr proteins. *Mol. Plant Microbe Interact.* 25: 1219–29.

Sierro, N., Battey, J., Bovet, L., Liedschulte, V., Ouadi, S., Thomas, J., Broye, H., Laparra, H., Vuarnoz, A. and Lang, G. 2018. The impact of genome evolution on the allotetraploid Nicotiana rustica–an intriguing story of enhanced alkaloid production. *BMC genomics.* 19: 1–18.

Sierro, N., Battey, J.N., Ouadi, S., Bovet, L., Goepfert, S., Bakaher, N., Peitsch, M.C. and Ivanov, N.V. 2013. Reference genomes and transcriptomes of Nicotiana sylvestris and Nicotiana tomentosiformis. *Genome Biol.* 14: 1–17.

Thyssen, G., Svab, Z., and Maliga, P. (2012). Exceptional inheritance of plastids via pollen in Nicotiana sylvestris with no detectable paternal mitochondrial DNA in the progeny. PLANT J. 72: 84–88.

Todorovska, E. (2007). Retrotransposons and their role in plant—genome evolution. Biotechnol. *Biotechnol. Equip.* 21: 294–305.

Umeda, M., Ohtsubo, h., and Ohtsubo, E. (1991). Diversification of the rice Waxy gene by insertion of mobile DNA elements into introns. JPN J GENET. 66: 569–86.

Vicient, C.M., and Casacuberta, J.M. (2020). Additional ORFs in plant LTR-retrotransposons. Front. Plant Sci. 11: 555.

Vimont, N., Quah, F.X., Schöepfer, D.G., Roudier, F., Dirlewanger, E., Wigge, P.A. Wenden, B., and Cortijo, S. (2020). ChIP-seq and RNA-seq for complex and low-abundance tree buds reveal chromatin and expression co-dynamics during sweet cherry bud dormancy. *Tree Genet. Genomes.* 16: 1–18.

Wang, J. and Han, G. (2022). A Missing Link between Retrotransposons and Retroviruses. mBio. 13:e00187-22.

Waugh, R., Mclean, K., Flavell, A., Pearce, S., Kumar, A., Thomas, B., and Powell, W. (1997). Genetic distribution of Bare–1-like retrotransposable elements in the barley genome revealed by sequence-specific amplification polymorphisms (S-SAP). *Mol. Genet. Genom.* 253: 687–94.

Weiner, (A.M). 2002. SINEs and LINEs: the art of biting the hand that feeds you. *Curr. Opin. Cell Bio.* 14: 343–50.

Wenke, T., Döbel, T., Sörensen, T.R., Junghans, H., Weisshaar, B., and Schmidt, T. (2011). Targeted identification of short interspersed nuclear element families shows their widespread existence and extreme heterogeneity in plant genomes. *The Plant Cell.* 23: 3117–28.

Wessler, S.R. (1996). Plant retrotransposons: turned on by stress. *Curr. Biol.* 6: 959–61.

Wicker, T., Sabot, F., Hua-Van, A., Bennetzen, J.L., Capy, P., Chalhoub, B., Flavell, A., Leroy, P., Morgante M., and Panaud, O. (2007). A unified classification system for eukaryotic transposable elements. *Nat. Rev. Genet.* 8: 973–82.

Wright, D.A. and Voytas, D.F. 1998. Potential retroviruses in plants: Tat1 is related to a group of Arabidopsis thaliana Ty3/gypsy retrotransposons that encode envelope-like proteins. *Genetics.* 149: 703–15.

Xu, S., Brockmöller, T., Navarro-Quezada, A., Kuhl, H. Gase, K., Ling, Z., Zhou, W., Kreitzer, C., Stanke, M., and Tang, T. 2017. Wild tobacco genomes reveal the evolution of nicotine biosynthesis. *Proc. Natl. Acad. Sci.* 114: 6133–38.

Yoshioka, Y., Matsumoto, S., Kojima, S., Ohshima, K., Okada, N., and Machida, Y. 1993. Molecular characterization of a short interspersed repetitive element from tobacco that exhibits sequence homology to specific tRNAs. *Proceedings of the National Academy of Sciences.* 90: 6562–66.

Zimmerman, J.L. and Goldberg, R.B. 1977. DNA sequence organization in the genome of Nicotiana tabacum. Chromosoma. 59: 227–52.

12

Retrotransposons and Crop Improvement

Bushra Hafeez Kiani[1][*]

1. Introduction

Before Barbara McClintock (McClintock, 1950) made the revolutionary finding of transposable elements (TEs) in the middle of the 20th century, it was believed that the genetic information coding for a particular phenotypic expression was structured in a static, one-dimensional manner. Barbara McClintock was behind the discovery of mutable loci that were unmappable on variegate corn kernels, which challenged this largely accepted idea, and initially needed to draw more inquisitiveness. This phenomenon is understood as common functional genetic components that have mobility, such as the McClintock-discovered Ac/Ds system. Transposable elements have been found in every species that has been examined as of yet, and account for over eighty percent of the genomic sequences in organisms such as maize and barley (Schnable et al., 2009; Wendel et al., 2016).

A better understanding of transposable elements has given way to many pieces of research illuminating the pivotal functionality of TEs in developmental phases, responding to ecological stimuli, and driving evolutionary tendencies. This is a complete contradiction of the early perception of transposable elements, where it was referred to as parasitic or junk DNA (Ohno, 1972). Along with numerous examples from plants, notable discoveries from other species of life, such as humans, are rounding out the most recent picture of transposable elements as essential basic components of life (Chuong et al., 2017) (Fig. 1).

Breeders, particularly those working in the organic industry, are currently in a difficult position. Under rapidly shifting environmental conditions, they must create

---

[1] Department of Biological Sciences (Female Campus), International Islamic University, Islamabad, 44000, Pakistan.

[*] Corresponding author: bushrahafeez.kiani@gmail.com

novel crop varieties that can lay the groundwork for a growing global population's food security (Fischer, et al., 2014). A focus of interest for plant breeders has recently been a fresh perception in the realm of epigenetic memory and epigenomics (Gallusci et al., 2017), in addition to the knowledge garnered from recent important discoveries in genetics (Abberton et al., 2016). The understanding of transposable elements as potential genomic resources can be useful in overcoming agricultural issues that are on the rise. Transposable elements are located at the intersection of genetics and epigenetics (Paszkowski, 2015).

To date, genetic engineering has been required to access the epigenome (Mirouze and Vitte, 2014). It is essential to learn more about the regulatory processes of these components to utilize them for the breeding of plants. This chapter will focus on the importance of transposable elements in improving crops and plant breeding.

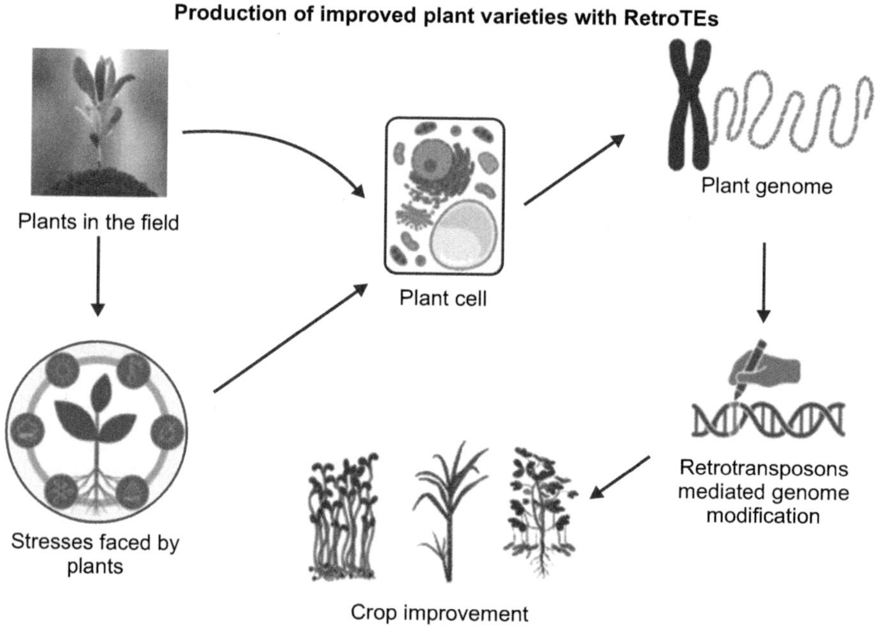

Fig. 1 Schematic representation of improvement of different crops by using RetroTEs strategies.

2. Importance of TEs in Crop Improvement

Although transposable elements have a lot of potential for use in crop breeding, recent technical developments have notably made it possible to monitor and mobilize class I retrotransposable elements. Retrotransposable elements, a potential group of elements with mobility that may be used for breeding, will be the main emphasis of this chapter in this context (Paszkowski, 2015). The phenotypic variety of different crops is, in fact, greatly influenced by continuing retrotransposition events (Vitte et al., 2014). Even as the underlying cause and implications of class I mobile elements in plants have been better understood, they have been frequently viewed as useful resources for advancing crop breeding (Mirouze and Vitte, 2014; Paszkowski, 2015). The selection of epigenetic variety, for example, brought on by the movement of TEs (Lisch, 2013) can be linked to important features like waxy phenotype in *Setaria italica* (Kawase et al., 2005) or apical dominance in Zea mays (Doebley et al., 1995).

Because of the complex mechanism of transposon silencing, there are only a small number of instances where the retrotransposition frequency was successfully raised in plants in a regulated fashion . However, there have been certain exceptions, such as when plants are subjected to severe pressures like tissue culture-induced dedifferentiation (Hirochika et al., 1996; Masuta et al., 2017) or in cases where substantial genomic stresses such as wide crossing take place (Wang et al., 2009).

Because transposable elements play a significant role in increasing genomic evolutionary tendencies their ability to move may be extremely beneficial in crop breeding. Active transposable elements have been utilized as mutagens for years by genetic engineering or by mobilizing those using tissue cultures. Mobilized transposable elements have the benefit over chemical mutagenesis in that they label the insertion site, making it possible to identify relevant genes. Depending on the plant species and variety, the dedifferentiation of explant cultures and their regeneration as new plants might be difficult. Due to these restrictions and the possible harm that transposable elements may bring to the host DNA, they have only been viewed negatively for a very long time. As a result, transposable elements have yet to be widely exploited (at least not intentionally) in plant breeding.

Transposable elements' role in causing alterations in epigenome and genome as a response to environmental cues should be studied to see if they could be controlled and mobilized in plants. The question of whether transposable elements function as an adaptation mechanism in plants might, potentially be tested in real time.

2.1 *Detection of Retrotransposition Events*

Effective detection of transposable elements before and after transposition is required for using them in plant breeding (Vitte et al., 2014). Effective traditional methods enable it to quantify the copy numbers of transposable elements (Southern blotting and qPCR) and the PCR-based "transposon display" identification of transposable element insertion loci (Ito et al., 2011). The field of research in transposable elements has substantially advanced thanks to improvements in downstream data processing and sequencing technology. The repeated nature of transposable elements causes detection limits of TE-associated structural variations, which are overcome in particular by the provision of longer sequencing runs (Debladis et al., 2017). The application of transposable elements in breeding programs will be made possible by recently created methods to specifically sequence the active mobilome which is a mobile genetic element in a cell (Lanciano et al., 2017; Thieme et al., 2017b). These methods also enable discussion of basic questions in transposable element biology and epigenetics.

2.2 Approaches for Mobilizing Transposable Elements for Crop Breeding

2.2.1 Genetic engineering

Certain transposable elements have now gone through modification and cloning in different plants because of the revelation that transposable elements may be mobilized in cell culture. Tnt1, for example, had been isolated in Tobacco (Grandbastien et al., 1989), is seen to be in the active form in heterologous plants including *Cucumis sativus* (Zhang et al., 2018), *Arabidopsis* (Lucas et al., 1995), and *Medicago truncatula* (D'Erfurth et al., 2003) It was discovered that these transgenic transposable elements were effective mutagens that enabled progressive mutant screenings.

Gene deletions associated with transposable element silencing are a different method for activating transposable elements in plants. Enzymatic proteins that are part of the upkeep or de novo DNA methylation of transposable elements were particularly effective in this regard through mutation. In Arabidopsis, it was demonstrated that the activation of the CACTA DNA transposon may be partially responsible for the developmental abnormalities seen in ddm1 mutants who have been self-crossed for many generations (Miura et al., 2001). Thereafter, several additional class 1 and class 2 transposable elements were discovered active in ddm1 (Tsukahara et al., 2009). The biological purpose of Arabidopsis plant-specific RNA polymerases IV and V was unknown for several years. In-depth molecular analyses made it abundantly evident how such polymerases helped RdDM target methylation in DNA at transposable elements. It was discovered later that crops with Pol IV or Pol V defects had mobile transposable elements in the presence of stressful stimuli (Ito et al., 2011) or through crossing such crops with met1 (Mirouze et al., 2009). This investigation of intracellular transposable elements in Arabidopsis silencing mutants was made possible by these reports.

In principle, inducing retrotransposition in crops by utilizing mutants deficient in transposable element silencing has potential. There are examples of the analogous mobilization of retrotransposons in plants. For example, when an H3K9 methyltransferase was knocked off in rice, the mobilization of the copia-like retrotransposons Tos17 improved (Ding et al., 2007). The limited number of mutant alleles is the reason retrotransposons use in plant breeding is currently constrained (Paszkowski, 2015), which may also have negative outcomes like those seen in Zea mays (Li et al., 2014) or Oryza sativa (Hu et al., 2014). However, recent significant developments in genetic modification and genome sequencing open more opportunities (Springer and Schmitz, 2017).

Along with targeted TE-silencer mutations employing TALENs, CRISPR-Cas9 systems, or ZFNs, particular "epigenome editing" techniques, such as those utilizing a Cas9 protein deficient in nucleases (dCAS9) coupled to a methyltransferase (Park et al., 2016) are advancing this domain. These techniques, meanwhile, require a transitory or steady alteration to be used. Genetic engineering varies greatly based on the species or type and may even be impossible. Additionally, the resulting plant crops might need to be legally classified as GMOs which may further influence the subsequent breeding applications.

2.2.2 *Temporary retrotransposons silencing inhibition*

By using several medications it is possible to temporarily disrupt the key enzymes in epigenetic silencing to boost retrotransposons movement in plants (Pecinka and Liu, 2014). The activation of retrotransposons using drugs does not need preliminary genomic information or complex transgenesis techniques because it targets highly conserved silencing pathways. Cytidine-analogs zebularine (Z) and 5-azacytidine (AZA) which act as DNMtase inhibitors were initially created as cancer-treating drugs (Lyko and Brown, 2005), but are now frequently employed to cause epigenetic alterations in plants (Griffin et al., 2016). These analogs have been integrated with DNA during cell replication. In bacteria, it has been demonstrated that the existence of cytidine analogs causes covalent (AZA) (Santi et al., 1983) or stable (Z) (Champion et al., 2010), leading to a reduction in active DNMtases and a successive DNA methylation loss.

Inhibiting DNMtases temporarily causes a decrease in methylation in all sequences in plants (Baubec et al., 2009; Griffin et al., 2016; Thieme et al., 2017b). The observed reduced DNA-demethylation for Z, which is thought to be because it is less toxic than AZA and more viable in water solutions (Cheng et al., 2004), led to the transcriptional release of many transposable elements, including the non-LTR retrotransposon LINE1-4 in Arabidopsis (Baubec et al., 2009). Plants treated with Z or AZA showed remarkably similar DNA methylation alterations, according to a genome-wide DNA methylation study. Notably, the DNA methylation levels of CACTA-like DNA transposable elements were significantly affected by both medications, with the largest effects (Griffin et al., 2016). The researchers were unable to locate any studies on the definite mobilization of transposable elements through the administration of epigenetic medicines on crops, even though these compounds effectively triggered the transcription of transposable elements.

The possibility of using epigenetic medications to activate transposable elements was explored to study transposable element bursts in crop wild-types and utilize them for plant production (Thieme et al., 2017b). As a reporter for the pharmacological treatments, the well-researched heat-stress responsive ONSEN retrotransposons were employed in Arabidopsis. Since both maintenance and RdDM-mediated methylation have silenced most transposable elements, it was intended to utilize medications that could impact both silencing pathways at different degrees (Matzke and Mosher, 2014; Yokthongwattana et al., 2010). The final step in the transposable element silencing pathway, DNA methyltransferases, is effectively inhibited by the aforementioned Z. To our knowledge, no medication has, however, been shown to effectively suppress RdDM.

The aim was to see if the RNA polymerase II (Pol II) enzyme complex could be involved in restricting transposable element mobility in light of claims that transcripts of RNA polymerase II (Pol II) could be behind the early extremely specialized signal targeting transposable elements (Cuerda-Gil and Slotkin, 2016). Crops that were partly deficient in Pol II (nrpb2-3) (Zheng et al., 2009) as well as a-amanitin (A) therapies (Lindell et al., 1970), a very powerful Pol II inhibitor was utilized to examine the role of Pol II in suppressing transposable element mobility. Through the use of medication combinations, it was discovered that the creation of ONSEN ecDNA could be effectively phenocopied by mixing Z and A. Researchers found over 75 new ONSEN insertions in plants that were descended from plants that had been subjected to heat, A, and Z stress.

As a result, the combination of A and Z and ONSEN therapy activated ONSEN and greatly amplified its presence in the wild-type background (Thieme et al., 2017b). From a molecular perspective, it can be postulated that Pol II is responsible for generating siRNAs by first creating antisense transcripts within the LTRs. These could target transposable element transcripts specifically in PTGS or cause the methylation of DNA at retrotransposon loci (Thieme et al., 2017b). A wide variety of phenotypes, including alterations in seed colors, chlorophyll levels, and blooming time, were brought about by the treatment that produced a significant increase in ONSEN copies. It was demonstrated that by the generalization of the brief blockage method of TE-silencing, such pharmacological therapies may also induce the mobility of transposable elements in rice. Hence, this shows that practically any plant can benefit from this technique to promote phenotypic variation (Thieme et al., 2017b).

It is important to keep in mind, when executing these treatments that they may also result in damage to DNA and epigenomic mutations that can be inherited which can cause abnormal phenotypes. Brassica rapa (Amoah et al., 2012) along with *Fragaria vesca* treatments with AZA have been linked to stable epigenetic alterations (Xu et al., 2016). Although these plants have quantitative variations in parameters like seed size and blooming time, genetic modifications, such as those caused by a putative AZA-induced mobilization of transposable elements, were not completely ruled out.

2.3 *Effects of Induced Transposition on Plant Phenotypes*

Mediated transposition has been employed for years to generate various phenotypes and find functional roles of genes, within the boundaries of plausibility (Gierl and Saedler, 1992; Naito et al., 2009). To achieve diversity in particular plants, the desired lineages have been either crossed with types that carry mobile transposable elements or those transposable elements have been exogenously introduced into plant cells. Class II DNA elements, such as Robertson's Mutator-Mu or Activator and Dissociation - Ac/Ds, were utilized in the majority of these instances (Settles, 2009). The fact that the altered loci are subsequently marked with a recognized TE, making it easier to identify the target gene, is a significant benefit of TE-mediated mutagenesis. The use of transposable element tagging techniques in forward genetic mutant screens, in conjunction with T-DNA insertions, has greatly aided the identification of multiple genes. When plants are difficult to transform, TE-mediated tagging is preferable over T-DNA insertions because a few successful transformation events can result in several individual transposable element insertions which could later be examined (Ramachandran and Sundaresan, 2001).

The previously underutilized endogenous retrotransposons for gene tagging have been made available through drug-induced mobilization of retrotransposons in wild-type Arabidopsis. Many phenotypic differences in Arabidopsis that might be brought on by the activation of ONSEN were observed. Researchers selected one line with additional ONSEN insertions having a highly distinct phenotype in the seeds to demonstrate that ONSEN did cause gene mutations. Testa on this line was transparent, which meant that they appeared yellow rather than dark brown as they are in the wild kind. They were able to locate and validate that the reported trait was caused by an insertion of ONSEN in TRANSPARENT TESTA 6 (TT6, AT3G51240) using a candidate gene method and genetic segregation patterns (Thieme et al., 2017b). This demonstrates that endogenous retrotransposons can now be used for forward genetic mutant screens in wild-type plants.

Researchers discovered several environmentally dependent changes in nonlinear functions, like blooming length and biomass output, which may represent the outcome of diverse genomic or epigenomics modifications brought forward by fresh ONSEN insertions (Lisch, 2013). Although researchers could definitively associate a homozygous ONSEN insertion on the TT6 locus to the observed transparent testa trait, it is still unknown if those transposable elements or prospective epigenetics or genotoxicity induced by drugs may be responsible for a few reported phenotypic traits. It has been demonstrated that Z-treatment might cause DNA degradation through strand formation in DNA duplication (Liu et al., 2015) as well as create class II CACTA-elements in Arabidopsis (Baubec et al., 2009; Griffin et al., 2016).

The possible removal of mobile DNA transposons along with inter- and intra-element recombination for either transposon type might theoretically result in genetic but also physiological changes across generations, which is crucial to note for trait stability (Vitte et al., 2007; Lisch, 2013; Bennetzen et al., 2014). Additionally, as previously indicated, significant retrotransposon copy number multiplications followed by filtering selections could result in physiological variance, as was demonstrated in wild *Brachypodium distachyon* and *Arabidopsis* lines (Quadrana et al., 2016; Stritt et al., 2018). The trait stability concern following an elicited burst of transposons in crops would be important for breeding and must be explored by researchers, even though no sudden significant increase nor reduction of ONSEN copy number across three inbreed generations of ONSEN high copy Arabidopsis plants (Thieme et al., 2017b).

2.4 Perspective: Transposable Element Mobilization in Crops and Applications in Breeding

As previously mentioned, transposon or epigenomic studies in model plants such as *Arabidopsis* may not always apply to agricultural crop varieties (Mirouze and

Vitte, 2014). Nevertheless, it was highly proposed that A and Z might both interact with TE-silencing in numerous plants due to highly conserved enzymatic proteins, particularly RNA-polymerase II associated with methylation of DNA in various plants (Nawrath et al., 1990) as well as eukaryotic organisms (Ream et al., 2009; Sweetser et al., 1987). In support of this hypothesis, researchers discovered that the dual therapies were responsible for activating the Houba retrotransposons in rice (Thieme et al., 2017b). It is also intriguing that it is indicated that Houba (Vitte et al., 2007), similar to ONSEN (Quadrana et al., 2016), demonstrated organic transposition in rice and Arabidopsis sequences. In rice, Houba is among the highest prevalent retrotransposons of the Copia family, and it has been operational for five million years (Wicker and Keller, 2007).

In light of the above examples, it may be assumed that plants usually have a silencing mechanism dependent on expression levels which relies on silencing signals generated by Pol II, even in genetically distinct species like Rice and Arabidopsis. Collectively, these findings imply that endogenous retrotransposons may be mobilized in almost all crops. Additionally, the scientific advancements in tracing transposons and their regulatory components which may be used to control transposon movement would encourage their usage in crop improvement (Paszkowski, 2015).

2.4.1 Breeding strategies

Transposons may be an effective tool for crop production due to their capacity to remodel genetic regulatory networks (Makarevitch et al., 2015), particularly when responding to stress. It has been demonstrated in previous studies that Arabidopsis plants with low numbers of siRNAs may use ONSEN retrotransposons which operate under thermal stress to make flanking genes respond to high-temperature induced stress (Ito et al., 2011). The prior stress observed in wild-type species is likely owing to stress-sensitive LTRs, which can also be linked to the impact on regulating transcription at insertion sites. It would be crucial to determine the degree to which certain groups of crop transposons show particular insertion biases. Therefore, one could anticipate scenarios in which these occurrences could benefit the plants and contribute to a quicker environmental response (Ito et al., 2016; Lisch, 2013; Makarevitch et al., 2015).

Although transposable elements were linked to adaptable capacities in species such as yeasts (Adams and Oeller, 1986), similar results are yet to be seen in crops. Researchers propose using the induced multiplication of endogenous retroelements for crop production, presuming that transposons can assist in the stress response. This technique is known as BUNGEE (breeding using jumping genes) or breeding utilizing inhibitors involved in retrotransposons silencing (Thieme et al., 2017b), as demonstrated in Fig. 2. Finding stress-responsive retrotransposons in the desired genetical context is the initial step in breeding stress-tolerant plants. A series of medications and stress therapies are used to mobilize such transposons if discovered. The resulting plantlets of the therapy could be further tested to identify stress-tolerant varieties (Fig. 2).

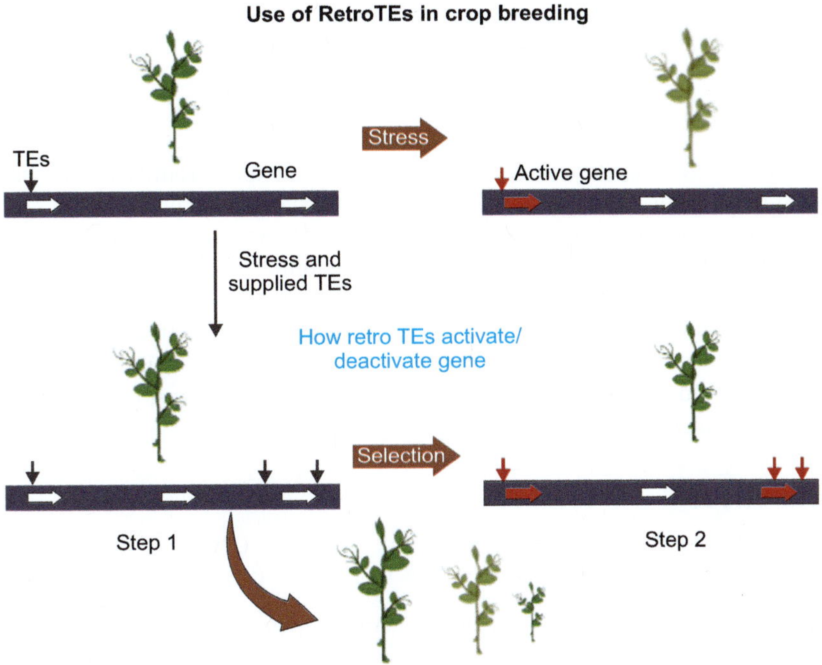

Fig. 2   A systematic RetroTEs methodology used in crop breeding

The above approach does have many benefits, including:

- Possibility of generating gain-of-function mutations.
- There is no requirement for previous information on the resistance mechanisms, in contrast with recombinant DNA technology and CRISPR-Cas9 technology.
- Desired genes responsible for a particular trait are marked with retrotransposon insertions, enabling quick detection.
- The copying and pasting method used by retrotransposons confers trait stability.
- A reference assembly is not necessary.

This technique has the drawback of producing mutations that may impact the production negatively. These lineages would be discarded during the breeding procedure, specifically during the selective stage. Another alternative is to use backcrossing to isolate undesired insertions of transposable elements. It could be intriguing to explore but also contrast cultivar-specific mobilomes and their responses to stressful conditions because retrotransposons-activity has been reported to vary among several genotypes of the same plant, as also observed in rice for Houba (Yuzbasioglu et al., 2016). The quick discovery of retrotransposons that respond to stress can be achieved through inhibitors used in transposon silencing (Thieme et al., 2017b). This may additionally be paired with selective gene editing as reported in cucumbers using Tnt1 (Zhang et al., 2018). According to (Hou et al.

2014), cisgenesis might be utilized for transferring transposable elements that respond to stress among various types of plants (Fig. 2).

Conclusion

From the above discussion, it can be revealed that the essential results of the *Arabidopsis* plant are backed by Houba-retrotransposons activation in rice. This opens new plant breeding opportunities considering the current agricultural constraints. The present advancements in epigenomics and transposable element biology, alongside innovative techniques in computational biology and genomic studies, are anticipated to increase the use of transposable elements in breeding strategies. Depending upon the desired plant species and objectives, various methods would be used to access epigenomic resources of transposable elements in genetic material.

References

Abberton, M., Batley, J., Bentley, A., Bryant, J., Cai, H., Cockram, J., et al. (2016). Global agricultural intensification during climate change: A role for genomics. *Plant Biotechnology Journal*, 14, 1095e1098.

Adams, J., and Oeller, P.W. (1986). Structure of evolving populations of Saccharomyces cerevisiae: Adaptive changes are frequently associated with sequence alterations involving mobile elements belonging to the Ty family. *Proceedings of the National Academy of Sciences*, 83, 7124.

Amoah, S., Kurup, S., Rodriguez Lopez, C.M., Welham, S.J., Powers, S.J., Hopkins, C.J., et al. (2012). A Hypomethylated population of Brassica rapa for forward and reverse Epi-genetics. *BMC Plant Biology*, 12, 193.

Baubec, T., Pecinka, A., Rozhon, W., and Mittelsten Scheid, O. (2009). Effective, homogeneous and transient interference with cytosine methylation in plant genomic DNA by zebularine. *The Plant Journal: For Cell and Molecular Biology*, 57, 542e554.

Bennetzen, J.L., and Wang, H. (2014). The contributions of transposable elements to the structure, function, and evolution of plant genomes. *Annual Review of Plant Biology*, 65, 505e530.

Champion, C., Guianvarc'h, D., Senamaud-Beaufort, C., Jurkowska, R.Z., Jeltsch, A., Ponger, L., et al. (2010). Mechanistic insights on the inhibition of c5 DNA methyltransferases by zebularine. PLoS One, 5, e12388.

Cheng, J.C., Weisenberger, D.J., Gonzales, F.A., Liang, G., Xu, G.-L., Hu, Y.-G., et al. (2004). Continuous zebularine treatment effectively sustains demethylation in human bladder cancer cells. *Molecular and Cellular Biology*, 24, 1270e1278.

Chuong, E.B., Elde, N.C., and Feschotte, C. (2017). Regulatory activities of transposable elements: From conflicts to benefits. Nature Reviews. Genetics, 18, 71e86. Clavijo, B. J., et al. (2017). An improved assembly and annotation of the allohexaploid wheat genome identifies complete families of agronomic genes and provides genomic evidence for chromosomal translocations. *Genome Research*, 27, 885e896.

Cuerda-Gil, D., and Slotkin, R.K. (2016). Non-canonical RNA-directed DNA methylation. *Nature Plants*, 2, 16163.

D'Erfurth, I., Cosson, V., Eschstruth, A., Lucas, H., Kondorosi, A., and Ratet, P. (2003). Efficient transposition of the Tnt1 tobacco retrotransposon in the model legume Medicago truncatula. *The Plant Journal*, 34, 95e106.

Debladis, E., Llauro, C., Carpentier, M.C., Mirouze, M., and Panaud, O. (2017). Detection of active transposable elements in Arabidopsis thaliana using Oxford Nanopore Sequencing technology. *BMC Genomics*, 18, 537.

Ding, Y., Wang, X., Su, L., Zhai, J., Cao, S., Zhang, D., et al. (2007). SDG714, a histone H3K9 methyltransferase, is involved in Tos17 DNA methylation and transposition in rice. *The Plant Cell Online*, 19, 9e22.

Doebley, J., Stec, A., and Gustus, C. (1995). Teosinte Branched1 and the origin of maize: Evidence for epistasis and the evolution of dominance. *Genetics*, 141, 333e346.

Fischer, R.A., Byerlee, D., and Edmeades, G.O. (2014). Crop yields and global food security: Will yield increase continue to feed the world? Canberra: Australian Centre for International Agricultural Research.

Gallusci, P., Dai, Z., Genard, M., Gauffretau, A., Leblanc-Fournier, N., Richard-Molard, C., et al. (2017). Epigenetics for plant improvement: Current knowledge and modeling avenues. *Trends in Plant Science*, 22, 610e623.

Gierl, A., and Saedler, H. (1992). Plant-transposable elements and gene tagging. In R.A. Schilperoort, & L. Dure (Eds.), *10 Years plant molecular biology* (pp. 39e49). Dordrecht: Springer Netherlands.

Grandbastien, M.-A., Spielmann, A. and Caboche, M. (1989). Tnt1, a mobile retroviral-like transposable element of tobacco isolated by plant cell genetics. *Nature*, 337, 376e380.

Griffin, P.T., Niederhuth, C.E., and Schmitz, R.J. (2016). A comparative analysis of 5-azacytidine- and zebularine-induced DNA demethylation. G3 (Bethesda), 6, 2773e2780.

Hirochika, H., Sugimoto, K., Otsuki, Y., Tsugawa, H., and Kanda, M. (1996). Retrotransposons of rice involved in mutations induced by tissue culture. Proceedings of the National Academy of Sciences of the United States of America, 93, 7783e7788.

Hu, L., Li, N., Xu, C., Zhong, S., Lin, X., Yang, J., et al. (2014). Mutation of a major CG methylase in rice causes genome-wide hypomethylation, dysregulated genome expression, and seedling lethality. Proceedings of the National Academy of Sciences of the United States of America, 111, 10642e10647.

Ito, H., Gaubert, H., Bucher, E., Mirouze, M., Vaillant, I., and Paszkowski, J. (2011). A siRNA pathway prevents transgenerational retrotransposition in plants subjected to stress. Nature, 472, 115e119.

Ito, H., Kim, J.M., Matsunaga, W., Saze, H., Matsui, A., Endo, T.A., et al. (2016). A stress-activated transposon in arabidopsis induces transgenerational abscisic acid insensitivity. *Scientific Reports*, 6, 23181.

Kawase, M., Fukunaga, K., and Kato, K. (2005). Diverse origins of waxy foxtail millet crops in East and Southeast Asia mediated by multiple transposable element insertions. *Molecular Genetics and Genomics*, 274, 131e140.

Lanciano, S., Carpentier, M.C., Llauro, C., Jobet, E., Robakowska-Hyzorek, D., Lasserre, E., et al. (2017). Sequencing the extrachromosomal circular mobilome reveals retrotransposon activity in plants. *PLoS Genetics*, 13, e1006630.

Li, Q., Eichten, S.R., Hermanson, P.J., Zaunbrecher, V.M., Song, J., Wendt, J., et al. (2014). Genetic perturbation of the maize methylome. *The Plant Cell Online*, 26, 4602e4616.

Lindell, T.J., Weinberg, F., Morris, P.W., Roeder, R.G., and Rutter, W.J. (1970). Specific inhibition of nuclear RNA polymerase II by a-amanitin. Science, 170, 447e449.

Lisch, D. (2013). How important are transposons for plant evolution? *Nature Reviews. Genetics*, 14, 49e61.

Liu, C.H., Finke, A., Diaz, M., Rozhon, W., Poppenberger, B., Baubec, T., et al. (2015). Repair of DNA damage induced by the cytidine analog zebularine requires ATR and ATM in arabidopsis. *The Plant Cell Online*, 27, 1788e1800.

Lucas, H., Feuerbach, F., Kunert, K., Grandbastien, M.A., and Caboche, M. (1995). RNAmediated transposition of the tobacco retrotransposon Tnt1 in Arabidopsis thaliana. *The EMBO Journal*, 14, 2364e2373.

Lyko, F., and Brown, R. (2005). DNA methyltransferase inhibitors and the development of epigenetic cancer therapies. *Journal of the National Cancer Institute*, 97, 1498e1506.

Makarevitch, I., Waters, A.J., West, P.T., Stitzer, M., Hirsch, C.N., Ross-Ibarra, J., et al. (2015). Transposable elements contribute to activation of maize genes in response to abiotic stress. PLoS Genetics, 11, e1004915.

Masuta, Y., Nozawa, K., Takagi, H., Yaegashi, H., Tanaka, K., Ito, T., et al. (2017). Inducible transposition of a heat-activated retrotransposon in tissue culture. *Plant & Cell Physiology*, 58, 375e384.

Matzke, M.A., and Mosher, R.A. (2014). RNA-directed DNA methylation: An epigenetic pathway of increasing complexity. Nature Reviews. *Genetics*, 15, 394e408.

McClintock, B. (1950). The origin and behavior of mutable loci in maize. Proceedings of the National Academy of Sciences of the United States of America, 36, 344e355.

Mirouze, M., Reinders, J., Bucher, E., Nishimura, T., Schneeberger, K., Ossowski, S., et al. (2009). Selective epigenetic control of retrotransposition in Arabidopsis. *Nature*, 461, 427e430.

Mirouze, M., and Vitte, C. (2014). Transposable elements, a treasure trove to decipher epigenetic variation: Insights from arabidopsis and crop epigenomes. *Journal of Experimental Botany*, 65, 2801e2812.

Miura, A., Yonebayashi, S., Watanabe, K., Toyama, T., Shimada, H., and Kakutani, T. (2001). Mobilization of transposons by a mutation abolishing full DNA methylation in Arabidopsis. *Nature*, 411, 212e214.

Naito, K., Zhang, F., Tsukiyama, T., Saito, H., Hancock, C.N., Richardson, A.O., et al. (2009). Unexpected consequences of a sudden and massive transposon amplification on rice gene expression. *Nature*, 461, 1130e1134.

Nawrath, C., Schell, J., and Koncz, C. (1990). Homologous domains of the largest subunit of eucaryotic RNA polymerase II are conserved in plants. *Molecular and General Genetics MGG*, 223, 65e75.

Ohno, S. (1972). So much "junk" DNA in our genome. *Brookhaven Symposia in Biology*, 23, 366e370.

Park, M., Keung, A.J., and Khalil, A.S. (2016). The epigenome: The next substrate for engineering. *Genome Biology*, 17, 183.

Paszkowski, J. (2015). Controlled activation of retrotransposition for plant breeding. *Current Opinion in Biotechnology*, 32C, 200e206.

Pecinka, A., and Liu, C.H. (2014). Drugs for plant chromosome and chromatin research. *Cytogenetic and Genome Research*, 143, 51e59.

Quadrana, L., Bortolini Silveira, A., Mayhew, G.F., LeBlanc, C., Martienssen, R.A., Jeddeloh, J.A., et al. (2016). The Arabidopsis thaliana mobilome and its impact at the species level. Elife, 5, e15716.

Ramachandran, S., and Sundaresan, V. (2001). Transposons as tools for functional genomics. *Plant Physiology and Biochemistry*, 39, 243e252.

Ream, T.S., Haag, J.R., Wierzbicki, A.T., Nicora, C.D., Norbeck, A.D., Zhu, J.K., et al. (2009). Subunit compositions of the RNA-silencing enzymes Pol IV and Pol V reveal their origins as specialized forms of RNA polymerase II. *Molecular Cell*, 33, 192e203.

Santi, D.V., Garrett, C.E., and Barr, P.J. (1983). On the mechanism of inhibition of DNA cytosine methyltransferases by cytosine analogs. Cell, 33, 9e10.

Schnable, P.S., Ware, D., Fulton, R.S., Stein, J.C., Wei, F., Pasternak, S., et al. (2009). The B73 maize genome: Complexity, diversity, and dynamics. *Science*, 326, 1112e1115.

Settles, A.M. (2009). Transposon tagging and reverse genetics. In A.L. Kriz, and B.A. Larkins (Eds.), Molecular genetic approaches to maize improvement (pp. 143e159). Berlin, Heidelberg: Springer.

Springer, N.M., and Schmitz, R.J. (2017). Exploiting induced and natural epigenetic variation for crop improvement. *Nature Reviews. Genetics*, 18, 563e575.

Stritt, C., Gordon, S.P., Wicker, T., Vogel, J.P., and Roulin, A.C. (2018). Recent activity in expanding populations and purifying selection have shaped transposable element landscapes across natural accessions of the mediterranean grass Brachypodium distachyon. *Genome Biology and Evolution*, 10, 304e318.

Sweetser, D., Nonet, M., and Young, R.A. (1987). Prokaryotic and eukaryotic RNA polymerases have homologous core subunits. Proceedings of the National Academy of Sciences of the United States of America, 84, 1192e1196.

Thieme, M., Lanciano, S., Balzergue, S., Daccord, N., Mirouze, M., and Bucher, E. (2017b). Inhibition of RNA polymerase II allows controlled mobilisation of retrotransposons for plant breeding. *Genome Biology*, 18, 134.

Tsukahara, S., Kobayashi, A., Kawabe, A., Mathieu, O., Miura, A., and Kakutani, T. (2009). Bursts of retrotransposition reproduced in Arabidopsis. *Nature*, 461, 423e426.

Vitte, C., Fustier, M.A., Alix, K., and Tenaillon, M.I. (2014). The bright side of transposons in crop evolution. *Brief Functional Genomics*, 13, 276e295.

Vitte, C., Panaud, O., and Quesneville, H. (2007). LTR retrotransposons in rice (Oryza sativa, L.): Recent burst amplifications followed by rapid DNA loss. *BMC Genomics*, 8, 218.

Wang, H., Chai, Y., Chu, X., Zhao, Y., Wu, Y., Zhao, J., et al. (2009). Molecular characterization of a rice mutator-phenotype derived from an incompatible cross-pollination reveals transgenerational mobilization of multiple transposable elements and extensive epigenetic instability. *BMC Plant Biol*, 9, 63.

Wendel, J.F., Jackson, S.A., Meyers, B.C., and Wing, R.A. (2016). Evolution of plant genome architecture. *Genome Biology*, 17, 37.

Wicker, T., and Keller, B. (2007). Genome-wide comparative analysis of copia retrotransposons in Triticeae, rice, and Arabidopsis reveals conserved ancient evolutionary lineages and distinct dynamics of individual copia families. *Genome Research*, 17, 1072e1081.

Xu, J., Tanino, K.K., Horner, K.N., and Robinson, S.J. (2016). Quantitative trait variation is revealed in a novel hypomethylated population of woodland strawberry (Fragaria vesca). *BMC Plant Biol*, 16, 240.

Yokthongwattana, C., Bucher, E., Caikovski, M., Vaillant, I., Nicolet, J., Mittelsten Scheid, O., et al. (2010). MOM1 and Pol-IV/V interactions regulate the intensity and specificity of transcriptional gene silencing. *The EMBO Journal*, 29, 340e351.

Yuzbasioglu, G., Yilmaz, S., Marakli, S., and Gozukirmizi, N. (2016). Analysis of Hopi/Osr27 and Houba/Tos5/Osr13 retrotransposons in rice. *Biotechnology & Biotechnological Equipment*, 30, 213e218.

Zhang, Q., Du, H., Lv, D., Xiao, T., Pan, J., He, H., et al. (2018). Efficient transposition of the retrotransposon Tnt1 in cucumber (*Cucumis sativus* L.). *Horticultural Plant Journal*, 4, 111e116.

Zheng, B., Wang, Z., Li, S., Yu, B., Liu, J.Y., and Chen, X. (2009). Intergenic transcription by RNA polymerase II coordinates Pol IV and Pol V in siRNA-directed transcriptional gene silencing in Arabidopsis. *Genes & Development*, 23, 2850e2860.

13

Retrotransposons and Plant Stress Responses

Alvina Gul[1][*] and *Noor ul huda*[2]

1. Introduction

Retrotransposons are DNA sequences that can replicate and move within the genome of a cell. They are a type of transposable element, which refers to a genetic sequence that can change its position within the genome. Unlike other transposable elements, retrotransposons replicate via an RNA intermediate, which is reverse transcribed into DNA and inserted into a new location in the genome. This process is called retro transposition (Bondar et al., 2022). Retrotransposons are "jumping genes" because of their ability to move within the genome. They can cause genetic changes and rearrangements, leading to new gene functions or altering existing ones. The mobility of retrotransposons can result in genome expansion, disrupting gene regulation and causing mutations. In plants, retrotransposons can play a role in stress responses and contribute to the evolution of the species. It is estimated that retrotransposons make up a significant portion of the plant genome, ranging from 50–80% in different species. Despite their potentially harmful effects, retrotransposons also play important roles in the regulation of gene expression and in shaping the genetic diversity of plants. This makes the study of retrotransposons crucial in understanding plant biology and the mechanisms of plant stress responses (Boyko and Kovalchuk, 2008).

[1] Associate Professor, Atta-ur-Rahman School of Applied Biosciences, National University of Sciences & Technology, Islamabad.

[2] Atta-ur-Rahman School of Applied Biosciences, National University of Sciences & Technology, Islamabad, noorulhudarao27@gmail.com

[*] Corresponding author: alvina_gul@yahoo.com

Retrotransposons play a crucial role in the evolution and adaptation of plants. They are responsible for a large portion of the plant genome and have been shown to have significant effects on gene expression and regulation. They also provide a source of genetic diversity and adaptation. Retrotransposons are known to respond to various environmental stressors and can affect gene expression patterns in plant stress responses. This has led to an increasing interest in understanding the mechanisms by which retrotransposons influence plant stress responses, and how they may be harnessed to improve the stress tolerance of crop plants. In addition, retrotransposons have been found to impact the structure and organization of chromosomes, which in turn affects processes such as meiosis, DNA repair, and gene expression. They also play a role in shaping the diversity of the plant gene pool, as they can insert into different regions of the genome and influence gene expression, leading to the creation of new alleles and traits. The abundance and dynamic nature of retrotransposons in the plant genome make them an important area of study for researchers interested in understanding the evolution of plants and their ability to adapt to changing environmental conditions. Retrotransposons are important in plant biology for several reasons. Firstly, they are a major component of the plant genome, accounting for a significant portion of the total DNA content in many species. In some cases, retrotransposons can make up over 80% of the genome, making them a dominant feature of the plant genetic landscape. This high level of representation in the genome has several important consequences, as discussed in this section (Bui and/Grandbastien, 2012).

Secondly, retrotransposons are important in shaping the structure and organization of the plant genome. They can cause chromosomal rearrangements, leading to changes in gene expression, as well as affecting processes such as meiosis, DNA repair, and gene duplication. Retrotransposons can also influence the distribution and spacing of genes within the genome, which can have effects on gene regulation and expression. For example, the insertion of a retrotransposon into the promoter region of a gene can lead to changes in its expression pattern, potentially leading to the development of new traits (Cardoso et al., 2022).

Thirdly, retrotransposons play a significant role in plant evolution and adaptation. By inserting into new locations in the genome, retrotransposons can create new alleles and traits and have been shown to play a role in the evolution of new species. For example, the insertion of a retrotransposon into a gene can lead to the creation of a new allele, which can be subject to selection and further evolution. This dynamic nature of retrotransposons means they can provide a source of genetic diversity and adaptation, allowing plants to evolve in response to changing environmental conditions.

Fourthly, retrotransposons play a role in plant stress responses, including responses to environmental stressors such as drought, cold, and heat. In response to stress, plants have been shown to activate specific retrotransposons, which can result in changes to the expression patterns of target genes. For example, the activation of certain retrotransposons has been linked to the upregulation of stress-responsive genes, leading to improved stress tolerance. Understanding the mechanisms by which retrotransposons are activated in response to stress, and the genes that they regulate is an important area of research with significant potential applications in the development of more stress-tolerant crop plants, which will be discussed in greater detail in this chapter (Cardoso et al., 2022).

Finally, retrotransposons provide a useful tool for researchers studying gene function and regulation. By using reverse genetics techniques, it is possible to specifically target and inactivate retrotransposons, providing a way to study the effects of changes in gene expression and regulation on plant development and growth. In this way, retrotransposons can be used to understand the role of specific genes in plant biology, including the regulation of stress responses.

In short, retrotransposons are an important and dynamic component of the plant genome, playing a significant role in shaping the structure and organization of the genome, influencing evolution and adaptation, and affecting plant stress responses. Understanding the mechanisms by which retrotransposons operate in plants is crucial for advancing our knowledge of plant biology and developing new approaches for improving the stress tolerance of crops (Cerbin et al., 2022).

2.   Role of Retrotransposons in Plants

Transposons are DNA sequences capable of changing their position in the genome, being a source of mutations and chromosomal rearrangements capable of altering cell identity, regulating other genes, and modifying the size of the genome, among others. They are part of the mobilome (mobile genetic elements that can move within a genome or between different genomes) along with plasmids, bacteriophages (or phages), and self-assembling molecular parasites (such as certain introns or homing endonucleases). Its distribution is practically ubiquitous in eukaryotic genomes. However, retrotransposons are mobile RNA intermediates (Chang et al., 2020).

Its discovery dates to the middle of the last century, thanks to Barbara McClintock's work on chromosome breakage in maize. After perfecting the staining techniques for maize chromosomes and identifying the linkage groups of its genome, he achieved the first experimental evidence of the idea proposed by Thomas H. Morgan two decades earlier: genes were physically positioned on chromosomes. Consequently, it was a possible description of the phenomena of crossing-over or genetic recombination.

During the 1940s, she observed that several maize varieties had frequent chromosomal breaks on chromosome 9, all at a particular position. These breaks are chromosomal abnormalities that involve DNA breaks and that can generate translocations, inversions, or deletions of sequences in the chromosome, causing unstable phenotypes (mottled, pigmented, or depigmented corn kernels on the corn cob) due to the interaction between a transposon and a gene that acts as a transcriptional regulator of the biosynthesis of phlobaphene (purple or red pigment) in aleurone. McClintock described a series of genetic factors (or controlling elements) associated with this characteristic break: Ds (Dissociation) located at the site of the chromosome break; and As (Activator) critical for activating chromosome

9 breakage at the Ds locus. Finally, he concluded that the speckled pigmentation patterns were the result of instability due to the transposition of the Ds locus outside the gene in which it is inserted and that Ac and Ds were mobile genetic elements in the plant genome (Duk et al., 2022).

McClintock's results and hypotheses were not well received by the scientific community at the time. Specifically, the idea of the transposition of genetic information did not fit with the conceptual framework of the genetics of the 40s: firstly, it contradicted the notion of the "constant genome", demonstrating that cells have systems capable of restructuring their genome; secondly, this finding challenges the unitary concept of a gene, so that a genetic locus is no longer a series of indivisible alternative alleles, but rather a complex mosaic that can be modified by the addition and removal of specific genetic elements. Finally, his work challenged the concept of the locus as an autonomous unit since the insertion of controlling elements from the same family into two physically separate loci caused them to express themselves under the same control (Galindo-González et al., 2017).

The importance of McClintock's discovery would not be appreciated until the advent of the first studies on retrotransposons in bacteria. The discovery of the Mu phage and its ability to insert its prophage into the Escherichia coli chromosome causing mutations. Added to this is the finding of insertion sequences (IS) after different studies on the regulation of gene expression of the gal and lac operons by (Grandbastien et al., 2005), which were key events to understand the importance of these elements in bacterial genomic reorganization and DNA transfer. Other authors showed that the IS would have a function like the controlling elements already described in maize by McClintock.

In this sense, studies on conjugation and bacterial transduction established the necessary conceptual bases to understand the genome as a fluid and complex entity. With the advancement of DNA and RNA analysis techniques and the discovery of the involvement of mobile genetic elements in evolution, McClintock's work was finally recognized, receiving the Nobel Prize in 1983 "for his discovery of mobile genetic elements" among other awards.

Regarding the classification and systematics of transposons, his study begins with the work carried out by (Orozco-Arias et al., 2019) based mainly on the proposed classification in Saccharomyces cerevisiae, Drosophila melanogaster, and human models.

The currently most accepted classification arose in 2007, after the PAG (Plant and Animal Genome) Conference in San Diego, California, whose main objective was to define a consensus for the classification of all eukaryotic transposable elements and, with it, the definition of criteria consisting of the characterization of the main superfamilies and families as well as the proposal of a nominal system. This is because transposons can be classified into two classes based on their transposition mechanism, which in turn are divided into subclasses (or orders) according to their chromosomal integration mechanism. These subclasses are divided into subgroups (or superfamilies) that are generally found in a broad spectrum of organisms but share a monophyletic origin. Finally, the most detailed level of classification is the families or subfamilies, which can be defined as groups of elements that are highly related to each other and that are descendants of a single ancestral unit, that is, they are divided based on their phylogeny (Grandbastien et al., 1997).

Class I elements (also called retrotransposons) consist of an RNA molecule that encodes proteins structurally homologous to the retroviral proteins gag and pol, enabling their reverse transcription resulting in a cDNA copy that integrates elsewhere in the genome. This process is generally referred to as a "copy and paste" mechanism, which can increase the number of copies of the element. The next level of classification is based on the absence or presence of LTRs (elements flanked by long terminal repeats): LTR retrotransposons, whose chromosomal integration occurs in a strand break and transfer reaction, catalyzed by an integrase, and non-LTR retrotransposons (which include SINEs and LINEs, or short and long scattered nuclear elements, respectively) that possess a mechanism of chromosome integration coupled to reverse transcription, called target-mediated reverse transcription (Horváth et al., 2017).

On the other hand, class II elements (also called DNA transposons) transpose without an RNA intermediary and according to a mechanism analogous to a "cut and paste" process, in which repair of the initial DNA site is carried out using the sister or homologous chromatid as a standard, so that no net excision is detected. Within this class, they are divided according to the absence or presence of TIRs (inverted terminal repeats) that serve as cleavage and reintegration sites when recognized by a transposase encoded by the element itself. The subclass lacking TIRs is characterized by a process of transposition and replication that does not involve a double-strand break, but rather a single-strand displacement. It is worth mentioning that the order Helitron is included in this subclass due to the absence of an intermediary RNA, not due to phylogeny, and that it involves a rolling circle replication model (Hou et al., 2019).

Each class of transposons contains autonomous and non-autonomous elements: autonomous elements possess Open Reading Frames (ORFs) that encode products required for transposition, while non-autonomous elements have no significant coding capacity. but they retain target sequences that are recognized by products of autonomous elements, carrying out their rearrangement. The integration of practically all the transposable elements results in the duplication of a sequence at the insertion site, being variable in size and/or sequence between the different superfamilies and families.

Higher plants are organisms with great flexibility in their growth and development, capable of rapidly and continuously adapting to environmental changes, especially during their post-embryonic development. To withstand biotic and abiotic stress conditions, plants require genetic variation: from single nucleotide polymorphisms to the duplication of their entire genome together with the contribution of epigenetic variation (Ito, 2022).

That said, transposons (which can make up more than 70% of the genome, as in *Zea mays*) are an obvious source of mutations and whose activation can affect the structure and function of the genome, as well as modify the regulation and expression of genes at different levels or alter their epigenetic state. Likewise, they can generate new genetic information, mobilize genes from the host genome, and regulate the response at different levels to abiotic and biotic stress. Therefore, they contribute to the generation of new and interesting phenotypes (stress tolerance, fruit color, and shape, among others), their important role in evolution being evident.

Activation of these elements under stress conditions was first recognized in 1985 in *Zea mays*, in response to barley streak mosaic virus (BSMV) infection, and since then numerous studies have shown not only its ability to activate under these conditions but also the possibility of granting memory after adaptation to stressful conditions.

In this way, retrotransposons can facilitate adaptation to stress and evolution by inserting into and modifying the genome of an organism. This can create genetic variation that may be beneficial in response to environmental stressors, allowing for natural selection to act on these new traits. In plants, retrotransposons can also modify gene expression in response to stimuli such as abiotic stress by altering the epigenetic landscape of the genome, leading to changes in chromatin structure and the activation or repression of genes. This can result in physiological and morphological changes in different plant organs, allowing for adaptation to a changing environment (Kinoshita and Seki, 2014).

3. Retrotransposons and Plant Stress Responses

3.1 *Effects of Retrotransposons on Plant Stress Tolerance*

Plants are exposed to a wide range of environmental stressors that can impact their growth, development, and productivity. These stressors can include drought, extreme temperatures, high salt concentrations, heavy metal toxicity, and pathogens, among others. To survive and reproduce under these challenging conditions, plants have evolved complex stress response mechanisms that allow them to maintain homeostasis and protect themselves from damage.

Plant stress responses can be broadly divided into two main categories: morphological and physiological. Morphological responses include changes to the plant structure, such as the development of deeper roots to search for water in times of drought, or the formation of protective tissues in response to herbivore attack. Physiological responses include changes to metabolic and biochemical processes, such as the regulation of water uptake, the activation of antioxidants, and the production of stress-responsive hormones such as abscisic acid (ABA) and salicylic acid (SA).

One of the key physiological responses to stress in plants is the regulation of water balance. During drought, for example, plants close their stomata to conserve water and reduce transpiration, which can lead to a decrease in photosynthesis. Additionally, plants can increase their root biomass and search for water deeper in the soil, to access new sources of moisture (Kumar and Bennetzen, 1999).

Another important physiological response to stress in plants is the activation of antioxidants, which help to protect the plant from oxidative damage caused by stress. Antioxidants are compounds that scavenge reactive oxygen species (ROS) generated as a byproduct of cellular metabolism. Under normal conditions, these ROS are scavenged by the plant's antioxidant defense system. However, during stress, the levels of ROS can increase dramatically, leading to oxidative damage to cellular components such as lipids, proteins, and DNA. To counter this damage, plants increase the expression of genes encoding antioxidant enzymes such as superoxide dismutases (SODs), catalases (CATs), and peroxidases (POXs).

In addition to changes in water balance and antioxidant activity, plants also activate stress-responsive hormones such as abscisic acid (ABA) and salicylic acid (SA). ABA is produced in response to a wide range of stressors and acts to regulate several stress-responsive genes, including those involved in water uptake and metabolism. ABA also plays a role in the regulation of seed germination and the development of stomata, among other processes (Liu and He, 2020).

Salicylic acid, on the other hand, is produced in response to biotic stressors such as pathogen infection and plays a key role in the plant's systemic acquired resistance (SAR) mechanism. SAR is a mechanism by which plants can mount a rapid and effective defense response to subsequent pathogen attacks, by activating a network of defense-related genes.

In addition to the molecular responses outlined above, plants also employ other mechanisms to cope with stress, including changes to gene expression patterns. For example, stress can cause the activation of specific retrotransposons, which in turn can lead to changes in the expression patterns of target genes. This can result in the upregulation of stress-responsive genes, leading to improved stress tolerance (Matsui et al., 2013).

Therefore, plants have evolved complex and multi-faceted stress response mechanisms to cope with environmental stressors. These responses can include changes to plant morphology, regulation of water balance and antioxidant activity, activation of stress-responsive hormones, and changes to gene expression patterns, among others. Understanding the mechanisms by which plants respond to stress is crucial for improving our understanding of plant biology and for developing new approaches for improving the stress tolerance of crops.

3.2 *Retrotransposons and Adaptation to Stress*

One of the primary ways that retrotransposons facilitate adaptation to stress is through the creation of genetic variation. Retrotransposons can insert into new locations within the genome, causing mutations and altering gene expression. These mutations and altered gene expression can lead to novel traits that may be beneficial in response to environmental stressors. This is known as the "TE-driven evolution" hypothesis.

Research in *Arabidopsis thaliana*, a model plant species, has provided evidence for the TE-driven evolution hypothesis. The Copia retrotransposon family in

Arabidopsis was found to be induced by drought stress. The insertion of Copia elements was also found to increase genetic diversity and facilitate the adaptation of Arabidopsis to drought stress. Similarly, a study on the genome of the desert poplar tree, *Populus euphratica*, found that retrotransposon activity was higher in the desert-adapted ecotype compared to the non-desert adapted ecotype. The authors suggest that the increased retrotransposon activity in the desert-adapted ecotype may be contributing to its ability to tolerate arid conditions (Matsui et al., 2013).

Retrotransposons can also contribute to adaptation by regulating gene expression. Retrotransposons can be inserted into regulatory regions of genes, leading to changes in gene expression. In some cases, retrotransposons may also provide new regulatory elements to genes. This can lead to changes in the expression of genes that are beneficial in response to environmental stressors.

An example of retrotransposon-mediated regulation of gene expression in response to stress comes from a study on grapevine (*Vitis vinifera*). The study found that a Copia retrotransposon family was induced by heat stress and was associated with the upregulation of several heat shock protein genes. Heat shock proteins are known to protect cells from damage caused by heat stress, so the upregulation of these genes likely contributed to the heat tolerance of grapevine (Mirouze and Paszkowski, 2011)

3.3   *Retrotransposons and Abiotic Stress Tolerance in Plant Organs*

Abiotic stresses, such as drought, salinity, and extreme temperatures, can have a significant impact on plant growth and development. Retrotransposons have been implicated in abiotic stress tolerance in various plant organs, including roots, leaves, and seeds.

Roots

Kinoshita and Seki (2014) investigated the role of retrotransposons in the response of rice roots to salt stress. The authors found that a Gypsy retrotransposon family was upregulated in response to salt stress in roots. The upregulation of this retrotransposon was associated with changes in the expression of genes involved in ion transport and stress response, and the authors suggest that it may be contributing to the salt tolerance of rice roots.

Leaves

Wang et al. (2022) studied the response of leaves of the desert plant Haloxylon ammodendron to drought stress. The authors found that several retrotransposon families were induced in response to drought stress and that this was associated with changes in the expression of genes involved in stress response and carbohydrate metabolism. The authors suggest that retrotransposons may be contributing to the drought tolerance of Haloxylon ammodendron leaves.

Seeds

A study by Zhao et al., (2016) investigated the role of retrotransposons in the response of wheat seeds to heat stress. The authors found that several retrotransposon families were upregulated in response to heat stress and that this was associated with changes in the expression of genes involved in stress response and seed development. The authors suggest that retrotransposons may be contributing to the heat tolerance of wheat seeds.

These studies provide evidence for the role of retrotransposons in abiotic stress tolerance in different plant organs and suggest that retrotransposons may be a valuable target for improving stress tolerance in crop plants.

3.4 Mechanisms of Retrotransposon Activation during Stress

Retrotransposons are present in most eukaryotic genomes, including plants, and are typically repressed to prevent their uncontrolled proliferation. However, they can be activated under certain conditions, such as stress, leading to their mobilization and potential impact on gene expression and genome structure. In this chapter, the mechanisms of retrotransposon activation during stress in plants will be discussed.

The mechanisms by which retrotransposons are activated during stress are complex and can involve a range of factors, including epigenetic modifications, changes in chromatin structure, and the activity of trans-acting factors such as transcription factors and small RNAs (Liu and He, 2020). The following are the key mechanisms that have been implicated in retrotransposon activation during stress in plants:

3.4.1 Epigenetic modifications

Epigenetic modifications are chemical changes to the DNA or associated proteins that can affect gene expression without altering the underlying DNA sequence. These modifications include DNA methylation, histone modifications, and small RNA-mediated gene silencing. In plants, DNA methylation is a key epigenetic mechanism for retrotransposon silencing, and changes in DNA methylation patterns have been linked to retrotransposon activation during stress (Sahu et al., 2013).

Under normal conditions, retrotransposons are usually heavily methylated, which helps to suppress their expression and activity. However, stress conditions can lead to changes in DNA methylation patterns that can activate retrotransposons. For example, a study by (Matsui et al., 2013) found that exposure of Arabidopsis plants to heat stress led to a decrease in DNA methylation in the promoters of several retrotransposons. This decrease in DNA methylation was associated with increased expression of the retrotransposons, as well as increased mobility of the retrotransposon RNA intermediates.

Similarly, a study by (Vanderschuren et al., 2022) found that exposure of maize plants to heat stress led to a decrease in DNA methylation in the promoter of a retrotransposon called Opie-2. This decrease in DNA methylation was associated with increased expression of the retrotransposon and transposition events, suggesting that changes in DNA methylation can activate retrotransposons during stress.

3.4.2 Changes in chromatin structure

Changes in chromatin structure can also play a role in retrotransposon activation during stress. Chromatin is the complex of DNA and associated proteins that make up the structure of the genome, and changes in chromatin structure can affect gene expression by altering access to the DNA sequence.

In plants, retrotransposons are usually silenced by a repressive chromatin structure that involves the formation of heterochromatin. Heterochromatin is a compact, tightly packed form of chromatin that is associated with gene silencing, and it is thought to help prevent retrotransposon expression and activity. However, stress conditions can lead to changes in chromatin structure that can activate retrotransposons (Nakaminami, Matsui, Shinozaki, and Seki, 2012).

For example, a study (Ito, 2022) found that exposure of rice plants to cold stress led to changes in the chromatin structure of the retrotransposon Tos17. The authors found that the heterochromatic marks on the Tos17 retrotransposon were decreased in response to cold stress, leading to increased expression of the retrotransposon. The authors suggest that changes in chromatin structure may be an important mechanism for retrotransposon activation during stress.

3.4.3 Trans-acting factors

Trans-acting factors, such as transcription factors and small RNAs, can play a significant role in the activation of retrotransposons during stress in plants. Transcription factors are proteins that bind to specific DNA sequences and control the rate of gene expression. Small RNAs are non-coding RNA molecules that can act as regulators of gene expression by binding to and modulating the activity of other RNA molecules.

1. *Transcription factors*

Several transcription factors have been implicated in the regulation of retrotransposons during stress in plants. One example is the Heat Shock Factor (HSF) family of transcription factors, which are known to regulate the expression of genes involved in the response to heat stress. HSFs can also bind to and regulate the expression of retrotransposons, leading to their activation during stress.

For example, a study by (Negi et al., 2016) found that exposure of Arabidopsis plants to heat stress led to the activation of several retrotransposons, including Ta3 and Athila, which were both upregulated in response to heat stress. The authors found that the upregulation of these retrotransposons was dependent on the activity of HSFs and that HSFs could bind to the promoters of these retrotransposons and activate their expression.

Another example is the MYB transcription factor family, which has been shown to play a role in the regulation of retrotransposons during abiotic stress in rice. A study by (Ohama et al., 2017) found that overexpression of the MYB transcription factor OsMYB2P-1 in rice plants led to the upregulation of several retrotransposons, including Tos17 and LTR-Retron2. The authors suggest that OsMYB2P-1 may regulate the expression of these retrotransposons by binding to their promoters and activating their transcription.

2. *Small RNAs*

Small RNAs can also play a significant role in the regulation of retrotransposons during stress in plants. Small RNAs are non-coding RNA molecules that can regulate gene expression by binding to and modulating the activity of other RNA molecules.

One example is the small RNA pathway involving the RNA-dependent RNA polymerase (RDR) and Dicer-like (DCL) proteins, which are involved in the generation of small interfering RNAs (siRNAs) that can induce RNA silencing. This pathway is known to play a role in the regulation of retrotransposons in plants, and changes in small RNA levels have been linked to retrotransposon activation during stress.

For example, a study by (Orozco-Arias et al., 2019) found that exposure of Arabidopsis plants to drought stress led to an increase in the levels of 24-nt siRNAs, which are known to be involved in the regulation of retrotransposons. The authors found that this increase in siRNA levels was associated with a decrease in the expression of several retrotransposons, suggesting that small RNAs may play a role in the suppression of retrotransposons during stress.

Another example is the role of microRNAs (miRNAs) in the regulation of retrotransposons during stress. MiRNAs are small RNA molecules that can regulate gene expression by binding to target mRNAs and inducing their degradation or translational repression.

A study found that exposure of maize plants to salt stress led to changes in the expression of several miRNAs, including miR156, miR167, and miR319. The authors found that these miRNAs could target and regulate the expression of several retrotransposons, including the LTR-Retron2 retrotransposon (Orozco-Arias et al., 2019).

Classic examples of epigenetic regulation of transposons under abiotic stress include the ONSEN element and the NAC gene. High temperature stress activates the ONSEN retrotransposon in *Arabidopsis thaliana*, through heat shock TFs that bind to the element sequence. Experiments with mutants deficient in the generation of small interfering RNAs or siRNAs demonstrated that these molecules are involved in its regulation, instead of the RdDM pathway. The NAC gene in maize encodes TFs that regulate numerous processes in plants, including the stress response. Genome-wide association studies (GWAS) showed a relationship between plants with drought-sensitive polymorphisms and the insertion of an MITE element in the promoter of an NAC transcription factor, as well as high levels of DNA methylation from the pathway. RdDM implied lower gene expression and worse drought tolerance (Paszkowski, 2015).

4.   Retrotransposons in Plant Genomes

Retrotransposons are abundant in plant genomes, and their activity has contributed significantly to the diversity and evolution of these genomes. Retrotransposons can be divided into two main groups: long terminal repeat (LTR) retrotransposons and non-LTR retrotransposons. LTR retrotransposons are further divided into Gypsy and Copia families, while non-LTR retrotransposons include LINEs (long interspersed nuclear elements) and SINEs (short interspersed nuclear elements).

In plant genomes, retrotransposons can constitute a significant proportion of the genome. For example, in the genome of maize (*Zea mays*), retrotransposons make up around 85% of the total genome size. Similarly, in the genome of rice (Oryza sativa), retrotransposons make up around 44% of the total genome size (Qu et al., 2023).

4.1   Structure of Retrotransposons in Plant Genomes

Retrotransposons are characterized by the presence of reverse transcriptase (RT) and integrase (IN) domains. The RT domain is responsible for the synthesis of a cDNA copy of the retrotransposon RNA, while the IN domain mediates the integration of the cDNA copy into a new genomic location.

LTR retrotransposons have a characteristic structure consisting of two LTRs, which flank an internal coding region that encodes the RT and IN domains. Non-LTR retrotransposons have a more variable structure but typically contain the RT and IN domains, as well as other conserved domains such as the endonuclease (EN) domain, which is involved in the cleavage of target DNA during integration (Sharma et al., 2022).

4.2   Diversity of Retrotransposons in Plant Genomes

The diversity of retrotransposons in plant genomes is vast, with different families and subfamilies exhibiting distinct patterns of distribution and activity. For example, in the genome of maize, the Copia retrotransposon family is more abundant than the Gypsy family, while in the genome of rice, the Gypsy family is more abundant than the Copia family.

Different families and subfamilies of retrotransposons can also exhibit differences in their activity and regulation. For example, the Tnt1 retrotransposon family in tobacco (Nicotiana tabacum) is transcriptionally active under normal growth conditions but is suppressed by small RNAs during stress. In contrast, the Tos17 retrotransposon family in rice is transcriptionally repressed under normal growth conditions but can be activated by stress (Singh et al., 2022)

In addition to differences in activity and regulation, retrotransposons can also exhibit differences in their sequence and structure. For example, the LTRs of different retrotransposon families can vary in length and sequence and may contain regulatory elements that influence the expression of the retrotransposon. The coding regions of retrotransposons can also vary in length and sequence and may contain additional domains that confer functional diversity.

Retrotransposons are a major component of plant genomes, and their activity has contributed significantly to the diversity and evolution of these genomes. Different families and subfamilies of retrotransposons exhibit distinct patterns of distribution, activity, and regulation, and can also exhibit differences in their sequence (Sudalaimuthuasari et al., 2022).

Once the concept of genetic mobility introduced by McClintock was accepted in the scientific community, retrotransposons were initially classified as "parasitic" or "selfish" DNA, because they contain sufficient genetic information to replicate and/or transpose into the genome., being able to compromise the genomic stability of the host and cause negative effects. However, systematic analyzes in eukaryotic genomes have revealed its role in the regulation of gene expression and evolution.

There are currently three main hypotheses about the function of these elements in the genome:

1. The emergence of epigenetic processes in evolution as defense mechanisms against transposable elements and viruses, suggests a mutual coadaptation process between the host genome and transposons.
2. Transposons as main agents of evolutionary innovation and phenotypic plasticity.
3. The existence of transposable elements as "genomic control modules", is a vision similar to the "Controller Element" hypothesis already postulated by McClintock.

These transposons form the largest fraction of "junk DNA" in the genome, in which there are segments of DNA without a relevant regulatory function for the organism or without a region that encodes proteins in the first instance. Likewise, transposons are ubiquitous elements, having been found in all organisms analyzed with few exceptions such as *Plasmodium falciparum*. Furthermore, there is a positive correlation between the content of transposable elements and the size of the genome, although the abundance of the different types of transposons in different organisms is highly variable. In some higher plants, such as *Arabidopsis thaliana*, the section of the genome made up of transposable elements is exceptionally small (only 14%) while in other angiosperms with large genomes (such as the genus Lilium with between 30 and 40 million kilobases) exceeds 90% (Sun et al., 2022).

Regarding autonomous class I elements, LTR transposons are the most abundant in the plant genome. They are divided into five superfamilies, two of which are present in plants: the Ty1-Copia and Ty3-Gypsy elements, which in turn are divided into multiple families. As the initials of their name indicate, LTR transposons are characterized by the presence of long terminal repeats, ranging in size from 100 base pairs (as Tos17 elements) to more than 5 kilobases (Sukkula elements) that are also flanked by insertion targets called TSDs (Target Site Duplication).

The genes gag (encoding a protein capsid) and pol (encoding the enzymatic machinery for reverse transcription and integration into the host genome), share structural homology with retrovirus genes, and the main difference between the Ty1-Copy and Ty3 elements -Gypsy is the physical position of the coding domains of the pol gene. Sequences belonging to the Copia group have been found in unicellular algae, bryophytes, gymnosperms, and angiosperms. On the other hand, Gypsy-type elements have been reported in various plants of agronomic interest such as corn, tomato, rice, and pineapple, as well as in other angiosperms and gymnosperms (Sunkar, 2010).

Other LTR transposons are non-autonomous elements, such as TRIMs (miniature retrotransposons with terminal repeats) and LARDs (long derivatives of retrotransposons). The former was discovered for the first time in an intron of the Solanum tuberosum urease gene and the latter, in barley. Non-autonomous elements are unable to encode the proteins necessary for their rearrangement, depending on other autonomous elements that recognize their LTR sequences for their subsequent mobilization. These two groups are differentiated by their size: the sequence of the TRIMs elements occupies from hundreds of bases to 4 kilobases and that of the LARDs always exceeds 4 kilobases. In plants, they have been described in various groups, from fungi and algae to various families of higher plants, with special diversification in Brassicaceae, although their small size makes their annotation difficult.

DIRS (Dictyostelium Intermediate Repeat Sequence) elements represent an order of retrotransposons present in most organisms, including plants and fungi and their structure denotes a different integration mechanism. to the rest of retrotransposons, but that has not yet been described.

Regarding non-LTR transposons, they are generally rare elements in the plant genome compared to LTR transposons. Even so, they are characterized by being flanked by TSDs and ending in a poly-A tail. These elements are subdivided into LINEs and SINEs (Todorovska, 2007).

LINEs have two open reading frames and are considered autonomous elements since they encode the machinery necessary for retro transposition. In contrast, SINEs are mainly composed of tRNA and rRNA, possessing boxes A and B in their sequence that serve as promoters for RNA polymerase III. However, they lack an open reading frame, depending on the LINEs machinery for their mobilization in the genome, so they are non-autonomous elements. SINEs have been found in fungi and higher plants such as tobacco, rice, and Brassica species.

PLE elements (Penelope-1 ike Elements) are another group of transposons with a wide distribution (fungi, metazoans, conifers, among others) and are considered a separate group from the two traditional classes of transposons, due to structural differences in their coding regions. In 2016, bioinformatic analyses concluded that these elements are the result of horizontal gene transfer from arthropods to conifers. Some exceptional examples due to their abundance in plants are LINE Del2 with up to 250,000 copies in Lilium (Leeton and Smyth, 1993), and SINE TS with more than 50,000 copies in tobacco (Vanderschuren et al., 2022).

As regards class II elements (or DNA transposons) they form a small mobilome portion of the plant genome compared to retrotransposons, due to the nature of their conservative (or "cut and paste") mechanism of transposition. They are divided according to the absence or presence of TIR sequences: the TIR transposons (which encode a transposase that acts as a DNA-binding protein and as an endonuclease, apart from being flanked by TIR sequences) and non-TIR transposons, both of which can be autonomous and non-autonomous. A notable example is the Ac/Ds system described by McClintock in which Ac is the autonomous element capable of activating Ds, or the MITEs (miniature transposable elements with inverted repetition, non-autonomous) that constitute an exception in regarding the proportion of class II elements, such as mPing in rice and mPIF in maize whose copy number reaches 6,000 (Wang et al., 2022).

Class II transposons also include elements with unusual replication mechanisms, without double-strand breaks, such as Helitrons (which involve replication in a rolling circle) or Mavericks (whose cleavage is mediated by single-strand breaks even though they contain sequences IRR). Helitrons can be considered autonomous or non-autonomous depending on whether they contain the Rep/Hel domain, which encodes a protein with helicase function (Hel) and rolling circle replication initiator (Rep) (Wheeler, 2013). These elements do not produce TSDs during transposition and are widely distributed in plants, fungi, and animals. Maverick elements are distinguished from the rest in the set of proteins that they require for their rearrangement, among which are DNA polymerase, retroviral integrase, cysteine-protease, and ATPase, in addition to their ability to sequester gene fragments of the host genome. Although they have not been isolated from higher plants, they are found in fungi, such as in the genus Glomus.

5. Importance of Retrotransposons in Abiotic Stress Tolerance and Phenotypic Plasticity

One of the most interesting roles of retrotransposons is their involvement in abiotic stress tolerance and phenotypic plasticity. Abiotic stress, such as drought, salinity, extreme temperatures, or heavy metal toxicity, can affect plant growth, development, and productivity. Retrotransposons are activated under stress conditions, and their activity can lead to genetic and epigenetic changes that can confer stress tolerance and allow for phenotypic plasticity.

For example, retrotransposon activity can cause DNA methylation changes, which can alter the expression of stress-responsive genes. Retrotransposon activity can also lead to the generation of new genetic variants, which can provide a pool of adaptive diversity that can be selected under stress conditions. Additionally, retrotransposons can act as stress sensors, responding to environmental cues and modulating gene expression in response to stress (Yushkova and Moskalev, 2023).

Overall, retrotransposons play an important role in abiotic stress tolerance and phenotypic plasticity by generating genetic and epigenetic diversity and modulating gene expression in response to environmental cues. Understanding the mechanisms by which retrotransposons contribute to stress responses can have important implications for improving crop productivity and developing new strategies for plant breeding and genetic engineering.

Stress can be defined as any factor, of biotic origin (pathogens, in vitro tissue culture, fungal infections, etc.) or abiotic (excess or deficiency of light, water, salts, temperature, nutrients, etc.) that decreases the rate of one or several physiological processes below the maximum rate that the plant would maintain under more favorable conditions. As a result, any of these external factors or a combination of these can negatively affect the growth, development, and productivity of plants. In the same way that stress damages the plant at different levels, it can generate a response that manifests itself at multiple levels of organization (molecular, histological, anatomical, and morphological), such as adjustments in the cell membrane system, changes in the cell wall architecture, transcriptional modifications, or metabolic responses. The responses of plants to abiotic stress are decisive for their survival given their sessile nature, which makes them face high levels of radiation, salinity, drought, hypoxia, or heavy metals, among others (Zhao et al., 2016).

Activation of retrotransposons under biotic and abiotic stress in plants is a reported and studied phenomenon, initially discovered in maize under biotic stress (mosaic virus infection). Specifically, the retro transposition of the Bs1 element in the coding region of the Adh1 gene, which encodes the enzyme alcohol dehydrogenase 1, was reported, causing infected plants capable of generating pollen grains resistant to allyl alcohol. The hypothesis that the activation of transposable elements in response to stress was initially proposed by Barbara McClintock and was confirmed thanks to advances in sequencing techniques, which made it possible to discover how some families of transposons contain stress response elements. or SREs in their promoters. Consequently, retrotransposons are not only capable of being activated by external factors such as abiotic and biotic stress but also of conferring a response capacity to these factors in coding genes close to the insertion or integration site (Wheeler, 2013).

The clearest role of retrotransposons in gene regulation is their mobilization and insertion into regulatory regions (such as promoters), which facilitate stress-inducible gene expression. A classic example is the activation under stress due to low temperatures of the Ruby Myb gene (activator of anthocyanin synthesis and responsible for the red color of the fruit in low temperature conditions) in the red or blood orange (Citrus x sinensis), which is mediated by the insertion of the Tcs1 retrotransposon, which acquires a second copy (Tcs2) of the same element in reverse orientation in Chinese varieties of this plant. On the other hand, the reference genome of rice has dozens of mPing elements, although the cultivar EG4 (cultivar Gimbozu) contains more than a thousand of these elements that, to this day, are still active, conferring response induction to saline stress or low temperatures at different genes. This demonstrates how transposable elements can produce phenotypic variation in a few generations, responding to stresses that cause strong selective pressure (Grandbastien, 1998).

The mutagenic nature of transposons makes them capable of creating genetic variations at insertion loci, which can be neutral, fatal, or beneficial to the plant genome. This fact can translate into the emergence of tolerant or sensitive phenotypes to different types of abiotic stress. For example, tolerance to metals such as aluminum (a characteristic root trait) in sorghum plants (Sorghum spp.) is associated with the insertion of these elements. Specifically, the main difference between aluminum-sensitive and tolerant cultivars in sorghum is the presence of a Tourist element in the AltSB gene sequence, which encodes an aluminum-inducible citrate transporter. Aluminum-tolerant plants show a tissue-specific expression in the root thanks to the activation of this MITE-type element. Likewise, its deleterious effect on certain plant stress response genes has also been reported: the comparison of the soybean cultivars Tiefeng 8 (tolerant to high salinity levels) and 85-140 (parent plant Tiefeng 8, sensitive to high levels of salinity) showed that a truncated transcript of the GmSALT3 gene product was produced in the sensitive cultivar, due to the disruption caused by a Ty1-Copia type retrotransposon in the third exon of the gene. The GmSALT3 gene encodes a protein located in the endoplasmic reticulum that contributes to sodium exclusion in soybean sprouts, whose product is truncated in cultivar 85–140.

LTR elements are the most abundant in plants and are involved in different responses to abiotic stress. Experiments under salt and light stress conditions reported the activation of the Ttd1a element in Triticum durum, which has a promoter with structural similarities to the transcriptional activation domains of plant defense response genes. In *Arabidopsis thaliana* seedlings, the activation of the At2G06045 element was demonstrated under different stress conditions (saline, osmotic stress, cold, heat and ABA (abscisic acid) treatments), causing the transcriptional induction of response genes to these stresses that encode TF or LEA proteins, among others. Likewise, wounds, salt stress and the culture of lemon (Citrus limon) leaf cells activated the CLCoy1 element, which, given its presence in different species of the genus Citrus, it is postulated that it may have been one of the main drivers of genomic variability during its domestication. Regarding class II elements, it is worth mentioning the MITE elements and their relationship with the stress response in rice. High doses of gamma radiation, low temperatures, and high salinity generate the rearrangement and amplification of mPing elements with high selectivity, avoiding exonic regions (Mukherjee et al., 2022).

Another way in which retrotransposons can generate new phenotypes is through exaptation. From an evolutionary point of view, exaptation is the subversion of a naturally selected set of adaptive features to achieve a new function. The sequences of the transposable elements can be directly exapted by specific phenotypic functions in the plant, and there are some examples in plants that, although they are scarce, play a critical role in plant growth and development. The TFs (transcription factors) FAR1 and FHY3 are essential in phytochrome PHYA signaling and were domesticated from MULE-like elements (Mutator-like elements). Similarly, the *Arabidopsis thaliana* DAY SLEEPER gene, which encodes a developmentally essential TF, is derived from class II hAt elements, with marked structural similarity. Another example is the MUG1-8 gene, also derived from the MULE superfamily, which plays an important role in floral development and reproductive fitness.

All the examples mentioned above demonstrate the decisive role those retro-transposable elements play in producing new genotypes and phenotypes in response to environmental perturbations. These elements are capable of, through sophisticated and diverse mechanisms, generating a great source of heritable intra- and inter-specific genomic variation, being considered today as important contributors to environmental adaptation (Singh et al., 2022).

6. Potential of Retro Transposable Elements in Crop Improvement Programs

Crop productivity is vulnerable to climate variability, as well as climate change associated with increases in temperature, CO_2, and changing rainfall patterns. In this context, the acquisition of phenotypic and genotypic diversity is crucial since poor adaptive plasticity can seriously harm crop species. Therefore, retrotransposons have gained great importance, suggesting the development of applied techniques based on induced and controlled transposition for crop improvement. These techniques could potentially create and fine-tune distinct regulatory networks induced by abiotic stress, as well as generate new epigenetic marks. However, they face a difficulty: the large fraction occupied by transposable elements in some crop plants (such as maize) implies extensive mobilization under stress conditions and therefore determines that the mechanisms for their control will require a deep knowledge of the environmental, epigenetic, and evolutionary triggers involved in the transpositional control of the different element families (Ito, 2022).

Another area where the study of transposable elements can assist in crop improvement programs is from molecular markers, gaining special importance in stress-induced chromosomal rearrangements, a common phenomenon in polyploid crops. Most of these studies have been carried out under biotic stress conditions:

1. IRAP (Inter Retrotransposon Amplification Polymorphism). This technique uses the LTR sequences as binding sites for the PCR primers, producing amplification of the region between the two LTR sequences.
2. REMAP (Retrotransposon-Microsatellite Amplified Polymorphism). The REMAP technique allows amplifying regions located between an LTR region and a microsatellite and is often used in conjunction with IRAP.
3. SSAP (Sequence Specific Amplification Polymorphism). This is a technique that uses a classic PCR primer and another adapter, which binds to the restriction fragments generated in the DNA sample thanks to prior digestion with restriction enzymes.
4. iPBS (inter-Priming Binding Site). A technique that is based on the recognition of the PBS sequence (universal 18 nucleotide sequence in autonomous and non-autonomous LTR retrotransposons) using a single primer.

WRGS (Whole-Genome Resequencing) has also been used together with KASPar (Kompetitive Allele-Specific Polymerase chain reaction) to analyze the GmSALT3 gene in more than a hundred soybean cultivars, demonstrating that salinity-resistant phenotypes contained a high content of transposon-associated polymorphisms. The KASPar technique allows genotyping of single nucleotide polymorphisms based on allele-competitive PCR that can detect allelic variations from fluorescent signals.

These techniques (together with other classic ones in genomics, such as PCR) directly support diversity studies or genotypic-phenotypic mapping, and make transposons highly important molecular markers, especially in studies of abiotic stress, since they are usually close to each other or within the sequence of stress response genes.

Mutagenesis by radiation exposure is a technique used in crop improvement programs, capable of generating mutations on a global genome scale. During this state of "genomic shock," it has been shown how the transposition of transposable elements is one of the most notable events during these experiments. Some groups have reported this phenomenon in obtaining sugarcane mutants tolerant to high sugar concentrations or high salt concentrations, suggesting a new approach in these programs.

Apart from what has been mentioned, all these techniques can provide new promoters, enhancers, and silencers, especially in the genes close to the transposon insertion sites, allowing the unmasking of new regulatory mechanisms not yet known in plants (Cerbin et al., 2022).

Conclusions

Retrotransposable elements are ubiquitous DNA sequences in eukaryotes capable of changing their position in the genome, giving rise to mutations, transcriptional modifications, chromosomal rearrangements, increases in genome size, or unusual expression patterns, among others. After being considered "junk DNA" and "selfish DNA", studies in the last two decades have provided a body of solid evidence that demonstrates how they generate heritable genotypic and phenotypic diversity, which translates into an important role in environmental adaptation and evolution.

These elements are widely distributed in plants and, through different mechanisms such as gene regulation, alternative splicing patterns, or the generation of polymorphisms and mutations, they can reconfigure the plant's transcriptional circuits under abiotic stress, as well as shape the architecture of the plant genome, ultimately leading to plant speciation and evolution. The plant genome has epigenetic mechanisms that repress the activity of retrotransposons, although, under abiotic stress conditions, these mechanisms are relaxed through the propagation of epigenetic marks to nearby loci and through the production of RNAs that contribute to regulating the epigenetic state of the host. Upon recovery, the patterns of retrotransposon silencing, and methylation are restored, demonstrating important dynamics in plant epigenetics during conditions prevalent in plants by their sessile nature.

Recent studies have demonstrated that retrotransposons play a crucial role in plant stress responses by generating genetic and epigenetic diversity and modulating gene expression in response to environmental cues. Abiotic stress, such as drought, salinity, extreme temperatures, or heavy metal toxicity, can lead to the activation of retrotransposons, causing genetic and epigenetic changes that confer stress tolerance and allow for phenotypic plasticity.

The rapid phenotypic plasticity provided by retrotransposons from an epigenetic perspective may be key in the search for resilient phenotypes in the current context of climate change that contemporary agriculture is facing. Studies of crosses and genetic maps with transposons demonstrate their potential in breeding programs in crop plants, although more attention should be paid to the molecular mechanisms that mediate their response to abiotic stress, to develop new hypotheses and new experiments.

In conclusion, the study of retrotransposons has shed light on the fascinating and dynamic mechanisms of plant stress responses. The integration of retrotransposons into the regulatory network of plants not only provides a pool of adaptive diversity but also offers a new perspective on the evolution of plant genomes and the plasticity of plant phenotypes. Further research on the role of retrotransposons in plant stress responses is needed to fully harness their potential for sustainable agriculture in the face of climate change and environmental challenges.

References

Bondar, E.I., Feranchuk, S.I., Miroshnikova, K.A., Sharov, V.V., Kuzmin, D.A., Oreshkova, N.V., and Krutovsky, K.V. (2022). Annotation of Siberian Larch (Larix sibirica Ledeb.) Nuclear Genome—One of the Most Cold-Resistant Tree Species in the Only Deciduous GENUS in Pinaceae. *Plants, 11*(15), 2062.

Bundock, P. and Hooykaas, P. (2005). An Arabidopsis hAT-like transposase is essential for plant development. Nature, 436(7048), 282–284. https://www.nature.com/articles/nature03667

Boyko, A., and Kovalchuk, I. (2008). Epigenetic control of plant stress response. *Environmental and molecular mutagenesis, 49*(1), 61–72.

Bui, Q., and Grandbastien, M.-A. (2012). LTR retrotransposons as controlling elements of genome response to stress? *Plant transposable elements: Impact on genome structure and function*, 273–96.

Cardoso, H., Campos, C., Grzebelus, D., Egas, C., and Peixe, A. (2022). Understanding the Role of PIN Auxin Carrier Genes under Biotic and Abiotic Stresses in Olea europaea L. *Biology, 11*(7), 1040.

Cerbin, S., Ou, S., Li, Y., Sun, Y., and Jiang, N. (2022). Distinct composition and amplification dynamics of transposable elements in sacred lotus (Nelumbo nucifera Gaertn.). *The Plant Journal, 112*(1), 172–92.

Chang, Y.N., Zhu, C., Jiang, J., Zhang, H., Zhu, J.K., and Duan, C.G. (2020). Epigenetic regulation in plant abiotic stress responses. *Journal of Integrative Plant Biology, 62*(5), 563–80.

Duk, M., Kanapin, A., Samsonova, A., Rozhmina, T., and Samsonova, M. (2022). *Structural variation in flax Linum usitatissimum L. genome.* Paper presented at the Bioinformatics of Genome Regulation and Structure/Systems Biology (BGRS/SB-2022).

Galindo-González, L., Mhiri, C., Deyholos, M.K., and Grandbastien, M.-A. (2017). LTR-retrotransposons in plants: Engines of evolution. *Gene, 626*, 14–25.

Grandbastien, M.-A. (1998). Activation of plant retrotransposons under stress conditions. *Trends in plant science, 3*(5), 181–87.

Grandbastien, M.-A., Audeon, C., Bonnivard, E., Casacuberta, J., Chalhoub, B., Costa, A.-P., . . . Poncet, C. (2005). Stress activation and genomic impact of Tnt1 retrotransposons in Solanaceae. *Cytogenetic and genome research, 110*(1– 4), 229–41.

Grandbastien, M.-A., Lucas, H., Morel, J.-B., Mhiri, C., Vernhettes, S., and Casacuberta, J.M. (1997). The expression of the tobacco Tnt1 retrotransposon is linked to plant defense responses. *Evolution and Impact of Transposable Elements*, 241–52.

Horváth, V., Merenciano, M., and González, J. (2017). Revisiting the relationship between transposable elements and the eukaryotic stress response. *Trends in Genetics, 33*(11), 832–41.

Hou, J., Lu, D., Mason, A.S., Li, B., Xiao, M., An, S., and Fu, D. (2019). Non-coding RNAs and transposable elements in plant genomes: emergence, regulatory mechanisms and roles in plant development and stress responses. *Planta, 250*, 23–40.

Ito, H. (2022). Environmental stress and transposons in plants. *Genes & Genetic Systems, 97*(4), 169–75.

Kinoshita, T., and Seki, M. (2014). Epigenetic memory for stress response and adaptation in plants. *Plant and Cell Physiology, 55*(11), 1859–63.

Kumar, A., and Bennetzen, J.L. (1999). Plant retrotransposons. *Annual review of genetics, 33*(1), 479–32.

Leeton, P.R. and Smyth, D.R. (1993). An abundant LINE-like element amplified in the genome of Lilium speciosum. Molecular and General Genetics MGG, 237, 97–104. https://riull.ull.es/xmlui/bitstream/handle/915/21706/Transposones%20y%20su%20relevancia%20en%20el%20estres%20abiotico%20en%20plantas.pdf?sequence=1&isAllowed=y

Liu, J. and He, Z. (2020). Small DNA methylation, big player in plant abiotic stress responses and memory. *Frontiers in plant science, 11*, 595603.

Liu, J. and He, Z. (2020). Small DNA methylation, big player in plant abiotic stress responses and memory. *Frontiers in Plant Science*, 11, 595603.

Matsui, A., Nguyen, A.H., Nakaminami, K., and Seki, M. (2013). Arabidopsis non-coding RNA regulation in abiotic stress responses. *International journal of molecular sciences, 14*(11), 22642–54.

Mirouze, M., and Paszkowski, J. (2011). Epigenetic contribution to stress adaptation in plants. *Current opinion in plant biology, 14*(3), 267–74.

Mukherjee, S., Sharma, D., and Upadhyaya, K.C. (2022). Differential transcriptional activation of copia family of different plant retrotransposons. *Journal of Plant Biochemistry and Biotechnology, 31*(4), 915–24.

Nakaminami, K., Matsui, A., Shinozaki, K., and Seki, M. (2012). RNA regulation in plant abiotic stress responses. *Biochimica et Biophysica Acta (BBA)-Gene Regulatory Mechanisms, 1819*(2), 149–53.

Negi, P., Rai, A. N., and Suprasanna, P. (2016). Moving through the stressed genome: emerging regulatory roles for transposons in plant stress response. *Frontiers in plant science, 7*, 1448.

Ohama, N., Sato, H., Shinozaki, K., and Yamaguchi-Shinozaki, K. (2017). Transcriptional regulatory network of plant heat stress response. *Trends in plant science, 22*(1), 53–65.

Orozco-Arias, S., Isaza, G., and Guyot, R. (2019). Retrotransposons in Plant Genomes: Structure, Identification, and Classification through Bioinformatics and Machine Learning. *International journal of molecular sciences, 20*(15), 3837. Retrieved from https://www.mdpi.com/1422-0067/20/15/3837.

Paszkowski, J. (2015). Controlled activation of retrotransposition for plant breeding. *Current Opinion in Biotechnology, 32*, 200–206.

Qu, Y., Shang, X., Zeng, Z., Yu, Y., Bian, G., Wang, W., . . . Wang, Q. (2023). Whole-genome Duplication Reshaped Adaptive Evolution in A Relict Plant Species, Cyclocarya paliurus. *Genomics, Proteomics & Bioinformatics*.

Sahu, P.P., Pandey, G., Sharma, N., Puranik, S., Muthamilarasan, M., and Prasad, M. (2013). Epigenetic mechanisms of plant stress responses and adaptation. *Plant cell reports, 32*, 1151–59.

Sharma, M., Kumar, P., Verma, V., Sharma, R., Bhargava, B., and Irfan, M. (2022). Understanding plant stress memory response for abiotic stress resilience: Molecular insights and prospects. *Plant Physiology and Biochemistry, 179*, 10–24. Retrieved from https://www.sciencedirect.com/science/article/pii/S0981942822001164. doi:https://doi.org/10.1016/j.plaphy.2022.03.004

Singh, A., Goswami, S., Vinutha, T., Jain, R., Ramesh, S., and Praveen, S. (2022). Retrotransposons-based genetic regulation underlies the cellular response to two genetically diverse viral infections in tomato. *Physiological and Molecular Plant Pathology, 120*, 101839.

Sudalaimuthuasari, N., Ali, R., Kottackal, M., Rafi, M., Al Nuaimi, M., Kundu, B., . . . Balan, J. (2022). The Genome of the Mimosoid Legume Prosopis cineraria, a Desert Tree. *International journal of molecular sciences, 23*(15), 8503.

Sun, C., Xie, Y.H., Li, Z., Liu, Y.J., Sun, X.M., Li, J.J., . . . Zhang, S. G. (2022). The Larix kaempferi genome reveals new insights into wood properties. *Journal of Integrative Plant Biology, 64*(7), 1364–73.

Sunkar, R. (2010). *MicroRNAs with macro-effects on plant stress responses.* Paper presented at the Seminars in cell & developmental biology.

Todorovska, E. (2007). Retrotransposons and their role in plant–genome evolution. *Biotechnology & Biotechnological Equipment, 21*(3), 294–305.

Vanderschuren, H., Zaidi, S.S.-e.-A., Shakir, S., Mehta, D., Nguyen, V., and Gutzat, R. (2022). Profiling of eccDNAs in Arabidopsis indicates shift in population of TE-derived eccDNAs in response to stress, cellular state, and epigenetic processes.

Wang, M., Zhang, L., Tong, S., Jiang, D., and Fu, Z. (2022). Chromosome-level genome assembly of a xerophytic plant, Haloxylon ammodendron. *DNA Research, 29*(2), dsac006.

Wheeler, B.S. (2013). Small RNAs, big impact: small RNA pathways in transposon control and their effect on the host stress response. *Chromosome research, 21*, 587–600.

Yushkova, E., and Moskalev, A. (2023). Transposable elements and their role in aging. *Ageing Research Reviews*, 101881.

Zhao, J., He, Q., Chen, G., Wang, L., and Jin, B. (2016). Regulation of non-coding RNAs in heat stress responses of plants. *Frontiers in plant science, 7*, 1213.

Index